Archimedes

New Studies in the History and Philosophy of Science and Technology

Volume 71

Archimedes has three fundamental goals: to further the integration of the histories of science and technology with one another; to investigate the technical, social and practical histories of specific developments in science and technology; and finally, where possible and desirable, to bring the histories of science and technology into closer contact with the philosophy of science.

The series is interested in receiving book proposals that treat the history of any of the sciences, ranging from biology through physics, all aspects of the history of technology, broadly construed, as well as historically-engaged philosophy of science or technology. Taken as a whole, Archimedes will be of interest to historians, philosophers, and scientists, as well as to those in business and industry who seek to understand how science and industry have come to be so strongly linked.

Submission / Instructions for Authors and Editors: The series editors aim to make a first decision within one month of submission. In case of a positive first decision the work will be provisionally contracted: the final decision about publication will depend upon the result of the anonymous peer-review of the complete manuscript. The series editors aim to have the work peer-reviewed within 3 months after submission of the complete manuscript.

The series editors discourage the submission of manuscripts that contain reprints of previously published material and of manuscripts that are below 150 printed pages (75,000 words). For inquiries and submission of proposals prospective authors can contact one of the editors:

Editor: JED Z. BUCHWALD, [Buchwald@caltech.edu]
Associate Editors:
Mathematics: Jeremy Gray, [jeremy.gray@open.ac.uk]
19th-20th Century Physical Sciences: Tilman Sauer, [tsauer@uni-mainz.de]
Biology: Sharon Kingsland, [sharon@jhu.edu]
Biology: Manfred Laubichler, [manfred.laubichler@asu.edu]
Please find on the top right side of our webpage a link to our *Book Proposal Form.*

Jutta Schickore • William R. Newman
Editors

Elusive Phenomena, Unwieldy Things

Historical Perspectives on Experimental Control

 Springer

Editors

Jutta Schickore
Department of History and Philosophy
of Science and Medicine
Indiana University
Bloomington, IN, USA

William R. Newman
Department of History and Philosophy
of Science and Medicine
Indiana University
Bloomington, IN, USA

ISSN 1385-0180   ISSN 2215-0064   (electronic)
Archimedes
ISBN 978-3-031-52956-6   ISBN 978-3-031-52954-2   (eBook)
https://doi.org/10.1007/978-3-031-52954-2

This work was supported by the Andrew W. Mellon Foundation and Indiana University.

This Springer imprint is published by the registered company Springer Nature Switzerland AG
The registered company address is: Gewerbestrasse 11, 6330 Cham, Switzerland

Paper in this product is recyclable.

Contents

1 Introduction: Practices, Strategies, and Methodologies of Experimental Control in Historical Perspective 1
Jutta Schickore

2 Christoph Scheiner's *The Eye, that is, The Foundation of Optics* (1619): The Role of Contrived Experience at the Intersection of Psychology and Mathematics 21
Tawrin Baker

3 One Myrtle Proves Nothing: Repeated Comparative Experiments and the Growing Awareness of the Difficulty of Conducting Conclusive Experiments 55
Caterina Schürch

4 Controlling Induction: Practices and Reflections in David Brewster's Optical Studies 105
Friedrich Steinle

5 Carl Stumpf and Control Groups 125
Julia Kursell

6 A "Careful Examination of All Kind of Phenomena": Methodology and Psychical Research at the End of the Nineteenth Century 149
Claudia Cristalli

7 Controlling Nature in the Lab and Beyond: Methodological Predicaments in Nineteenth-Century Botany 179
Kärin Nickelsen

8 Controlling the Unobservable: Experimental Strategies and Hypotheses in Discovering the Causal Origin of Brownian Movement 209
Klodian Coko

**9   From the Determination of the Ohm to the Discovery of Argon:
      Lord Rayleigh's Strategies of Experimental Control**   243
      Vasiliki Christopoulou and Theodore Arabatzis

**10   Controlling Away the Phenomenon: Maze Research
      and the Nature of Learning**   269
      Evan Arnet

**11   Controlling Animals: Carl von Heß, Karl von Frisch,
      and the Study of Color Vision in Fish**   291
      Christoph Hoffmann

Contributors

Theodore Arabatzis Department of History and Philosophy of Science, School of Science, National and Kapodistrian University of Athens, University Campus, Ano Ilisia, Athens, Greece

Evan Arnet Indiana University Bloomington, Bloomington, IN, USA

Tawrin Baker South Bend, IN, USA

Vasiliki Christopoulou Department of History and Philosophy of Science, School of Science, National and Kapodistrian University of Athens, University Campus, Ano Ilisia, Athens, Greece

Klodian Coko Philosophy Department, Ben-Gurion University of the Negev, Beersheba, Israel

Claudia Cristalli Department of Philosophy, University of Tilburg, Tilburg, Netherlands

Christoph Hoffmann Department of Cultural and Science Studies, University of Lucerne, Lucerne, Switzerland

Julia Kursell Faculty of Humanities, University of Amsterdam, Amsterdam, Netherlands

William R. Newman Department of History and Philosophy of Science and Medicine, Indiana University, Bloomington, IN, USA

Kärin Nickelsen Ludwig Maximilian University Munich, Munich, Germany

Jutta Schickore Department of History and Philosophy of Science and Medicine, Indiana University, Bloomington, IN, USA

Caterina Schürch Institute of History and Philosophy of Science, Technology, and Literature, Technical University Berlin, Berlin, Germany

Friedrich Steinle Institute of History and Philosophy of Science, Technology, and Literature, Technical University Berlin, Berlin, Germany

Chapter 1
Introduction: Practices, Strategies, and Methodologies of Experimental Control in Historical Perspective

Jutta Schickore

Control is the hallmark of scientific experimentation. If an experiment is deemed to be lacking in control, it is unlikely to gain traction in the scientific community; arguably, an uncontrolled intervention is not even a genuine experiment. Today, scientific articles routinely mention controls and handbooks and instruction manuals on methods in the life sciences call for controlled experiments. Evaluating the appropriateness of controls is a core element of successful peer-review.

But despite its centrality to modern scientific inquiry, many foundational and historical questions about experimental control remain open. Experimental practice has been studied for decades, but only few analyses of scientific control practices in experimentation exist,[1] with almost nothing written on controlled experimentation in the *longue durée*.[2] We know little about changing expectations for well-controlled experiments or about different kinds of control, experimenters' interpretations of control, or reasons given for applying controls. There is not even consensus about whether experimental control is an ancient, early modern, or Enlightenment concept, or whether it is a more recent feature of scientific inquiry.[3] This is, in part, because the concepts "control," "control experiment," and "controlled experiment"

---

[1] This is slowly changing, see Guettinger (2019); Sullivan (2022); Guettinger (2019); Desjardins et al. (2023).

[2] Only the randomized controlled trial has been studied historically and systematically. See Marks (1997, Chap. 5); Worrall (2007); Cartwright (2007); Keating and Cambrosio (2012). For the control group and (double) blind tests, see Kaptchuk (1998); Strong and Frederick (1999, including further references); Dehue (2005); Holman (2020).

[3] For a variety of views, see, for instance, McCartney (1942); Beniger (1986); Levin (2000a, 13–14); Amici (2001).

J. Schickore (✉)
Department of History and Philosophy of Science and Medicine, Indiana University, Bloomington, IN, USA
e-mail: jschicko@indiana.edu

© The Author(s) 2024
J. Schickore, W. R. Newman (eds.), *Elusive Phenomena, Unwieldy Things*, Archimedes 71, https://doi.org/10.1007/978-3-031-52954-2_1

are polysemous, like "replication" or "significance." In addition, methodological concepts for experimental practice have until recently received comparatively little scholarly attention.

"Control" has been studied mostly as a broader cultural phenomenon in the Western world. Cultural histories of control focus on ideologies and technologies for governing people, procedures, or systems of machines (Levin 2000b; Derksen 2017). Historical studies of control and science have shown how cultural currents, for better or worse, transformed scientific practices into more rigorous endeavors. Historians of science have noted the increasing importance in science of a quantifying spirit (Frängsmyr et al. 1990) and the values of precision (Wise 1995). They have examined the influence on science of tools such as statistics (Porter 1995; Gigerenzer et al. 1989) and surveillance devices (Foucault 1975, 1979), as well as bureaucratic procedures such as record-keeping, double bookkeeping, and accounting. These authors have argued that institutional changes in science, such as the rise of the university and urban research laboratories, have helped to standardize scientific practice and make it more exact (Tuchman 1993; Dierig 2006). Eighteenth-century sciences of state promoted record-keeping, accounting, and statistical assessments of experimental data (Seppel and Tribe 2017). Nineteenth- and twentieth-century physics and engineering helped to create automated feedback control mechanisms (Bennett 1993), intertwined control and communication systems (Wiener 1948), and "networks of power" (Hughes 1983). They also brought about catastrophic failure of control, as in failed aerospace missions, plane crashes, and collapsing bridges (Schlager 1995). Industrial and technological advancements allowed researchers to engineer the development of living organisms and human heredity (Pauly 1987; Paul 1995), to standardize living things as model organisms for experiments (Rader 2004), and to measure human performance (Rabinbach 1990). The twentieth-century nexus of military, industry, and information technologies enabled wide-ranging control over data and information flow (Galison 2010; Franklin 2015).

Of course, broader socio-political and cultural developments such as industrialization, the institutionalization of university research laboratories, and the expansion of bureaucracies and state administration are impactful. These developments change how practices of research, recording, and record-keeping are organized, as so many authors have demonstrated. But they do not fully determine experimental designs or experimenters' views on what is considered good and well-controlled or deficient and poorly controlled experimental practice.

This volume shifts the focus from broader socio-political and cultural contexts of control onto practitioners' methodological strategies of inquiry and experimental design. While acknowledging that broader cultural forces do affect control practices, we contend that these forces only partially shape experimental design and strategy. We identify additional social dimensions of experimental control. On the one hand, identifying experimental conditions, confounders, and solutions to technical problems in experimental design takes time, and unfolds by the activities of multiple individuals or groups. On the other hand, whether an experiment counts as "sufficiently" or even "fully" controlled is not entirely decided by the experimenters themselves, nor can the question be settled by comparing actual experimentation with an abstract standard of the ideal controlled experiment.[4] The adequacy of control critically depends on the social interactions and negotiations among experimenters and their various interlocutors; as such, the issue is open to revisiting, revision, and renegotiation.

To capture the complicated and multilayered history of experimental control, it is useful to distinguish control strategies, control practices, and methodological ideas about experimental control. *Control strategies* are general designs and plans to follow in an experiment, like the comparison of an intervention target with a control. *Control practices* are the concrete actions by which experimenters implement control strategies in particular contexts. These contexts comprise all the resources available to the experimenters, including materials, tools, techniques, local expertise, and institutional opportunities. *Methodological ideas* are the broader notions of how to study nature and everything in it. They are contained in accounts of control strategies and practices, as the practitioners themselves give them.[5]

Contributions to this volume deal with the details of experimental control practices, as well as with the expectations and perceived obstacles for experimental designs. The chapters are also sensitive to long-term developments of control strategies and methodological ideas. We provide a set of focused studies on control practices, strategies, and ideas that, together, cover a period of more than 300 years, with glimpses back to antiquity and forward to the late twentieth century. We contend that the long-term perspective is productive for understanding experimental methodologies and experimental control in particular.[6] The chapters offer several examples of how control practices using those strategies and ideas are shaped by local contexts—material-technical, conceptual, and social. Together, they illustrate that control strategies and methodological ideas often remain stable for a long time and change only gradually.

To study controlled experimentation from a historical perspective, we must distinguish at least two notions of control. The first is a broad sense of control as "managing," "restraining," or "keeping everything stable except the target system to be intervened upon." This notion primarily but not exclusively concerns the experiment's material side—the objects, the setting and environment, and the tools, as

---

[4] Two classic studies of how experimenters sought to "control" their audiences are Shapin and Schaffer (1985) and Geison (1995, especially Chap. 5).

[5] These ideas are also articulated in the philosophy of science, of course. In this volume, however, we are concerned mostly with practicing experimenters' working philosophies.

[6] Some historians have strong reservations about long-term histories "lining up unconnected look-alikes through the ages" (Dehue 2005, 2), or "ahistorical narratives" comparing, for instance, early modern and Victorian experiments "merely because of superficial similarity 'in the use of controls'" (Strick 2000, 5, commenting on spontaneous generation experiments). Our volume shows that it is possible to write long-term histories without comparing apples to oranges.

well as the guided manipulation or intentional intervention in an otherwise stable situation to see what will happen.[7]

In an uncontrolled situation, experimenters cannot determine the changes resulting from their interventions. To extract information from unwieldy experimental situations, they must standardize instruments and experimental targets and hold fixed the experimental background conditions. They ought also to be free of preconceived opinions and other sources of influence. Experimenters seek to make the experimental setting and background as stable and rigorous as possible because effects, both expected and novel, appear most distinctly against a stable background.[8] Generally, then, we can consider any aspect of experimental practice from the perspective of control; a key question is how experimenters identify what must be controlled in concrete contexts and how they achieve that control.

There is also a narrower notion of control, referring to comparative experimental designs.[9] It primarily but not exclusively concerns the experiment's epistemic side, or the conditions required for the experiment to generate knowledge. Modern scientists typically associate with "control experiment" a particular experimental strategy or design, namely the comparison to a control case. An experimental intervention is compared with a baseline; the target system of the intervention is compared with a similar target system that, unlike the experimental object, was not intervened on (the "control mouse," say, which did not receive treatment). This strategy encapsulates the requirements for an experiment to be informative about cause-effect relations.[10]

In the narrow sense, comparison to a baseline is needed to find out whether it really was the manipulation of *this* particular variable that made a difference to the experimental outcome.[11] Of course, the more similar the experimental situations are, the more informative the comparisons will be. Making informative comparisons thus requires control practices in the broader sense explained above, to ensure that the two experimental settings are stable, save the intervention.

We should avoid confusing the emergence of terms such as "control experiment" and "experimental control" in the scientific literature with the emergence of explicit discussions about control practices and strategies. The terms "control experiment," "controlled experiment," and "experimental control" are recent terms. Google Ngram shows a steep increase for "control experiment" in the last decades of the nineteenth century in English, French, and German-language scientific literature. Of course, Ngram is not a rigorous tracker for word usage, but based on its data, we

---

[7] These distinctions are inspired by one of the few systematic studies of controlled experimentation, Edwin Boring's "The Nature and History of Experimental Control" (Boring 1954).

[8] This insight underlies Ludwik Fleck's and Thomas Kuhn's accounts of scientific change.

[9] Comparison, Boring noted, "appears in all experimentation because a discoverable fact is a difference or a relation, and a discovered datum has significance only as it is related to a frame of reference, to a relatum" (Boring 1954, 589).

[10] For the epistemic ideal underlying this design, the "perfectly controlled experiment," see Guala (2005, 65–69).

[11] I keep this characterization vague because I do not want to commit to a specific philosophical understanding of causality here.

can safely assume that control practices were common long before the term spread in scientific writing.[12] As our volume demonstrates, discussions about stable experiments antedate the appearance of the term "control" in this literature. Concerns about the adequate management of experimental settings were voiced as soon as experimentation became widespread. Robert Boyle, for one, published two famous essays on "unsucceeding" experiments, where he discussed the obstacles posed by impure chemicals, the variability of body parts in different corpses, and other issues threatening experimental success (Boyle 1999a, b).

The history of experimental control, then, encompasses four distinct yet related strands. The first is the historical development of control practices to stabilize and standardize experimental conditions. The second is the emergence and career of the comparative design in experimentation, understood as a way of generating and securing knowledge of cause-effect relations. The third involves the unfolding, both in philosophy of science and in the sciences themselves, of methodological discussions on control practices and designs in experimental practice. The fourth is the history of the term "(experimental) control."

This volume concerns itself most with the first three strands. We do not systematically explore the history of the term "control;"[13] in fact, several contributions discuss research from before the late nineteenth century. However, precisely because control practices and strategies predate the term "control" in scientific literature, we keep terminological questions in mind as we analyze past experimental reports and methodological discussions. We pay careful attention to the terms past practitioners did use, whatever they were, to describe, explain, and defend control practices and strategies.

The contributions here examine how control practices and comparative designs developed, and include past accounts of critiques and defenses for these practices. Control is a multifaceted and elusive concept, and our volume reflects this. We have not attempted to reduce our discussion to a single definition of "control." Although this introduction provides some points of orientation for analyzing control practices and strategies, each contributor further explains the concept for specific experimental contexts. The chapters range over different fields, from botany and vision studies, ecology and plant physiology, human physiology and psychology to animal behavior and experimental physics. They cover a period from the early seventeenth to the twentieth century. They examine experiments with complex and sometimes unwieldy objects and elusive phenomena. Chapters deal with studies on learning and judgment; color blindness in animals; auditory perceptions of tones, pitch, and vowel sounds; irregular movements; psychic forces; unobservable elements; and the best "photogenic climate" for promoting photosynthesis. Experiments on such objects and phenomena are hard to design, stabilize, and carry out, and they are often controversial. For this reason, they showcase questions and reflections on control in science particularly well.

---

[12] Technical terms such as "positive" and "negative" control are even more recent (and outside the timeframe of our volume). They are also poorly understood.

[13] For a brief overview of historical definitions of control, see Levin (2000a, 21–31).

The very practice of creating and maintaining a stable experimental situation is old, arguably as old as experimental intervention itself. Over time, experimenters learn what must be managed and tracked in experimental contexts; they seek to localize the phenomena of interest as well as the elements of the experimental setting in order to make interventions more exact. Gradually they develop new tools to do this. Precision instruments, elaborate recording devices, and other technologies available in the last century or two can assist with these tasks. The history of research laboratories can be written as the history of efforts to create highly controlled research environments. Nineteenth-century physicists worked at night or retreated to the lab basement to escape city noise, vibrations from trams, and exuberant students (Hoffmann 2001). Today's scientists turn to specialized construction companies when they need "clean rooms" for research.[14] All-metal or all-plastic labs are built for research into the impacts of micro-plastics on materials and tissues or on radiation, respectively. Particle physicists dive to recover radiation-free lead from ancient shipwrecks to prevent contaminating their measurements.

Such materials and technologies often make it easier to keep an experimental situation stable and to track interesting changes.[15] At the same time, however, closer analysis of actual episodes shows that advancements in instrumentation, impressive as they may appear in hindsight, do not guarantee improved control. In fact, obtaining control often becomes *more* difficult, not least because researchers must learn the instruments' proper functioning. "The more finely a method of investigation operates, the more complicated the devices used must be," as Carl Stumpf noted (1926, 8).[16]

Moreover, the history of control is a history of efforts—and efforts can fail. Implementing control strategies often fails, as even the experimenters themselves sometimes admit. Our volume illustrates how difficult it can be to manage an experimental setting, how resourceful some experimenters were in their management, and how they sometimes failed to achieve it despite intense effort. Claudia Cristalli's researchers of psychic phenomena walk the line between controlling the psychic powers of the "percipients" in their experiments, and preventing them from sensing any phantasms at all. Christoph Hoffmann's study of color blindness in fish shows how experimenters dealt with the tricky problem of controlling animals' behavior. Experimenters found different solutions, both difficult to implement and neither completely satisfying. One option was to train the fish—much more challenging to do than training, say, a dog or rat. The other was to design the experimental setting in such a way that the "normal" behavior of the fish was taken into account when the behavior of interest was elicited. But what is the "normal" behavior of fish? And how can it be accommodated in the unnatural environment of a laboratory fish tank?

---

[14] See Holbrook (2009).

[15] See, e.g., Kuch et al. (2020).

[16] The quotation is drawn from Kursell's chapter in this volume.

Other contributions illustrate how experimenters approached the creation and monitoring of an experimental setting. They discuss the multifaceted nature of the associated problems and the obstacles the experimenters had to overcome when attempting to stabilize unwieldy things, such as the irregular movements of microscopic parts, the germination, sprouting, and growth of plants, and auditory perceptions. The contributions describe the solutions they found to these problems. Experimenters tried their best to identify the smallest details of the experimental settings deemed relevant, and sometimes invented remarkably elaborate contraptions to keep them stable.

Caterina Schürch depicts the curious machines with which eighteenth-century plant physiologists tried to electrify plants and seeds with precise doses of electricity. Kärin Nickelsen shows how the nineteenth-century plant physiologist Julius Wiesner designed an artificial environment for his plants: double-walled glass jars, with the space between the walls filled with a solution of iodine in carbon disulphide. Because this liquid layer absorbed all visible light but heat rays, Wiesner could examine the impact of those rays on plant growth. Julia Kursell describes the giant arrangement of tubes Carl Stumpf erected to compare how his experimental subjects perceived natural and machine-generated vowels. She notes that, according to Stumpf, the increased finesse of experimental tasks required ever more complex experimental devices. Cristalli shows how Faraday, attempting to stop participants in table-turning experiments from making involuntary movements, designed a device consisting of a stack of cardboard sheets, arranged like a voltaic pile, with pellets of wax in between. The device would be placed between the hands of the séance participants and the tabletop. The sheets were arranged and marked in such a way that their displacement would indicate hand movements prior to the table's movement.

These devices often astonish with their ingenuity, but the point is that they are the material realizations of what experimenters recognized as the relevant conditions and potential confounders for their experiments. They are therefore purpose-dependent, as Kursell notes; at the same time, they both constitute and constrain the generation of experimental knowledge. Cristalli's, Schürch's, Nickelsen's and Evan Arnet's chapters demonstrate this constraint: over time, views about what factors to manipulate, keep fixed, or monitor in controlled experiments might change considerably, even within a single research tradition. While Faraday built tools to control his subjects' involuntary movements, his American colleague and erstwhile admirer Robert Hare turned to designing machines that would prevent voluntary movements in psychic experiments—in other words, to prevent fraud.

Schürch's account illustrates a most dramatic change of focus. After decades of carefully controlled experimentation, which supported the view that electrification promotes plant growth, Jean Ingen-Houz showed, using the same control strategies, that it was not electricity but differences in light intensities that affected the plants. He thus re-oriented the entire research program of plant growth, rendering previously "well-controlled" experiments uncontrolled.

Similarly, in maze research on animal learning, later investigators critiqued their predecessors for stabilizing—"controlling for"—the very phenomenon they should have studied, as Arnet's work illustrates. Nickelsen shows how control practices in photosynthesis research changed fundamentally as the experiments moved from the laboratory to the field. As she observes, the changes were not just practical—measuring natural light is harder than measuring laboratory light—but also conceptual. What mattered was no longer just "daylight," but a complex set of factors consisting of the specific light individual plant parts received, intensity fluctuations during the day and the season, and so forth. Klodian Coko charts another kind of reorientation in his study of research on Brownian movement. Using the strategy of comparative experimentation, nineteenth-century researchers tried to establish what could and could not be the cause of Brownian movement. Later in the century, Brownian movement itself became evidence for a new kinematic-molecular theory of matter, which changed the understanding of rigor and experimentation.

Several chapters also direct attention to the fact that many experimenters were explicitly concerned with developing coping strategies for "limited beings" (Wimsatt 2007) in sub-optimal situations. Researchers faced challenges not only because background factors were difficult or too numerous to monitor, but also because those factors were not immediately observable. Remarkably, the physicist Lord Rayleigh devoted several of his public-facing remarks to the theme of "deficient rigor." As Vasiliki Christopoulou and Theodore Arabatzis point out, for Rayleigh, the pursuit of absolute ("mathematical") rigor could even be detrimental to progress in physics. It was in this situation that experimenters insisted on using two or more different experimental techniques to check if both converged on the same outcomes, as detailed in the contributions by Christopoulou and Arabatzis and by Coko.

Notably, experimenters developed strategies to guard against entirely *unknown* influences on their experiments. The notion that natural phenomena in an experiment might occur and not occur in unforeseeable ways is centuries old. The metaphysical interpretation of this notion has changed dramatically over time (Hacking 1984, 1990), but there was wide and long-standing agreement about how to address it: namely, through multiple repetitions of experimental trials. Both the early seventeenth-century experimenter Scheiner and the late nineteenth-century experimenter Rayleigh gave the idea of multiple repetitions an important role in rigorous experimentation, if for different reasons.

In an early essay on medical experience, the ancient physician and anatomist Galen discussed the possibility that what is seen only once in a patient may not be a regular occurrence, and thus may not be worthy of acceptance and belief. Galen suggested this point in the middle of his attempt to demonstrate that medical practice is not just *logos*, but also experience.[17] As part of the argument, Galen alluded to the instability of memory and also noted that medicines work sometimes but not always (Galen 1944). In clinical medicine, at least, one single drug test might not produce reliable results, because "some things are frequent and some are rare"

---

[17] Much of the text rebuts the sorites argument, according to which it is impossible to clarify the notion of seeing something "very many times" (see Galen 1944, 124–25). For a reconstruction of the argument, see (Kupreeva 2022).

(Galen 1944, 113). It must therefore be repeated several times, and even then, it may not tell us what is usually the case.[18] Ibn Sīnā (Avicenna) expressed a similar idea in a proposal for rules of drug testing, albeit with a positive spin. He wrote that "the effect of the drug should be the same in all cases or, at least, in most. If that is not the case, the effect is then accidental, because things that occur naturally are always or mostly consistent" (Nasser et al. 2009, 80).

In the early modern period, we encounter this idea frequently, now also in discussions about experimentation beyond drug testing in clinical medicine. Repeating experimental trials several times, indeed "very many times," became an imperative for rigorous experimentation—in this way, unknown or contingent and accidental influences on experiments could be avoided.[19] In later centuries it was to become a hallmark of rigorous experimentation that a trial be done more than once or on large samples.[20] However, as Schürch's chapter shows, the appropriate number of repetitions remained contested.

Scholars looking for the "first" control experiment in the history of scientific inquiry typically assume, but in most cases tacitly, the narrower notion of "control" as comparative trial. They have found quite early examples for comparative designs in experimental practice. These examples often come from medicine, where it is both vitally and commercially important to discover the efficacy of certain drugs and treatments. The reputation of a practitioner depended on the treatments' success.

For example, historian of statistics Stephen Stigler finds an instance of comparative experimentation in the Old Testament, in the Book of Daniel (around 164 BCE). Servants on a vegetarian diet are compared with children who eat "the king's meat": "And at the end of ten days their countenances appeared fairer and fatter in flesh than all the children which did eat the portion of the king's meat" (Daniel 1:5–16).[21]

A passage by Athenaeus (200 CE) describes how some convicted criminals had been thrown among asps and survived. It turned out that they had been given lemons prior to their punishment. The next day a piece of lemon was given to one convict

---

[18] For the Aristotelian notion of the memory of many instances, see Bayer (1997). For its application in the scholastic-mathematical tradition, see Dear (1991).

[19] On repetition and "many, many" trials, see Schickore (2017, chapters 1–3).

[20] A popular passage by Karl Popper expresses this idea: "Every experimental physicist knows those surprising and inexplicable apparent 'effects' which in his laboratory can perhaps even be reproduced for some time, but which finally disappear without trace. Of course, no physicist would say in such a case that he had made a scientific discovery (though he might try to rearrange his experiments so as to make the effect reproducible). Indeed the scientifically significant physical effect may be defined as that which can be regularly reproduced by anyone who carries out the appropriate experiment in the way prescribed. No serious physicist would offer for publication, as a scientific discovery, any such 'occult effect,' as I propose to call it—one for whose reproduction he could give no instructions. The 'discovery' would be only too soon rejected as chimerical, simply because attempts to test it would lead to negative results. (It follows that any controversy over the question whether events which are in principle unrepeatable and unique ever do occur cannot be decided by science: it would be a metaphysical controversy)" (Popper 2002, 23–24).

[21] This example is also quoted on the website of the Institute for Creation Research as a model for sound experimental design (Treece 1990).

but not to another. The one who ate the lemon survived the bites, the other died instantly.[22] The pseudo-Galenic treatise on theriac describes a trial with a similar design, whereby two birds would be poisoned and only one given an antidote (Leigh 2013). The trial tests the efficacy of medicines: if both animals survived, the tested antidote was recognized to be ineffectual. That experiment was again reported in the Middle Ages, notably by Bernard Gordon (McVaugh 2009).

Another famous ancient example is the legend of Pythagoras. As the story goes, he observed that most combinations of blacksmiths' hammers generated a harmonious sound when striking anvils at the same time, while some did not. Pythagoras discovered that harmonious sounds were produced by those hammers whose masses were simple ratios of each other, while other hammers made dissonant noises when struck simultaneously. Notably, Ptolemy later criticized the Pythagorean experiment because, to him, it *lacked* control (Zhmud 2012, 307).

The Pythagorean case is interesting. It clearly has a comparative component, inspecting the sound of hammers whose masses were simple ratios of each other and that of other hammers. But in the historiography of science it does not serve as an example of an early "control experiment." In fact, the ancient texts have too little information to determine whether it was consciously performed as an experiment compared with a control, whether Pythagoras simply varied the setup, or whether he arrived at his conclusions by observing different blacksmiths at work.

Conscious and explicit implementation of comparative designs appears to become more common in seventeenth- and eighteenth-century experimental practice. In his studies on the generation of insects, Francesco Redi famously compared samples of organic materials—"a snake, some fish, some eels of the Arno, and a slice of milk-fed veal in four large, wide-mouthed flasks" (Redi 1909, 33)—kept in open and closed containers. The samples were periodically inspected for traces of life. No life developed in closed containers, which Redi took as evidence against the spontaneous generation of maggots from putrefying flesh. Here, the comparative design demonstrates a cause-effect relation through the comparison with a "control." Redi showed that maggots in open containers were generated by flies' eggs.[23]

The case of spontaneous generation research illustrates particularly well why it is useful to distinguish between comparative design strategies and a broader notion of control as management of the experimental setting. Redi's experimental research was not decisive, and after him many other experimenters investigated spontaneous generation. They all contested each other's experiments and many argued that their opponents had not properly maintained the experimental settings; they also argued that they themselves really had taken the necessary precautions to do so. John T. Needham, for instance, claimed that he could demonstrate the spontaneous generation of animalcules in infusions. He told his readers that he had "neglected no

---

[22] Deipnosophists or Banquet of the Learned, 3.84 d-f:2. The reference is from McCartney (1942, 5–6).

[23] For details on Redi's experiments, see Parke (2014). Historians of biology as well as science educators regularly cite Redi's experiments on spontaneous generation as "the first control experiments."

Precaution, even as far as to heat violently in hot Allies the Body of the Phial; that if any thing existed, even in that little Portion of Air which filled up the Neck, it might be destroy'd, and lose its productive Faculty" (Needham 1748, 638). Notably, he did not report a comparison with a vial that had not been heated in fire. It may have been superfluous to him, because it was obvious that animalcules would appear in it, as so often had been observed. The debates continued throughout the nineteenth century. Experimental designs and interpretations for possible contaminants varied, but the comparative strategy generally remained the strategy of choice.[24] As Schürch's contribution shows, in the decades around 1800, experimenters across Western Europe advocated comparative experimental designs.

Reports of comparative trials can be found in many fields, from agriculture to clinical medicine.[25] A notable but little-studied example is steeping experiments (Pastorino 2022). A comparative experiment by Francis Bacon served as a template for many subsequent experiments on the effects of plant growth when steeping seeds in various fluids.

Our volume illustrates comparative trial designs in plant physiology, physics, animal behavior studies, and psychology. The episodes exemplify both the conscientious application of these strategies and the obstacles experimenters faced as they attempted to realize well-controlled comparative trials.

The earliest pre-modern reports of experimental trials and comparative designs contain little express discussion on control practices and strategies. There are exceptions, of course, especially in medical contexts. I already noted Galen's writings, and we know that medieval scholars such as Ibn Sīnā developed rules for drug testing (Crombie 1952). Mostly, however, comparative designs were simply described and rarely justified; there was little explicit concern with managing the details of experimental settings. When ancient and medieval authors noted the drug test on two birds, they surely meant to show a test to support the drug's efficacy, but the argument for the comparative approach often remained implicit. In modern scientific writing, by contrast, we sometimes find detailed discussions and justifications of experimental designs—in controversies about experimental results, in debates about the status of heterodox scientific fields such as research on psychic phenomena, and in situations of uncertainty.

In this volume, Tawrin Baker's chapter on Scheiner and Christopoulou and Arabatzis's chapter on Rayleigh epitomize both the scarcity and the abundance of practitioners' discourse on their control practices and strategies. Scheiner demonstrated to his readers how experimentation could serve as a legitimate check on a theory of vision. He did not expound or defend methodological ideas in detail, although he did focus attention on the process of experimentation. Words and

---

[24] In his well-known book on Pasteur, Gerald Geison drew on Pasteur's experiments with infusions to show that the negotiations of what does and does not count as a properly controlled experiment in the spontaneous generation debates turned into battles motivated by political and religious concerns. Geison argues that Pasteur effectively "controlled" his audiences (Geison 1995).

[25] Bertoloni Meli (2009) describes many other comparative experiments from the early modern period. See also Schickore (2021).

pictures conveyed the experimental setups. Scheiner instructed his readers to make certain experiences and experiments; he discussed the implications for the theory of vision. However, as Baker notes, several issues remained open, such as how often an experiment should be repeated or how one ought to deal with discrepancies. Christopoulou and Arabatzis's chapter on Rayleigh shows that late-nineteenth-century scientists wrote not only about the details of their experiments but also about experimental control. Experimenters drew attention to how they had re-designed instruments to make their measurements more precise and how they had employed additional instruments to check the quality of their measurements. They often insisted on using two measurement methods to guard against error.

We still know little about the unfolding of methodological discussions in the centuries after Scheiner's appeal to a variety of experiences and experiments and Boyle's musings on unwieldy, "uncontrolled" experimental settings and about the practices appropriate for managing and extracting knowledge from these settings. Little is known about the emergence of explicit methodologies for comparative trials. According to some scholars, notably Edwin Boring, it was not until the mid-nineteenth century that we find such explicit methodologies. Boring associated the first methodology of comparative experimental designs with a philosophical text, John Stuart Mill's *System of Logic* (Boring 1954). While the contributions to our volume do not tell a comprehensive history of methodological accounts on experimental control, they do suggest that it would be misleading to identify Mill as the sole originator and principal representative of these accounts.[26] As Schürch's, Coko's and Nickelsen's chapters demonstrate, Mill was one of several early-nineteenth-century commentators on science who urged investigators to keep background conditions constant across trials, to "analyze" the background into different experimental conditions, and to compare the effects of interventions in one setting to another setting left untouched. But a broader history of these developments would still be desirable.

Our volume also shows that reflections about and justifications of control strategies predate modern philosophies of science. From Schürch's study of late-eighteenth-century plant physiology we learn that, prior to Mill, practitioners not only called for rigorous and properly managed interventions, but also did much more: they reflected on control practices as validation procedures and debated their relative merits, practicality, and limitations. They observed that, to be instructive, comparisons must be made on sufficiently similar experimental subjects in similar situations. At times they disagreed about whether they or their colleagues had done enough to control their experiments. They criticized each other for not making comparative trials, for not controlling the right thing, or for not repeating a trial often enough.

The content of these debates and reflections tells us something about the experimenters' own understanding of methodological issues concerning control, rigor,

---

[26] Our volume focuses on practitioners' methodological accounts. However, even in philosophy of science, Mill had predecessors in this regard: Dugald Stewart and John Herschel, for instance, cover territory very similar to Mill's four methods of experimental inquiry.

reliability, certainty, and failure in experimentation. Christopoulou and Arabatzis's and Coko's chapters illustrate this. As many contributors show, satisfactory control of an experiment is, in the end, an intersubjective, iterative achievement. Schürch and Christopoulou and Arabatzis note that experimenters such as Ingen-Housz and Rayleigh call upon others to check the results they themselves had obtained and to contribute additional experiments.[27] Cristalli charts the decades-long negotiations and re-negotiations among physicists, chemists, and psychologists on experimental practices deemed adequate to study psychic phenomena. The experimenters understood that their projects' success depended on "controlling" their interlocutors as well.[28]

This volume does not aim to replace earlier systematic discussions in history and philosophy of science on these issues, such as those on epistemological strategies of experimentation (Allan Franklin), tests for error (Deborah Mayo), representing and intervening (Ian Hacking), and how experiments end (Peter Galison). Our volume complements them. In fact, our discussions overlap with these approaches as we trace the history of controls while keeping epistemological strategies of experimentation in mind. We do contend that re-directing attention to control practices, control strategies, and practitioners' accounts thereof illuminates new aspects of the history of experimental practices.

Control strategies and practices can be viewed as long-term and short-term methodological commitments, along the lines suggested by Peter Galison (1987). Arnet's contribution to this volume uses this approach. Material and conceptual organizations of experiments vary, as do the identification of target systems, conditions, and confounders. The tools for stabilizing them change as well and are often (but by no means always!) local, context-specific, and relatively short-lived. Modern technologies allow for creative and sometimes intricate solutions to the problems of stabilization, standardization, and tracking. Yet the strategies have long been in place.

Control strategies are persistent. Even in the most complicated settings and with the most elusive phenomena, experimenters try to implement established control strategies as best they can, as shown in Schürch's study of plant electrification, Coko's discussion of experiments on Brownian movement, Cristalli's study of psychic experiments, Kursell's work on elusive auditory judgments, and Nickelsen's discussion of plant physiology. Experimenters look for experimental conditions and confounding factors; they vary them to weigh their influence on experimental processes; they probe for error (Mayo 1996); they make their interventions less "fat-handed" (Woodward 2008); they compare situations meant to be similar and assess robustness, presupposing the no-miracle argument (Hacking 1985). At the same time, they develop specific, contextual implementations for these strategies, and they do not always agree on whether a particular implementation is effective.

---

[27] For another example of appeals to the community in the struggle to identify the causes of blue milk, see Schickore (2023, 37).

[28] See Schürch's discussion of Ingen-Housz in this volume, for example.

In doing all this, experimenters face both technical and conceptual challenges. It may take a long time to harness experimental conditions, identify potential confounders, and find suitable techniques for doing so. Solutions to control problems will typically remain less than ideal. Hoffmann's contribution demonstrates this fragility in control procedures. In debates about spontaneous generation, it took centuries to refine the tools to prevent contaminations from reaching the materials under investigation, and every new tool generated new issues for further exploration. Along the way, the understanding changed regarding the causes, conditions, and potential modifying factors and confounders. New technical challenges arose as a result.

Several chapters show that the implementation of control strategies may *generate* entirely new technical and conceptual problems for the experimenter, or even produce "surplus findings," as Kursell writes.[29] Nickelsen, for instance, tracks changes in both the conceptualization and the logistics of managing background conditions for experiments on the influence of light on plant growth. Christopoulou and Arabatzis suggest that disturbances in physics experiments could become research topics in their own right. Arnet's work also brings into relief the problematic implications of an over-emphasis on rigor and control. Early mazes were designed as simple systems of tracks in order to minimize environmental cues. But for a more complete understanding of animal learning, later researchers re-introduced precisely those same environmental features. The early mazes embodied a regime of control that stripped animals of certain sensory and environmental cues. Those mazes, however, excluded exactly those features that later researchers thought essential to advanced rodent learning.[30]

Finally, several chapters suggest that it is fruitful to think of experiments as "controls of inferences," because this perspective also brings out relevant methodological issues and their historical development. As Baker demonstrates, for early modern experimenters coming to grips with their Aristotelian heritage, the role of experiments in scientific inquiry was a crucial issue. In hindsight, studying how they managed this issue can also tell us something about Aristotle's own ideas on the role of experimentation in empirical inquiry. For eighteenth- and nineteenth-century inquirers, then, the question is not so much *whether* but *how,* exactly, experimentation and experimentally generated knowledge can help us to understand nature. Steinle, Coko, Nickelsen, Kursell, and Hoffmann show how intricate the question can be as experiments target unobservable phenomena. As these experiments involve increasingly complicated instruments, hypotheses, assumptions, chains of inferences, and interpretations, the challenges for experimenters increase accordingly.

---

[29] For another example of how control practices themselves become the object of study, see Landecker (2016).

[30] Researchers today have identified other areas of concern for over-emphasizing rigor and control. One example is over-standardized mice (Engber 2013), and these studies highlight the importance of balancing control with other demands on research design. In public health studies, researchers must overcome barriers for recruitment, attrition, and sample size, which may necessitate lowering the bar for rigor to gather any valuable information at all (Crosby et al. 2010). Thus, the implication of an over-emphasis on rigor may be epistemic, socio-political, or both.

We place practitioners' methodologies, experimental designs, strategies of inquiry, and practices of implementation in the center of our analyses. We thereby draw new trajectories and connections in the history of experimental inquiry. We identify lines of experimentation that sometimes turned into models of rigorous experimental design while other times being criticized. Bacon's steeping experiments with plant seeds, as analyzed by Pastorino, exemplify a specific kind of comparative experimentation. It would be applied again and again throughout the eighteenth century, not just in plant science but also in other scientific fields. Pythagoras' hammer experiments too were repeated, at least repeatedly reported, by several scholars prior to Galileo and Mersenne. In this case, the design was not a model but a point of critique for later scholars.

Our studies on control practices and on their discussion and justification have revealed other lineages and cross-fertilizations—among physics and psychology, physiology, botany and ethology, chemistry, medicine, agriculture, and philosophy. Control practices and strategies are contextual, in that the context determines what is controlled and how to achieve control. But control strategies and at times even control practices are not discipline-specific. The same strategies travel across disciplines, from physics to medicine and physiology to chemistry and back again. Several chapters suggest that the same methodological ideas and control strategies are advocated across national boundaries (see especially Schürch and Coko). Control strategies such as comparative designs and multiple repetitions are relatively stable across historical periods. But they may be justified in different ways at different times and may cease to be justified at all.

With our work, we hope to stimulate broader discussions about the longer-term history of rigorous experimentation: what are the strategies involved in it? And how do debates concerning well-designed experiments unfold in different fields and periods? By our effort we seek to clarify the roles of experimental strategies and methodologies as driving forces for scientific change, and as tools for determining what it means to do—or not to do—good science.

***

This volume (and its companion, a collection of essays on analysis and synthesis) originated in a Sawyer Seminar at Indiana University Bloomington titled "Rigor: Control, Analysis and Synthesis in Historical and Systematic Perspectives," which was funded by the Andrew W. Mellon Foundation. Mellon Sawyer Seminars are temporary research centers, gathering together faculty, postdoctoral fellows, and graduate students for in-depth study of a scholarly subject in reading groups, seminars, and workshops. As part of our activities, we organized two international conferences. They brought together scholars in history, philosophy, and social studies of science who examine historical and contemporary dimensions of rigor in experimental practice. The contributors to this volume participated in the second of the Sawyer conferences (March 2022) and reconvened a few months later for an authors' workshop, at which the draft chapters for this volume were intensely discussed.

Several institutions and individuals helped to make our work possible. We gratefully acknowledge the Mellon Foundation's generous financial support, and especially the Foundation's flexibility as we dealt with the challenges of pursuing collaborative scholarship during a pandemic. We are grateful to Director of Foundation Relations Cory Rutz at Indiana University's Office of the Vice President for Research, for his prompt and efficient assistance in administering the grant. The authors' workshop took place at the IU Europe Gateway (Berlin) and was funded by a combined grant from the IU College of Arts and Sciences and the College Arts and Humanities Institute. We very much appreciate this support. We are indebted to Jed Buchwald for including our work in the *Archimedes* series, and to Chris Wilby for his efforts in moving the publication along. A big thank you to our department manager Dana Berg (Department of History and Philosophy of Science and Medicine at IU), office assistant Maggie Herms (IU HPSC), and Andrea Adam Moore (IU Europe Gateway), all of whom helped to organize our conferences and workshops. Finally, we warmly thank the many participants at the two conferences and at the various other Sawyer events for their valuable input, comments, questions, and critique.

References

Amici, Raffaele Roncalli. 2001. The History of Italian Parasitology. *Veterinary Parasitology* 98: 3–30.

Bayer, Greg. 1997. Coming to Know Principles in Posterior Analytics II 19. *Apeiron* 30: 109–142.

Beniger, James. 1986. *The Control Revolution. Technology and Economic Origins of the Information Society*. Cambridge, MA: Harvard University Press.

Bennett, S. 1993. *A History of Control Engineering*. Wiltshire: Redwood Books.

Bertoloni Meli, Domenico. 2009. A Lofty Mountain, Putrefying Flesh, Styptic Water, and Germinating Seeds. In *The Accademia del Cimento and its European Context*, ed. Marco Beretta, Antonio Clericuzio, and Larry Principe, 121–134. Sagamore Beach: Science History Publications.

Boring, Edwin Garrigues. 1954. The Nature and History of Experimental Control. *American Journal of Psychology* 67: 573–589.

Boyle, Robert. 1999a. The First Essay, of the Unsuccessfulness of Experiments. In *The Works of Robert Boyle*, ed. Michael Hunter and Edward B. Davies, 37–56. London: Pickering & Chatto. Original edition, 1661.

———. 1999b. The Second Essay, of Unsucceeding Experiments. In *The Works of Robert Boyle*, ed. Michael Hunter and Edward B. Davies, 57–82. London: Pickering & Chatto. Original edition, 1661.

Cartwright, Nancy. 2007. Are RCTs the Gold Standard? *BioSocieties* 2: 11–20.

Crosby, Richard, Laura F. Salazar, Ralph DiClemente, and Delia Lang. 2010. Balancing Rigor Against the Inherent Limitations of Investigating Hard-to-Reach Populations. *Health Education Research* 25: 1–5.

Dear, Peter. 1991. Narratives, Anecdotes, and Experiments: Turning Experience into Science in the Seventeenth Century. In *The Literary Structure of Scientific Argument: Historical Studies*, ed. Peter Dear, 135–163. Philadelphia: University of Pennsylvania Press.

Dehue, Trudy. 2005. History of the Control Group. In *Encyclopedia of Statistics in Behavioral Science*, ed. Brian S. Everitt and David C. Howell, 829–836. Chichester: Wiley.

Derksen, Maarten. 2017. *Histories of Human Engineering: Tact and Technology*. Cambridge: Cambridge University Press.

Desjardins, Eric, Derek Oswick, and Craig W. Fox. 2023. On the Ambivalence of Control in Experimental Investigation of Historically Contingent Processes. *Journal of the Philosophy of History* 17 (1): 130–153.

Dierig, Sven. 2006. *Wissenschaft in der Maschinenstadt. Emil Du Bois-Reymond und seine Laboratorien in Berlin*. Berlin: Wallstein Verlag.

Engber, Daniel. 2013. The Mouse Trap. How one Rodent Rules the Lab. Slate. http://www.slate.com/articles/health_and_science/the_mouse_trap/2011/11/the_mouse_trap.html.

Foucault, Michel. 1975. *The Birth of the Clinic: An Archaeology of Medical Perception*. New York: Vintage Books.

———. 1979. *Discipline and Punish: The Birth of the Prison*. New York: Vintage.

Frängsmyr, Tore, J.L. Heilbron, and Robin E. Rider. 1990. *The Quantifying Spirit in the Eighteenth Century*. Berkeley, CA: University of California Press.

Franklin, Seb. 2015. *Control: Digitality as Cultural Logic*. Cambridge, MA: MIT Press.

Galen. 1944. *Galen on Medical Experience*. First edition of the Arabic ed. New York [etc.]: Pub. for the trustees of the late Sir Henry Wellcome by the Oxford University Press.

Galison, Peter. 1987. *How Experiments End*. Chicago, IL: University of Chicago Press.

———. 2010. Secrecy in Three Acts. *Social Research: An International Quarterly* 77: 941–974.

Geison, Gerald. 1995. *The Private Science of Louis Pasteur*. Princeton, NJ: Princeton University Press.

Gigerenzer, Gerd, Zeno G. Swijtink, Theodore M. Porter, Lorraine Daston, John Beatty, and Lorenz Krüger. 1989. *The Empire of Chance: How Probability Changed Science and Everyday Life*. Cambridge: Cambridge University Press.

Guala, Francesco. 2005. *The Methodology of Experimental Economics*. Cambridge/New York: Cambridge University Press.

Guettinger, Stephan. 2019. A New Account of Replication in the Experimental Life Sciences. *Philosophy of Science* 86: 453–471.

Hacking, Ian. 1984. *The Emergence of Probability: A Philosophical Study of Early Ideas about Probability, Induction and Statistical Inference*. Cambridge: Cambridge University Press.

———. 1985. Do We See Through a Microscope? In *Images of Science*, ed. P.M. Churchland and C.A. Hooker, 132–152. Chicago and London.

———. 1990. *The Taming of Chance*. Cambridge: Cambridge University Press.

Hoffmann, Christoph. 2001. The Design of Disturbance: Physics Institutes and Physics Research in Germany, 1870–1910. *Perspectives on Science* 9: 173–195.

Holbrook, Daniel. 2009. Controlling Contamination: the Origins of Clean Room Technology. *History and Technology* 25 (3): 173–191. https://doi.org/10.1080/07341510903083203.

Holman, Bennett. 2020. Humbug, the Council of Pharmacy and Chemistry, and the Origin of "The Blind Test" of Therapeutic Efficacy. In *Uncertainty in Pharmacology: Epistemology, Methods and Decisions*, ed. Barbara Osimani and A. Lacaze, 397–416. Dordrecht: Springer.

Hughes, Thomas Parke. 1983. *Networks of Power: Electrification in Western Society, 1880–1930*. Baltimore, MD: Johns Hopkins University Press.

Kaptchuk, Ted. 1998. Intentional Ignorance: A History of Blind Assessment and Placebo Controls in Medicine. *Bulletin of the History of Medicine* 72: 389–433.

Keating, Peter, and Alberto Cambrosio. 2012. *Cancer on Trial*. Chicago, IL: University of Chicago Press.

Kuch, Declan, M. Kearnes, and K. Gulson. 2020. The Promise of Precision: Datafication in Medicine, Agriculture and Education. *Policy Studies* 41 (5): 527–546. https://doi.org/10.1080/01442872.2020.1724384.

Kupreeva, Inna. 2022. Galen's Empiricist Background: A Study of the Argument in On Medical Experience. In *Galen's Epistemology: Experience, Reason, and Method in Ancient Medicine*, ed. Matyáš Havrda and R.J. Hankinson, 32–78. Cambridge: Cambridge University Press.

Landecker, Hannah. 2016. It Is What It Eats: Chemically Defined Media and the History of Surrounds. *Studies in History and Philosophy of Biological and Biomedical Sciences* 57: 148–160.

Leigh, Robert Adam. 2013. *On Theriac to Piso, Attributed to Galen. A critical edition with translation and commentary*. Exeter: Department of Classics, University of Exeter.

Levin, Miriam. 2000a. Contexts of Control. In *Cultures of Control*, ed. Miriam Levin, 13–39. Amsterdam: Harwood Academic Publishers.

―――, ed. 2000b. *Cultures of Control*. Amsterdam: Harwood Academic Publishers.

Marks, Harry M. 1997. *The Progress of Experiment. Science and Therapeutic Reform in the United States, 1900–1990*. Cambridge: Cambridge University Press.

Mayo, Deborah G. 1996. *Error and the Growth of Experimental Knowledge*. Chicago: University of Chicago Press.

McCartney, Eugene S. 1942. A Control Experiment in Antiquity. *The Classical Weekly* 36: 5–6.

McVaugh, Michael. 2009. The 'Experience-Based' Medicine of the Thirteenth Century. In *Evidence and Interpretation in Studies on Early Science and Medicine*, ed. Edith Sylla and William R. Newman, 105–130. Leiden: Brill.

Nasser, M., A. Tibi, and E. Savage-Smith. 2009. Ibn Sīnā's Canon of Medicine: 11th Century Rules for Assessing the Effects of Drugs. *Journal of the Royal Society of Medicine* 102: 78–80.

Needham, John Turbervill. 1748. A Summary of Some Late Observations upon the Generation, Composition, and Decomposition of Animal and Vegetable Substances; Communicated in a Letter to Martin Folkes, Esq.; President of the Royal Society, by Mr. Tubervill Needham, Fellow of the Same Society. *Philosophical Transactions of the Royal Society of London* 45: 615–666.

Parke, Emily C. 2014. Flies from Meat and Wasps from Trees: Reevaluating Francesco Redi's Spontaneous Generation Experiments. *Studies in History and Philosophy of Biological and Biomedical Sciences* 45: 34–42.

Pastorino, Cesare. 2022. Francis Bacon's Controlled Experiments on Seed Steeping and Germination: Their Context, Circulation and Methodological Significance. Paper presented at the conference *Control Practices in Historical and Systematic Perspectives*, Indiana University Bloomington, March 2022.

Paul, Diane B. 1995. *Controlling Human Heredity*. Atlantic Highlands, NJ: Humanities Press International.

Pauly, Philip J. 1987. *Controlling Life. Jacques Loeb & the Engineering Ideal in Biology*. Oxford: Oxford University Press.

Popper, Karl. 2002. *The Logic of Scientific Discovery*. London and New York: Routledge.

Porter, Theodore M. 1995. *Trust in Numbers. The Pursuit of Objectivity in Science and Public Life*. Princeton, NJ: Princeton University Press.

Rabinbach, Anson. 1990. *The Human Motor: Energy, Fatigue, and the Origins of Modernity*. Berkeley, CA: University of California Press.

Rader, Karen. 2004. *Making Mice: Standardizing Animals for American Biomedical Research*. Princeton, NJ: Princeton University Press.

Redi, Francesco. 1909. *Experiments on the Generation of Insects*, 1688. Chicago, IL: Open Court.

Schickore, Jutta. 2017. *About Method: Experimenters, Snake Venom, and the History of Writing Scientifically*. Chicago, IL/London: University of Chicago Press.

―――. 2021. Methodological Ideas in Past Experimental Inquiry: Rigor Checks Around 1800. *Intellectual History Review*. https://doi.org/10.1080/17496977.2021.1974225.

―――. 2023. Peculiar Blue Spots: Evidence and Causes around 1800. In *Evidence: The Use and Misuse of Data*, ed. The American Philosophical Society, 31–55. Philadelphia, PA: American Philosophical Society.

Schlager, Neil. 1995. *Breakdown: Deadly Technological Disasters*. Dretroit, MI: Visible Ink Press.

Seppel, Marten, and Keith Tribe. 2017. *Cameralism in Practice: State Administration and Economy in Early Modern Europe*. Rochester, NY: Boydell & Brewer.

Shapin, Steven, and Simon Schaffer. 1985. *Leviathan and the Air-Pump. Hobbes, Boyle, and the Experimental Life*. Princeton, NJ: Princeton University Press.

Strick, James. 2000. *Sparks of Life. Darwinism and the Victorian Debates over Spontaneous Generation*. Cambridge, MA: Harvard University Press.

Strong, I.I.I., and C. Frederick. 1999. The History of the Double Blind Test and the Placebo. *Journal of Pharmacy and Pharmacology* 51: 237–238.

Sullivan, Jacqueline A. 2022. Novel Tool Development and the Dynamics of Control: The Rodent Touchscreen Operant Chamber as a Case Study. *Philosophy of Science* 89 (5): 1–19.

Treece, James W. 1990. Daniel and the Classic Experimental Design. Accessed January 29, 2023. https://www.icr.org/article/daniel-classic-experimental-design.

Tuchman, Arleen M. 1993. *Science, Medicine, and the State in Germany. The Case of Baden, 1815-1871.* New York: Oxford University Press.

Wiener, Norbert. 1948. *Cybernetics, or, Control and communication in the animal and the machine.* New York/Paris: J. Wiley, Hermann et Cie.

Wimsatt, William C. 2007. *Re-Engineering Philosophy for Limited Beings.* Cambridge, MA: Harvard University Press.

Wise, M. Norton, ed. 1995. *The Values of Precision.* Princeton, NJ: Princeton University Press.

Woodward, James. 2008. Cause and Explanation in Psychiatry: An Interventionist Perspective. In *Philosophical Issues in Psychiatry: Explanation, Phenomenology and Nosology*, ed. K. Kendler and J. Parnas, 132–184. Baltimore, MD: Johns Hopkins University Press.

Worrall, John. 2007. Why There's No Cause to Randomize. *British Journal for the Philosophy of Science* 58: 451–488.

Zhmud, Leonid. 2012. *Pythagoras and the Early Pythagoreans.* Oxford: Oxford University Press.

Jutta Schickore is Ruth N. Halls Professor of History and Philosophy of Science and Medicine at the Department of History and Philosophy of Science and Medicine, Indiana University (Bloomington). Her research interests include philosophical and scientific debates about scientific methods in past and present, particularly debates about (non)replicability, error, and negative results.

Chapter 2
Christoph Scheiner's *The Eye, that is, The Foundation of Optics* (1619): The Role of Contrived Experience at the Intersection of Psychology and Mathematics

Tawrin Baker

2.1 Introduction

This chapter examines the Jesuit polymath Christoph Scheiner's (1573–1650) 1619 work, *Oculus hoc est: fundamentum opticum*, or *The Eye, that is, the Foundation of Optics* (hereafter *Oculus*). I consider two broad issues from the history and philosophy of science.[1] The first has to do with the problem of establishing first principles in natural sciences based on experience and experiment. Early accounts in the history and philosophy of science, attempting to understand when and how modern experimental science arose, took the sixteenth and seventeenth centuries as a turning point: Aristotelian natural science, while at least nominally based on sense perception, supposedly neglected, discouraged, or was outright hostile to experimental investigation. The new science, developed by figures such as Francis Bacon, Galileo Galilei, and others, was said to be responsible for reforming natural philosophy and placing natural science on firm experimental foundations, and it did so by rejecting Aristotelianism. Although it is not false in every respect, this picture has proven inadequate, and recent attempts to understand Aristotelian contributions to seventeenth-century developments in natural science have helped to remedy this oversimplified account. More yet needs to be done, however.

The Jesuits have been of particular interest in this regard given their pedagogical influence in the seventeenth century, along with the sheer number of treatises they produced. Nevertheless, lingering assumptions about the supposed anti-experimentalism embedded in Aristotle's works—or Aristotelianism, however

---

[1] I would like to thank Jutta Schickore, William Newman, Julia Kursell, Cesare Pastorino, as well as the rest of the contributors to this volume for their helpful comments, corrections, and suggestions.

T. Baker (✉)
South Bend, IN, USA

© The Author(s) 2024
J. Schickore, W. R. Newman (eds.), *Elusive Phenomena, Unwieldy Things*,
Archimedes 71. https://doi.org/10.1007/978-3-031-52954-2_2

understood—have colored these accounts. Looking at Scheiner's *Oculus*, I argue that the concern with the problem of establishing the foundations of natural science on the basis of experiments and contrived experiences was not confined to the *novatores* such as Galileo; it appears to have been a general project. Furthermore, I argue that Aristotelianism *per se* presented no in-principle obstacles to the use of contrived experiences and experiments in establishing the first principles or axioms of a science. Finally, I suggest that Paduan Aristotelianism was an important influence on Scheiner. Although Ernst Cassirer (1906, 139), John Herman Randall (1940), and others following them have argued, controversially, that the Paduan professor in logic and natural philosophy Jacopo Zabarella was a key figure in the development of the "scientific method," I propose that Scheiner's use of experience and experiment to establish first principles in the science of optics was inspired more by anatomists such as Hieronymus Fabricius ab Aquapendente.

The second issue I touch on has to do with the problem of how to incorporate, into natural science, contrived first-person experiences and experiments, in particular judgments about visual phenomena such as color, distance, number, shape, and so on. Much of the history and philosophy of science, particularly for the early modern period, has focused on how experiments were used as evidence for or against theories together with the problem of establishing public facts via observation and experimentation. Such scholarship centers on the development of the concept of natural law and has made connections between concepts of public evidence in legal contexts and experimental evidence in scientific communities. In contrast, what we might call "self-perceptual" experiments do not fit this model of public evidence. These kinds of experiments or contrived experiences belong more to the history and philosophy of psychology than to physics. Indeed, from antiquity through the early modern period the discipline of optics was a demonstrative science that drew its principles in part from the science of the soul (what today we might classify as psychology), which was itself understood as a branch of, or topic within, natural philosophy. (As a so-called middle science, optics also drew its principles from geometry.) Prior to and largely during the seventeenth century, optics was primarily the science of seeing, not the science of light (Smith 2015). Examining Scheiner's *Oculus* with this background in mind thus contributes to questions of experimental rigor and control in the history of psychology, the senses, and the sensibles (e.g., light, color, and distance perception) in the *longue durée*. Scheiner's *Oculus* and other contemporaneous works, such as François de Aguilon's 1613 *Opticorum libri sex*, are part of a history that begins in antiquity and connects to key works in physiological optics by Hermann von Helmholtz and Ewald Hering in the nineteenth century.

To address both of these issues—how to establish first principles on the basis of experience and experiment, and how to deal with contrived first-person experiments—it is necessary to address anachronistic readings of Aristotle's *Posterior Analytics* and their influence on the historiography of the Scientific Revolution. The subject of the *Posterior Analytics* is demonstration, or "a deduction that produces knowledge." This work was carefully studied in Aristotelian (including Jesuit) education of the early modern era. It ends with a notoriously brief chapter (II.19) that offers an account of how the first principles of a demonstrative science are grasped. I follow several recent scholars who have argued that we should not understand the *Posterior Analytics*—particularly II.19—as a treatise on epistemology, i.e., an account of how to justify knowledge claims. Rather, II.19 is better understood as a general psychological *description*, beginning with sense perception and ending with the comprehension of universal first principles. Because each science had distinct methods for arriving at first principles, an account of how to justify the foundations of that science, according to Aristotle, belonged at the beginning of such treatises and not in a very general account such as the one given in the *Posterior Analytics*.

I argue that this reading is closer to the interpretation of the *Posterior Analytics* in the early modern period as well, particularly in light of the so-called "Aristotle Project" in late sixteenth-century Padua that revived Aristotle's science of animals and the animal soul. This view has major ramifications for understanding how authors, such as Scheiner, used experience and experiment in the process of grasping first principles. That process for Scheiner is not a matter of stripping away the particularities of any individual sense experience in order to arrive at the universal core of a sensation (i.e., it does not directly invoke a realist position with regard to universals); nor is it a matter of accessing *a priori* universal first principles in a Neoplatonic fashion. Finally, it is not a matter of invoking universal "common sense" observations as a starting point. Rather, Scheiner is influenced by the anatomical tradition according to which the reliability of one's determination of the properties, activities, and purposes of the universal anatomical part arises from the combination of skill at dissection, long experience, and deft engagement with the accounts of anatomical authorities, both ancient and contemporary. It involves being able to demonstrate one's anatomical findings, ideally in person, and to argue convincingly from those observations.

In short, in their attempts to understand how experience and experiment were incorporated into the sciences in the myriad ancient, medieval, and renaissance Aristotelianisms, historians and philosophers have looked in the wrong places. Many supposed problems that experiment raised for Aristotelian science simply disappear if we do not assume that historical actors took II.19 in Aristotle's *Posterior Analytics* as a work of epistemology; examining specific scientific treatises to assess how Aristotle and Aristotelians justified the foundations of their sciences is the better route. We can read Scheiner's *Oculus*, then, as a treatise on how, in a (broadly) Aristotelian treatment of vision science, one starts with sense perception, forms memories, gathers these memories to form experiences, and finally how from that state of *being experienced* or *having experience* one securely grasps first principles. This is precisely the title of his work: *The Eye, that is, the Foundation of Optics*. I show here that Scheiner understands the foundation for the science of optics to consist of dissecting the eye, performing experiments and generating contrived experiences to understand the actions of the eye (primarily sight, a complex action), and appealing to particular (and even unique or singular) observations and experiences of others. That is, early in the process of developing Aristotelian experience, Scheiner requires the reader to perform contrived experiences and experiments to become truly experienced, and being experienced includes having the written records of an exhaustive body of experience to hand. Only after all of this has been done can one combine these experiences with geometrical knowledge in order to grasp the first principles or axioms of optics. Notably, although he accepts Kepler's revolution in optics, which moved the site of sense perception from the lens to the retina, Scheiner retains the traditional axioms of optics. As Scheiner (1619, 124) states at the end of Book III part 1, and shows in Book III part 2, "All the axioms reported by Euclid's still hold in the strictest rigor."

The plan for this chapter is as follows. I first briefly review the historiography of early modern experience and experiment. Because he has addressed these issues in the most detail, is considered an authority on these matters, and is one of few scholars to have analyzed Scheiner's *Oculus* at length, I focus on Peter Dear's work. In this section, I also show the influence of Paduan anatomy on Scheiner. I next discuss Aristotle's *Posterior Analytics* and the issue of first principles in pre-modern and early modern optics, followed by a brief introduction to the science of optics at the beginning of the seventeenth century. Following this, I analyze the use of contrived experiences and experiments in Scheiner's *Oculus*.

2.2   Historiography

To begin, however, a short biography of Scheiner is needed; for more detail, see (Shea 2008) and (Daxecker 2004). Scheiner was born in Wald, a small village in Bavaria that was active in the Counter-Reformation, including a strong Jesuit presence. He was educated by Jesuits and joined the Order in 1595, studying mathematics and philosophy in Ingolstadt from 1600. He later also studied theology there, beginning in 1610. He is best known today for his observations of sunspots in 1611, and for his priority disputes with Galileo over their discovery as well as debates over their nature. He published many other works on mathematics and was an accomplished inventor as well, writing treatises on instruments he created including the pantograph. Another notable publication is his later *Rosa ursina sive sol*, which includes a detailed account of his sunspot observations, comparisons between the human eye and a telescope, a description of his helioscope, and many other astronomical observations and arguments. In this chapter, however, we are mainly concerned with his *Oculus*, which was first published in 1619 in Innsbruck, and reprinted in 1621 in Freiburg im Breisgau and in 1652 in London.

Questions surrounding experience and experiment were the bread and butter of the historiography of the Scientific Revolution as the disciplines of the history and philosophy of science were being formed in the first two-thirds of the twentieth century. Some specific narratives about experience and experiment—particularly the failure of Aristotelian science to use experiments to either ground or test claims—became commonplace in mid-twentieth century accounts. I pulled at random an early Scientific Revolution textbook from my shelf as an example; thus Charles Gillispie (1960, 12–13):

[Aristotelian science] started from experience apprehended by common sense, and moved through definition, classification, and deduction to logical demonstration. Its instrument was the syllogism rather than the experiment or the equation. […] For however congenial Aristotelian physics was to the self-knowledge of the minds that elaborated it, nature is not like that, not an enlargement of common sense arrangements, not an extension of consciousness and human purposes.

Much has been done since to challenge such blanket statements (Newman 2006; Ragland 2017), but Peter Dear's work is arguably the most influential attempt to understand the Jesuit Aristotelian approach to experience and experiment in the seventeenth century. His writings remain authoritative in some respects, and he is one of few to have analyzed Scheiner's *Oculus* in detail. I will therefore quote him at length. Dear (2006, 109) writes:

For Aristotle, a science of the physical world should, ideally, take the form of a logical deductive structure derived from incontestable basic statements or premises. The model for this was the structure of classical Greek geometry as exemplified in Euclid's *Elements*. […] In the case of sciences that concerned the natural world, however, such axioms could not be known by simple introspection. In those cases, the axioms had to be rooted in familiar and commonly accepted experience.

For Aristotelians, it seems, it is only communally agreed-upon experiences that can serve as the basis for a deductive science (Dear, 109, emphasis in original):

This kind of experience, therefore, was of universal behaviors rather than particulars: The sun *always* rises in the east; acorns *always* (barring accidents) grow into oak trees. Singular experiences … were more problematic because they could only subsequently be known by historical report, as something that had happened on a particular occasion. They were thus unfit to act as scientific axioms because they could not receive immediate free assent from all: Most people had not witnessed them.

He continues (110):

Aristotle's natural philosophy was especially concerned with 'final causes,' the purposes or ends toward which processes tended or that explained the conformation and capacities of something. […] Active interference, by setting up artificial conditions, would risk subverting the natural course of things, hence yielding misleading results: experimentation would be just such interference. […] To the extent that Aristotle's natural philosophy sought the final causes of things, and thereby to determine their natures, experimental science was therefore disallowed.

Several clear counterexamples to this last statement have come to light since 2006. Examples are the physician and chymist Daniel Sennert (Newman 2006, 86–125) and the physician and anatomist William Harvey (Lennox 2006, 5–26; Distelzweig 2013, 151–69; Goldberg 2016; Lennox 2017, 151–68).[2] Both were self-professed followers of Aristotle, something we see both in the intellectual content of their writings, and also, most importantly, in their methods for investigating nature (Klein 2014, 136–37; Ekholm 2011, 45–46). These included performing experiments as self-professed Aristotelians to discover material, efficient, formal and—especially

---

[2] On the early modern term chymistry, distinct from both medieval alchemy and modern chemistry, see Newman and Principe (1998).

in anatomy—final causes. Based on these figures alone, we see that the position held by Dear and others, namely that experience and experiment were somehow at odds with the search for Aristotelian final causes, is false, particularly for the science of animals.

Dear offers a complicated story of a difficult transformation within Aristotelianism, spearheaded by Scheiner and other Jesuits, to accommodate Aristotle's writings to experimentation and mathematical physics. In a seminal article on the Jesuit contribution to the role of experience and experiment in seventeenth-century investigations of nature, he describes a philosophical-methodological puzzle faced by the Jesuits as they attempted to incorporate experimentation and mathematics into natural philosophy (1987, 160):

> The employment of constructed experiences in the mathematical sciences threatened to violate not only the requirement that scientific premises be evident, but also the strict artificial/natural distinction at the heart of the Aristotelian world-view.

While Dear has perhaps tempered his opinion on this issue since, it is still clearly present in his writings 20 years later. Moreover, he expresses an attitude common since the mid-twentieth century at least. Even if it is no longer taken for granted by many early modern scholars, the narrative persists among non-specialists as a key component of the scientific revolution.[3]

But what, for Aristotelians, was the artificial/natural distinction? It was simply to identify whether the cause of some change—local motion, alteration, or substantial transformation—can be found in the innate capacities, or the nature, of the substance in question, or whether one ought to look outside that substance for a cause. Aristotle in the *Generation of Animals* (735a2–4) writes: "For the art is the starting-point and form of the product; only it exists in something else, whereas the movement of nature exists in the product itself, issuing from another nature which has the form in actuality."[4] Aristotle's example in the *Physics* is that of a wooden bed: if we plant a wooden bed frame and it starts to sprout on its own, it will grow into more wood, i.e., a tree, not into more beds (193b7–193b12). The wood has the nature of a tree, and if moved by that nature it changes insofar as it has the innate capacities of growth, nutrition, and reproduction, all of which are guided by the *telos* of becoming a flourishing tree. But the wood is made into a bed, with the ability to promote good sleep, by art—that is, by an external force moving it to ends external to the tree or wood itself. In another example used by Aristotle, if we suppose (impossibly) that an axe was a natural substance with a soul, then it would be able to achieve its final cause—chopping wood—owing to its own internal nature, rather than, as a product of art, only via an external mover. Natures are principles of motion

---

[3] The literature on this topic is extensive, and I do not delve too deeply into it here. For a sustained argument against the traditional account of the art/nature distinction, see Newman (2005), especially chapter 5. On the epistemology of early modern meteorology, to which the traditional account of Aristotle's art/nature distinction cannot do justice, see Martin (2011), especially chapter 1.

[4] All translations of Aristotle are from the Barnes edition (Aristotle 1984), which I cite only by Bekker number. Unless otherwise noted, translations of all other works are my own.

and rest, and they are in a sense ontologically primitive in Aristotelian philosophy. They describe inward impulses moving substances toward particular ends. On the other hand, the ends for artificial substances, and the activities needed to achieve these ends, come from outside.

Seventeenth century anti-Aristotelians called such natures "occult," and so too did some Aristotelians. But on a fundamental level it is difficult to see why the notion of Aristotelian natures would be an obstacle to experimental investigation— no more, that is, than experimental investigation within other philosophical and methodological frameworks. The most famous seventeenth-century proponent and theorizer of the experimental method, Francis Bacon, sought through his new logical tool for scientific investigation, the *Novum organum*, to discover the basic natures in Nature, natures arguably no less occult than those the Aristotelians posited. As Bacon famously wrote at the beginning of the *Novum*'s first book (Bacon 2004, 65): "As for works man can do nothing except bring natural bodies together or put them asunder; nature does the rest from within." Now Bacon thought that such basic natures were few and combined to form new natures in the way letters combine to form words; in contrast, Aristotle and most of his followers seemed happy to admit as many natures as there are species of plants and animals. Nevertheless, it is not obvious that the art/nature distinction in Bacon offers any more or fewer obstacles to experiment than the art/nature distinction held by the Aristotelians he attacked.[5] Dear and others argue that artificially constructed experiments or experiences do not reveal the normal course of nature. But although Aristotelians indeed held that one can only have a natural science of things that happen always or for the most part, this in no way entails that Aristotle or Aristotelians had any issue with experimentation *per se*. One must simply ensure that one's experiments shed light on what happens in nature always or for the most part—a requirement, it seems, for any investigator of nature.

It is difficult to find early modern Aristotelians cautioning against experimentation in general or discussing the problem of singular experiments. We do find a sort of example later in the century, by the philosopher Giovanni Maffei in Pisa writing to the Grand Duke Ferdinand II around 1670.[6] In defending Aristotle against the experiments of Galileo, he writes (Galluzzi 1995, 1329):

> [Note that] experience is fallible and dangerous, as Hippocrates holds, and that the intellect needs to correct the sense when it knows that [what is sensed] is not the case; I also say that in order for an evident proof to be drawn from experience, it is necessary that the effect experienced is known, time and time again, to be uniform, since one sensation of a particular effect is not enough from which to derive a universal proposition, but long observation is necessarily required, made up of many, many experiences, and from this used as a scientific foundation.[7]

---

[5] For a thorough discussion of the art/nature distinction in Bacon, see Newman (2005, 256–71).

[6] Thanks to William Newman for bringing this to my attention.

[7] "Alla prima dubitatione rispondo che l'esperienza è fallace e pericolosa al sentire d'Hippocrate e che deve correggersi dall'intelletto il senso quando conosce ciò che non è; dico inoltre che acciò che da quella si cavi una prova evidente è necessario che l'effetto sperimentato sia più e più volte

He likewise cautions against confusing natural causes, which operate spontaneously, with artificial ones, and warns against those who believe "that nature operates
similarly when left free, as when it is constrained and deprived of its natural ability
by art" (1329). His criterion, he mentions several times, is uniformity. Here, then,
we have a polemical attack by a Catholic natural philosopher against the
"Democritean" doctrine of Galileo, lamenting in the second half of the seventeenth
century that Aristotelian doctrine is not being taught properly in the schools, and
that Aristotelianism, even if imperfect, is being replaced by flimsy and empirically
unsupported philosophies: "To extract a universal proposition from many particulars, one needs the eyes of an Argus in quantity and of a Lynceus in perspicacity, and
such eyes are certainly not the eyes of the common intellect" (1329). What we see
in these remarks, however, is primarily that Aristotelians and *novatores* such as
Galileo had different opinions about which basic motions existed *by nature*; apart
from ad-hominem attacks on the carelessness of the *novatores*, it is hard to read
much else here. Singular experiments that cannot be replicated are always suspect;
determining whether an observed effect is genuinely attributable to nature or merely
an artifact of the experimental set-up is always a problem needing resolution.
Whether there are indeed such things as irreducible Aristotelian natures, or whether
those supposed natures are reducible to other (perhaps mechanical) causes, is a
separate issue.

I would argue that the contribution by Arnet to Chap. 10, this volume, for
instance—where twentieth-century attempts to study learning in rats led to interventions that arguably distorted the learning capacities of the rats—is, metaphysical
issues aside, perfectly comprehensible if reframed in Aristotelian terms. In Arnet's
study, artifice with the aim of "control" destroyed certain capacities of the rats in the
experiment, rendering them imperfect (as Aristotelians would frame the alteration)
and thereby unable to move toward their goals as rats typically do. Thus certain
experiments did not, in fact, reveal the aspect of the rat's nature that the experimenters believed they were investigating. I am not proposing we ought to reframe the
episode in this way; I am suggesting only that Aristotelianism was highly adaptable,
and that in many cases little is gained by setting it in opposition to modern science
by default.

Paduan anatomists around the turn of the seventeenth century offer good examples of early modern Aristotelians who engaged with experiment and contrived
experience, with Hieronymus Fabricius ab Aquapendente (1533–1619) being perhaps the most influential on vision and the eye (1600).[8] Paduan anatomists are worth
examining, moreover, because their treatises on the eye influenced Jesuit optics,
including Scheiner in his *Oculus* (1619, 20, 119), and because their revival of the

---

conosciuto uniforme, non bastando una sensatione d'un effetto particolare per cavarne poi una
propositione universale, ma si richiede necessariamente un'osservatione lunga, fatta in molte e
molte esperienze, e di quella valersene per fondamento scientifico."

[8] We also see a similar approach to anatomy in his student and eventual rival in Padua, Julius
Caserius (1609), as well as those trained in Padua around the turn of the century, such as
William Harvey.

"Aristotle Project" for studying animals and the universal animal soul contained explicit Aristotelian methodological precepts, specific to the study of animals, developed from Aristotle's *History of Animals, Parts of Animals,* and *On the Soul* (Cunningham 1985). To answer the question, "What is the capacity for sight in animals?" these anatomists give the history (*historia*), action (*actio*), and usefulness (*usus*) of the parts of the eye. This process begins by carefully dissecting as many different animals as possible, with as many individuals of each as possible. (These writers do not specify an ideal number of cases; more is simply better.) This experience provides the basis for the *historia*, also referred to as *structura* or *fabrica*, of one's investigation into the nature of vision; based on criticisms anatomists made, as well statements on why their own investigations ought to be trusted, we can say that developing anatomical *historia* demands meticulous dissection technique, careful observation using all five senses, and a thorough review and critique of all authorities that have investigated the eye.

Next, they look at the activity of vision itself, which necessitates investigations into the nature of light and color, whether rays might be emitted from the eye, which parts of the eye actively receive impressions from visible things, how the soul, spirit, or visual faculty receives or generates visual information, and so on. That is to say, the *actio* section contains arguments for a theory of vision along with a theory of light and color. Finally, they determine the *usus* or *utilitates* of the parts of the eye, a project which is also framed in terms of Aristotelian final causes. This account of the purposes of the parts—like the shape, size, situation, color, texture, refractive power, color, etc., of the crystalline lens vs. the vitreous humor vs. the aqueous humor—necessitates discussing rival theories of vision. By taking for granted that the parts of the eye do have final causes (they exist for the sake of vision), and by combining this view with a theory of light and color previously established in the *actio* section, the anatomist can be called on to perform experiments that provide evidence for or against various theories of vision.

Fabricius conducted simple, public experiments on how the crystalline lens, separated from the living animal, refracts light and focuses rays into a cone whose point would lie somewhere within the vitreous humor. These he performed to refute extramissionist (largely Galenic) theories of vision (1600, 102–3), and also to challenge certain aspects of perspectivist optics. The Aristotelian logician and natural philosopher Zabarella (who also wrote a commentary on Aristotle's *Posterior Analytics*) also saw these experiments, suggesting that they were performed publicly. In his natural philosophy textbook *De rebus naturalibus*, Zabarella wrote as a witness to a singular event: "I saw the crystalline separated from the other humors in a dissection of the eye, which when placed near a small lit candle was made to shine all over…" (1590, 632–33). Like Fabricius, he used these experiments to argue against Galenic theories of vision. Via his careful description of the sizes, shapes, and relative refractive powers of the parts of the eye, all discovered by the anatomist's knife, Fabricius also offered an empirical refutation of certain key aspects of the mathematical optics of the perspectivists, such as Alhazen, Witelo, and Peckham. These observations and experiments were performed with dead eyes that lacked the animating nature of a living eye—indicating that the eyes were not in their "natural" state. The experiments also relied on the assumption that everything in the body was fabricated for the sake of specific ends—i.e., that final causes exist, and indeed are prior to the body's matter—in order to argue against accounts of vision that rendered the shapes, sizes, temperaments, relative positions, etc. of certain parts of the eye purposeless. Thus, at this time, assuming that the parts of the body possessed final causes improved, rather than hindered, the effectiveness of Fabricius's and Zabarella's arguments, which were empirical and to some extent experimental (Baker 2019, 137–42).

Although the literary and investigative framework for Fabricius's project—*historia, actio, usus*—was largely Galenic, he followed Aristotle's account in the *Posterior Analytics* for how to arrive at first principles i.e., the faculties, capacities, or natures of the parts of the body (1600, unnumbered prefaces to *De visione* and *De voce*). Many other anatomists drew on Aristotle's *Posterior Analytics* as well, including, most famously, Fabricius's student William Harvey (Cunningham 1985; Goldberg 2012, 214–32; Distelzweig 2013, 13–151).

2.3   Aristotle's Posterior Analytics, First Principles, and Early Modern Mathematical Optics

It may still be the case that, as Dear and others claim, there is something about astronomy, optics, mechanics, and other sciences—sciences that combined mathematics and natural philosophy—that rendered experiment and contrived experience problematic for Aristotelians. To see why this is not the case we should review Aristotle's notoriously brief account in the *Posterior Analytics* of how to establish first principles in a science.

Aristotle held that "All teaching and all intellectual learning come about from already existing knowledge" (*Posterior Analytics,* 71a25–71a30). Much of the *Posterior Analytics* is concerned with scientific demonstration, or how new knowledge can be derived from previously existing knowledge, particularly in the natural sciences. Crucial to his account, therefore, is the problem of establishing *first principles*, which cannot be demonstrated from previous knowledge owing to the threat of infinite regress. Aristotle's solution comes in the final chapter of the second book. The puzzle about inquiry in the *Meno* is first summarized: if we have knowledge of first principles innately, then it seems absurd that we would not recognize that we have such knowledge, which is even more precise than knowledge arising from demonstration. If we do not have innate knowledge of first principles, then it seems impossible that we can acquire knowledge of them without preexisting knowledge, that is, it seems that knowledge would arise out of nothing. Aristotle resolves this puzzle with the potency-act distinction: we evidently have an innate capacity or potential for such knowledge, even if we do not have that specific piece of knowledge in actuality: "And *this* evidently belongs to all animals; for they have a connate discriminatory capacity, which is called perception" (99b35–100a2). (Here Aristotle is using "knowledge" very broadly. The sort of knowledge that an animal can potentially have depends on their other cognitive capacities, and so an animal without the capacity for memory can only "know" in the sense of directly perceiving particulars.) In human beings, sensation leads to memory and many memories lead to experience, which he says results in "the whole universal that has come to rest in the soul (the one apart from the many, whatever is one and the same in all those things)." This leads to skill, in the case of practical arts, or understanding (*nous*) of first principles in the case of speculative disciplines (100a10–100a14):

> Thus the states [of comprehending first principles] neither belong in us in a determinate form, nor come about from other states that are more cognitive; but they come about from perception—as in a battle when a rout occurs, if one man makes a stand another does and then another, until a position of strength is reached. And the soul is such as to be capable of undergoing this.

The primitives or first principles arise from a sort of induction. This account seems to be one provocation for Francis Bacon's criticism of the Aristotelian logic or method for investigating nature, which was that they fly too quickly to first principles ("axioms" in Bacon's terminology). But while Aristotle's remarks here are brief, there is no reason to conclude that all Aristotelians held that the process of grasping first principles is itself brief or uncomplicated. As Aristotle mentions time and again, becoming a person of experience is a long process.[9]

In the twentieth century, this passage, and the *Posterior Analytics* overall, were read as a treatise on epistemology: Aristotle was supposed to be giving an account of how knowledge claims are justified. But Aristotle's use of *episteme* has at least two senses here, as either a body of knowledge or as a cognitive state. Taking the second sense as primary, the *Posterior Analytics* can be read, not as an account of the justification of knowledge, but as an account of how demonstration works as a form of explanation (Salmieri et al. 2014, 2–3). Aristotle's notoriously brief treatment in book II chapter 19, then, can be seen as just a highly general account of the psychological act of acquiring first principles of a science, or a description of how we move from sensation to first principles; it does not offer an epistemological justification for the truth of those principles (Aydede 1998, 38–39). Justification is therefore not offloaded to *nous* in some mysterious way, and the act of intuition (i.e., of grasping universal first principles as the result of sufficient experience) is not, in this view, assumed to be infallible (Aydede 1998, 19). Rather, the justification of the first principles of a science requires a case-by-case approach, and such a justification properly belongs at the beginning of the treatises of each specific science (Salmieri et al. 2014, 33). This is arguably what we see in Aristotle's works, where the norms of inquiry, including the establishment of first principles, are specific to each domain (Lennox 2011, 23–46; 2021).

---

[9] E.g., 316a5–10, 980b26–981a12, 1142a12–21.

Aristotelians such as Scheiner, I argue, held that the process of acquiring the requisite experience to grasp first principles is also not simple or straightforward.[10] In the early modern period, whether we examine written statements by historical actors or reconstruct actual practices, experimentation hardly seems barred from this experience-gathering phase in natural science. This would include sciences that drew on principles of natural philosophy, such as optics.

2.4 The State and Scope of Optics Circa 1620

In the first half of the seventeenth century optics was still fundamentally a science about understanding and explaining first-person visual experience, rather than understanding light and image formation. That is, geometrical rays were investigated largely to make sense of first-person visual experiences. This was the understanding of optics at least until Kepler, as A. Mark Smith has shown (2015, ix):

> For the vast majority of its history, the science of optics was aimed primarily at explaining not light and its physical manifestations, but sight in all its aspects from physical and physiological causes to perceptual and cognitive effects. Consequently, light theory was not only regarded as subsidiary to sight theory but was actually accommodated to it.

Prior to Kepler, the lens or crystalline humor was believed to be the primary seat of visual sensation. Smith argues that Kepler's retinal theory of vision transformed (at least eventually) optics into its modern form. But even though Kepler's work contributed to this reconfiguration of optics' scope, I argue elsewhere (following Antoni Malet) that even as late as Descartes's 1637 *Dioprique* this inversion was not yet accomplished—that the goal of optics was still largely to understand first-person visual perception and not reflection, refraction, and image formation (Malet 2005; Baker 2016). Accordingly, to understand Scheiner's *Oculus* in context, we must understand the scope and aim of optics immediately after Kepler's radical work; it will also help to review Descartes's investigations for a fuller view of the approach. I focus here on the order of topics addressed so that we can see how an understanding of light, refraction, ocular anatomy, and the visual faculty came together in seventeenth-century accounts of vision.

From antiquity until the seventeenth century, optics, or *perspectiva*, was often referred to as a so-called "middle" science. It was subordinate to (and thus took its principles from) both natural philosophy and geometry. One goal of optics was to save the phenomena in the most literal sense of that phrase, i.e., "account for the appearances" (Smith 1981). The mathematical cone employed by Euclid and Ptolemy was used to do so (see Fig. 2.1). The postulates in Euclid's optics, for example, are the following (Euclid 1947, 357):

---

[10] On the notion of experience according to William Harvey, which is similar to what we find in Scheiner (not surprising given the Paduan influence on both), see Goldberg (2016).

Fig. 2.1 The visual cone model, showing the same size line appearing smaller if it is farther away from the observer (above), and larger as the observer moves closer to it (Euclid 1557, 8)

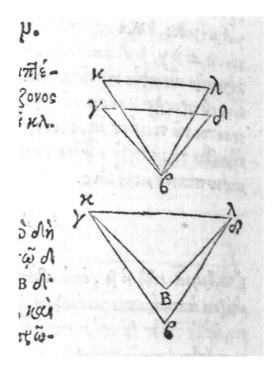

1. Let it be assumed that lines drawn directly from the eye pass through a space of great extent;
2. and that the form of the space included within our vision is a cone, with its apex in the eye and its base at the limits of our vision;
3. and that those things upon which the vision falls are seen, and that those things upon which the vision does not fall are not seen;
4. and that those things seen within a larger angle appear larger, and those seen within a smaller angle appear smaller, and those seen within equal angles appear to be of the same size;
5. and that things seen within the higher visual range appear higher, while those within the lower range appear lower;
6. and, similarly, that those seen within the visual range on the right appear on the right, while those within that on the left appear on the left;
7. but that things seen within several angles appear to be more clear.[11]

Euclid and Ptolemy both posited an extramitted visual cone, but after Ibn al-Haytham in the eleventh century most perspectivists followed him in accommodating the visual cone to Aristotle's more satisfactory physics according to which the

---

[11] Note that Burton labels these "definitions," but early modern editions refer to them as posits, suppositions, or axioms. For example, the influential early edition by Jean Pena (Euclid 1557, 4) labels them *posita*.

forms of light and color enter into the eye from without. This synthesis was accomplished, among other things, by positing a very specific, *a priori* geometrical account of the eye, and by locating the seat of visual perception in the crystalline humor. Early modern anatomical investigations of the eye that questioned this geometrical arrangement therefore threatened the visual theory of the perspectivists (Baker 2016).

In this period the order of topics presented any given optical treatise largely tracks the epistemic priorities of the author. The perspectivists begin with an account of light and color, followed by a qualitative account of refraction and an account of sight in direct vision; only after this are we given an account of the anatomy and physiology of the eye needed to accommodate the visual cone (Fig. 2.2). John Peckham's *Perspectiva communis*, the typical introduction to optics from the fourteenth through the sixteenth centuries, can be outlined thus (here I draw primarily from the 1504 Venice edition):

- Book I: Vision by direct rays

  - Propositions 1–27: The properties of light and color and their propagation
  - Propositions 15, 16: Qualitative account of refraction
  - Proposition 28: The manner of direct vision, namely,
  - "Sight occurs through lines of radiation directly [i.e., perpendicularly] incident upon the eye." (5v)

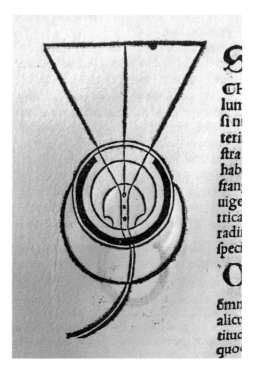

Fig. 2.2 The geometry of the eye accommodated to the intromitted visual cone (Peckham 1504, 6v)

– Propositions 29–46: Anatomy and physiology of the eye and the act of visual perception

"Visible things are grasped (*comprehensio*) by means of a pyramid of radiation; the certitude of apprehension (*certitudo apprehensionis*) however is made by the axis [of vision] being carried all over the visible." (6v)

– Propositions 47–54: Physical requirements for vision
– Propositions 55–79: Psychology of vision

- Book II: Vision by reflected rays
- Book III: Vision by refracted rays

The works of the other perspectivists—Alhazen, Witelo, and Bacon—give the same order as Peckham's book.

Drawing on recent anatomical investigations, Kepler introduced the retinal theory in his 1604 *Paralipomena ad Vitellionem, quibus astronomiae pars optica traditur*, or *Supplement to Witelo, in which the Optical Part of Astronomy is Given*. Along with Alhazen's *Optics*, Witelo's *Optics* was a standard advanced treatise prior to the seventeenth century. Kepler's argument for the retinal theory begins by establishing thirty-eight propositions about the nature of light and color. Although he claims that they are "among the principles in Euclid, Witelo, and others" (Kepler 2000, 20), their clear Neoplatonic basis would have been controversial for many at the time.[12]

In the second chapter he solves a long-standing problem of pinhole images. In the next he refutes Euclid, Witelo, and Alhazen on the formation of images in mirrors, and in so doing Kepler attempted to place catoptrics, or the mathematics of reflection, on a more secure footing (Goulding 2018). Chapter four tackles refraction in a thorough and sophisticated manner without, however, the benefit of the sine rule of refraction. Finally, in chapter five, he gives first the anatomy of the eye and only then the means of vision, namely, that vision occurs when an inverted image is cast upon the retina in the manner of a *camera obscura*. He argues for this claim using a mathematical account of the path of rays through a transparent sphere—i.e., the caustic of a sphere—then confirms the mathematics with experiments of light passing through spherical urinal flasks. Lastly he shows how these mathematical results demonstrate that the crystalline humor indeed refracts rays such that they bring innumerable cones of rays, sent out from each point in the thing seen and landing on every part of the cornea (there forming the bases of those cones), back to

---

[12] For example, in establishing the basis for his thirty-eight propositions on rays of light and color he writes, "The spherical is the archetype of light (and likewise of the world)" (Kepler 2000, 19). As one clear example of a controversial position on the nature of light itself, see Proposition 32, "Heat is a property of light" (39). For Kepler's Neoplatonic account of light, see Lindberg (1986). Kepler's theory of light cannot easily be captured here. The introduction and first chapter of his *Paralipomena* dedicate thirty-eight propositions (plus lengthy corollaries) to establishing his mathematical-physical-theological account of light and color, and this is followed by an attack on Aristotle's account of light, as Kepler interpreted it. The latter, ignoring entirely nearly 2000 years of commentary, would hardly have been convincing to scholastic Aristotelians. For a more detailed account of Kepler's Neoplatonism, influenced by Proclus in particular, see Michalik (2019).

single points on the retina. He does all this, Kepler says, without having performed or attended a dissection of the eye (Kepler 2000, 171).

Note the order of investigation: whereas Peckham and the other perspectivists offer the manner of vision first followed by the anatomy of the eye, Kepler places the anatomy of the eye prior to his determination of the manner of vision. He therefore derives his projection of a picture onto the retina in part from empirical investigations of the eye. He took his empirical account of the eye from anatomists and from them also drew his order of investigation—in particular, from Fabricius ab Aquapendente, via the latter's student Johannes Jessenius (Baker 2019, 141–42). This new order, in which the anatomy of the eye precedes and helps determine the manner of vision, was also followed by later writers, including Scheiner, Descartes, and the Jesuit mathematician and polymath François d'Aguilon (1613, 2–12).

Even contemporaries with the mathematical aptitude to understand Kepler's results might be skeptical of certain steps in his larger investigation. In his 1637 *Dioptrique*, on the other hand, Descartes takes a different approach. Like Kepler, Descartes begins with a treatise on light and argues that light and color are mechanical—a combined tendency of linear and rotational motion of the tiny globules comprising the second kind of matter in Descartes's physical world.[13] He next demonstrates the sine rule of refraction in discourse two; gives an abbreviated anatomy of the eye in discourse three; posits, in discourse four, an account of the senses in general, in which he discards the scholastic view on natural images; and then argues for the retinal theory of vision in discourse five. He accomplishes the last goal, however, by pointing to a simple experiment that allows one to see the inverted picture of the world cast on the retina: "if, taking the eye of a newly deceased man, or for lack of this, that of ox or some other large animal, you skillfully cut through the three coats that enclose it at the base…" (1637, 35). Here Descartes does not give a mathematical demonstration of the path of rays through the eye, but merely refers to the experience of seeing an inverted image on the back of a dissected eye. With this qualitative account of how the rays ought to refract within the eye, he then trusts that his reader will either perform the experiment or else assent to the scheme depicted in his famous diagram.

Notably, Scheiner himself made the same argument in his 1626 *Rosa ursina,* though he presented it as a witnessed experience rather than performance instructions. He also omitted a diagram, which was crucial for Descartes. Scheiner (1626, 110) writes:

> For the rest, that the crossing of rays is made before the image of the object is formed on the Retina yz was not only demonstrated by many exceedingly evident experiments and reasonings in my *Oculus*, but also in a human eye seen publicly here in Rome in the Jubilee Year [1625], where having removed the sclera from the base of the eye, the light of a candle sent through the pupil fell upon the Retinal tunic with crossed rays: which I have also shown to

---

[13] For Descartes's account of light and color, and the experimental basis for his description of the rainbow, see Buchwald (2008).

be true in the eyes of many brute animals. This anatomy of the eye was made by the Reverend Father Niccolò Zucchi in my presence, performed as a favor to me.[14]

Zucchi (1586–1670) was a Jesuit philosopher, astronomer, and mathematician. The experiment he helped perform likely occurred years before similar ones by Descartes. As Scheiner says, however, in 1619 he had recourse only to reasoning from more indirect, though "exceedingly evident," experiments.

Descartes was able to argue for the *camera obscura* model of vision rather easily, and did so in a work aimed at a more general audience, in part because the retinal theory was already making significant inroads. But, for the most part, this was not because Descartes's readers had wrestled with Kepler's difficult *Paralipomena*. Scheiner likely deserves a good deal of credit for converting people to the retinal theory, particularly given the pedagogical reach of the Jesuits. So too does Descartes's one-time collaborator and later critic, the physician Vopiscus Fortunatus Plempius, who advocated for an Aristotelized retinal theory in his medico-philosophical *Ophthalmographia* (1632, 172–74).

2.5 Scheiner on the Eye as the Foundation of Optics

Scheiner divides the *Oculus* into three books. Book I has two parts, the first containing an in-depth anatomy of the eye, although he says that he omits details relating to medicine and the functioning of the eye in general that do not pertain to the foundations of a mathematical theory of vision. In the second part of book I he writes,

> In the second part we report experiences (*experientiae*) as needed in these matters, so that from them we might establish the truth, and refute errors. Indeed, one true experience, as the Philosopher attests, is worth more than a thousand deceitful strings of sly reasoning.[15] (1)

In his preface describing book II he writes, "we examine the visual ray *formaliter*, first from the nature of refraction in general, then with respect to that which concerns the eye in particular" (*ii). By "formaliter" Scheiner means that the path of rays is treated without an account of the physical causes of refraction. Finally, in the preface he also describes the aim of book III: "The retinal tunic is established as the organ of vision, the visual angle is described in detail, and various objections, difficulties and curious questions are examined" (*iii).

---

[14] "Caeterum decussationem radiorum fieri antequam imago objecti in Retina yz effigietur, non tantum in Oculo meo multis evidentissimis experimentis atque rationibus demonstravi, sed etiam in oculo humano hic Rome anno Iubilaeo apertissimè vidi, ubi abrasi in fundo oculi sclerode, immissum candelae per pupillam lumen radiis decussatis in tunicam Retinam accidit: id quod in multis brutorum oculis saepius expertus eram. Facta est autem haec Oculi Anatome à in praesentia R. P. Nicolae Zucchi, in gratiam meam instituta."

[15] "Parte secunda experientias pro re nata adferimus: ut ex illis veritatem stabiliamus, refellamus errores. Una enim vera experientia, Philosopho teste, plus valet, quam mille rationum subtolarum fallaces argutiae."

The first part of book I, as mentioned, is a detailed anatomy of the eye, along with a new diagram or image of the eye containing significant innovations compared to previous diagrams (Raynaud 2020, 108). This material also includes an account of the physical causes of the construction of the eye, and a description of how to dissect it. As to the necessity of this first anatomical section—consisting of about twenty-eight pages—Scheiner says:

> The preconditions for beginning our work are not so much the Phenomena, but rather experiments drawn out with singular zeal, which are of two kinds: the one from the inspection of the eye; the other from the species of things perceived in the eye under certain conditions.[16] (Scheiner 1619, 1)

What Scheiner means by "phenomena" will be discussed later. He lists several reasons for the necessity of ocular anatomy: to determine both whether substances are continuous or distinct, which is necessary to identify places where rays refract; and to determine the degrees of transparency and opacity of the parts, their shapes, the density or rarity of the parts and the differences between them, the magnitude of the parts, and where they are located. All these factors affect the path of rays in the eye, and the foundation of optics just is the determination of the path of rays in the eye. The purpose of doing anatomy, therefore, is both to refute mistaken assumptions that earlier authors had relied on, and to establish a true natural-philosophical account of the act of seeing, from which basis mathematical propositions related to vision can be demonstrated. Scheiner, in fact, cites Fabricius's remarks in *De visione,* part 3 chapter 8, where the latter writes that a true anatomy of the eye should be used to establish the progression of rays through it, the angles of their refraction through the parts, and so on (ab Aquapendente 1600, 105). Note, however, that Scheiner says he came upon this passage only after he had begun his own anatomical investigations (Scheiner 1619, 20).

We might find it curious that Scheiner does not proceed by developing or drawing from the mathematical science of dioptrics first, in order to use those results to understand the path of visual rays in the eye and thus the manner of vision. From the middle of the seventeenth century we increasingly see this order, relegating the formation of images on the retina to a special case of the science of optics (understood as the mathematics of the reflection and refraction of light). For his part, Scheiner still conceives of optics as a middle science, a discipline that combines the principles of physics—particularly the science of the soul, which encompasses perception—with those of mathematics. His treatment of the eye and visual experience is an attempt to establish those areas of physics (including physiology and psychology) that will be relied on to generate the postulates or axioms of optics, and also to tell us what we need to do to establish and secure those parts of physics. The

---

[16] "Praecognita ad institutum nostrum non tam sunt Phaenomena, quam experimenta singulari studio hausta, eaque duplicis generis; altera ex oculi inspectione; altera e specierum a rebus aspectabilibus in oculum diffusarum consideratione desumpta." Dear renders the first sentence, "singular experiments derived from study," and while I find, with Dear, that Scheiner makes a distinction between "singular experiments" and general (and more universal) experiences, I don't quite see this distinction highlighted here.

geometrical foundations for optics, on the other hand, were relatively better established, although he does review some of this material in his treatise.

Scheiner's uses experiments with particular dissections of the eye to build a universal account of the human organ of vision. In this method Scheiner was inspired by the anatomical tradition—particularly out of Padua—which was, as we have seen, built on a synthesis of Aristotelian and Galenic approaches. The methodological norms related to control in contemporaneous anatomical treatises are largely implicit, but by teasing them out via a comparison of many such treatises (and by relying on some explicit statements, particularly from Paduan anatomists), we find that particular experiments or dissections ought to be repeated. The visual and tactile sensations given in those dissections were then to be compared with the writings (and perhaps visual depictions) of relevant authorities, until the investigator arrived at a state of secure understanding for the fabric of the universal part. Moreover, after Vesalius, images were increasingly important for anatomical knowledge, and one important result of Scheiner's anatomy is what he calls an *effigies*—an image or diagram—of the eye (see Fig. 2.3).

Scheiner makes several important corrections to prevailing ocular anatomies. The most obvious one, and that most noted by historians, is that the optic nerve enters the bulb nasally. All previous anatomists and perspectivists thought that it was in a direct line with the visual axis (D in Fig. 2.3), and this change impacts Scheiner's account of vision. He also argues that the corneal bulge, only recently noted by anatomists, is not spherical as commonly described, but either a parabolic or hyperbolic spheroid (8); in this he was almost certainly taking inspiration from Kepler's discussions of conic sections in the *Paralipomena* (Kepler 2000, 106–23, 183–87). Scheiner's crystalline humor (G) is somewhat more forward than most

Fig. 2.3 An effigies or diagram of the eye (Scheiner 1619, 17)

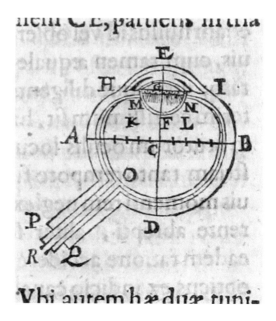

anatomists placed it as well. He says that both sides of the crystalline humor are portions of a sphere, the rear having a small radius, and thus more curvature, than the front (9); this contradicts Kepler's claim that the rear is a hyperbolic conoid (Kepler 2000, 179).[17] Despite adopting the retinal theory from Kepler, Scheiner's dissections led him to a more traditional description of the surfaces of the crystalline lens.

It should be noted that Scheiner first gives the names and a general description of the parts, then in a subsequent chapter enumerates the parts according to transparency and opacity. He then describes, in a geometric fashion, the eye's tunics, then its humors, and lastly explicates at length his image or diagram of the eye, which is in a sense a synthesis of the previous chapters.

Concerning his remarks on dissection procedure and on how he arrived at his anatomical account of the eye, he says that he leaves a more thorough investigation to the physicians (*medici*) and physicists (*physici*), and that "it is sufficient for us to investigate the number, size, shape, position, transparency, density, and similar characteristics of the parts of the eye" (25). This investigation would require an extremely sharp knife and several long needles to probe and secure the parts as needed. He also writes, "it is better to examine and pursue a single aspect of the eye with precision and certainty, rather than attempting to grasp everything all at once" (25). This goal implies a number of dissections and thus multiple eyes; it also contrasts with accounts in anatomical treatises of how to dissect eyes in public anatomies. Although human eyes are preferred, any animal eye will suffice, as Scheiner says that "the visual organ in the human eye is of the same kind (*species*) as that found in a bull or a horse" (26–27). Here he cites Fabricius's dissections of many different animals, made in order to understand the nature of vision generally. One can dissect either fresh (*crudus*) eyes or ones that have been tightened up (*constipatus*) somewhat in warm water. Each will reveal something different, although boiling the eye will ruin the anatomy. Likewise, eyes of the immediately deceased reveal different things compared to eyes that have dried out somewhat. For investigating the humors, only fresh eyes will do, and Scheiner gives instructions for how to empty the various chambers in order to measure the quantities of the humors. Finally, one should dissect in multiple ways, e.g., cutting transverse to the axis of vision, removing the tunics from the rear, from the front, and so on. In short, Scheiner's instructions are more detailed, and directed to more specific ends, than almost all contemporary accounts of ocular dissection.

All this fits easily into the well-trod epistemic path—roughly Aristotelian with Galenic influences—of gathering sensory experiences, storing them in memory, comparing them with the reports of others, and repeating until one *is experienced*, and from this is able to grasp the universals that are the starting points for the science in question. But Scheiner does not discuss just how much experience one needs to either securely arrive at universal knowledge of the relevant anatomical parts or how to resolve disagreements. He says merely that experience with many

---

[17] On the crystalline humor as a hyperbolic conoid, see Baker (2023, 138–40).

eyes is needed. He also says little about the problem of anatomical difference and individual variation. Moreover, he says that knowledge of the fabric of the eye will not alone reveal the nature of vision.

Scheiner goes beyond his predecessors in anatomy and optics in his meticulous measurements of the magnitude of the parts of the eye. While he does mention the problem of diversity and the difficulty of establishing precise accounts for the relative magnitudes of the parts, he concludes that if one measures a certain proportional magnitude in most cases, then the figures can be "firmly established" (12). His method for determining the curvatures of the various parts of the eye also involves careful and novel experimentation, though we will not cover that here.

In the second part of book I, Scheiner begins to use experiments to extend and develop the Aristotelian epistemic framework for grasping the first principles of a science: "Book I part 2, in 14 chapters, brings forth wonderful yet well-tested experiences on behalf of the teachings immediately following" (unnumbered index).[18] As Dear has observed, Scheiner makes a deliberate distinction between experiences (*experientiae*) and experiments (*experimenta*). Singular experiments lead to a general state of *having experiences* or *being experienced*. The *experimenta* are best understood in the context of *Posterior Analytics* II.19: sensations lead to memories, a sufficient collection of which result in experience. Again, we might say the experiments are events or singular sensations, while experience is something like a cognitive state. Experience without understanding is thus similar to the condition of an apprentice house-builder who knows how to do everything involved in house-building but does not yet know the *why* of it all.[19] An experienced person can therefore know or act in situations exactly like the particular ones they have experience with, but their efforts fail in new circumstances.

In the Middle Ages this issue was often discussed in the context of the problem of universals, thereby tying it to metaphysical questions—for instance, what is the ontological status of universals such that a series of particular experiences can cause universal knowledge to arise in the mind or soul? But Scheiner is not concerned with metaphysics and offers instead methodological and other practical solutions connected to the move from experience to understanding—imitating, it seems, Aristotle's approach to the problem in his treatises on individual sciences. Experiment, for Scheiner, thus becomes a way to further develop and refine one's cognitive state of having experience or being experienced, a refinement necessary to establish true first principles for a science.

To understand his methodology we can take the first two *experimenta*. His *experientia prima* concerns variations in the pupil, and he offers two *experimenta* to build one's experience of pupil variation. In the first, which is not new to him, he implores the reader to have someone look at a bright light and note how their pupil contracts; if this person looks away from this light, their pupil will dilate. In the

---

[18] "Liber 1, pars altera, miras sed exploratas pro subsequenti doctrina experientias capitibus 14 depromit."

[19] See 981a13–981b9.

second *experimentum* he asks the reader to look at a needle held at varying distances
from the eye, while another person examines their pupil and notes its changes. Hold
the needle a finger's breadth from the eye and then move it closer, and the pupil will
contract; move away once again and it will dilate. From this result, "a sensible difference will be seen by he, whom you employ as a witness to truth and friend of
philosophy" (30–31).[20] He mentions that no change in illumination is necessary to
cause this dilation and contraction. After this narrative description of the procedure,
he gives a more formal mathematical description, now written in the third person
and illustrated with a diagram: "Let there be an eye, ABC, pupil AB, which is aiming toward a remote grain D. If the grain approaches towards point E…", and so on
(see Fig. 2.4).

Observations about pupil dilation were used to support and attack theories of
vision since Galen at least, but Scheiner's explicit structuring—listing several particular experiments that are supposed to lead one to a more general experience of
how the pupil dilates—is perhaps new. His observation that the pupil contracts

Fig. 2.4 Experience 1,
experiment 2: a friend
observes your pupil as a
needle moves closer to and
further away from your eye
(Scheiner 1619, 31)

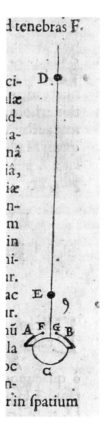

---

[20] "Ita ut crassitie digiti vix absit; una cum accessua aciculae ad oculum tuum claudetur pupilla
eiusdem, una cum recessa ab eodem aperitur, sensibili differentiae quod videbit is, quem veritatis
testem & philosophiae amicum adhibueris."

when close objects are moved closer appears to be novel as well. Scheiner here is building a curated stock of recorded experiences—starting with the easiest to gather—that he will later draw from to establish his first principles of optics.

While different, his arguments from pupil dilation were not unprecedented; Scheiner's second *experientia*, however, appears to be entirely new. He titles it "Things Seen through a Small Hole by Means of Crossed Rays" (32). Scheiner gives a sequence of contrived experiences that he directs his readers to have, but here he does not list a number of *experimenta* that build up to a generalized (though still narrow) *experientia*. This is because these experiences involve mere attention to one's direct, first-person perceptions under special conditions; indeed, he refers to the reader in the second person, imploring them to perform these *experimenta* and gather this body of *experientia* directly. This analysis supports the idea that he sees experience as an internal state, although one which can be shared by many people. "Experience" here is therefore best understood in the psychological sense from Aristotle's *Posterior Analytics*, rather than in the sense of evidence or shared public facts in the legal sense, as described by Barbara Shapiro (Shapiro 2000). Scheiner's contrived experiences are, we might say, private facts that are capable of being described and enumerated, and thus unproblematically referred to in demonstrations.

Scheiner directs his readers to take a *lamina* or thin sheet of opaque material (metal foil works well), which he designates DEFG, and which has a hole H. He says to peer through it to some object, which he labels IK (see Fig. 2.5). If another small opaque plate NO is placed between our eye and DEFG, and slowly moved leftwards across the hole from the right, we notice that our view of the body IK is obscured in the reverse direction—that is, we perceive point I disappearing before

Fig. 2.5 Observing IK through a small aperture as the aperture is gradually blocked by plate NO (Scheiner 1619, 31)

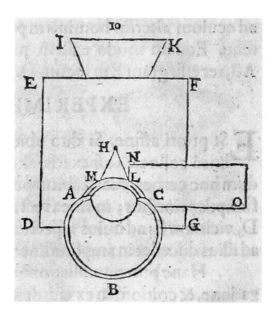

point K, and indeed before the hole itself is obscured (Scheiner 1619, 32). Placing the small plate NO behind the sheet DEFG, and once again moving it across the hole from right to left, we notice that this time point K is obscured before point I. Scheiner's marginal index for this section reads, "Decussatio radiorum demonstratur," or "Crossing of the rays is demonstrated" (32). He lists, moreover, seven *proprietates* or special characteristics arising from this illuminating experience (*lucentia experienta*). In addition to conclusions about the crossing of rays, these results include all sorts of things seen and deduced from the experience: if the hole is too big the rays do not cross; when the hole is small the things seen through it appear to be smaller, but are apprehended more precisely, distinctly, and accurately; if we move the small hole away from our eye, the thing seen through the hole appears smaller and becomes more indistinct; and so on.

There are eleven *experientiae* in book I part II. Experience five is titled, "With one eye it is possible to see, distinctly, the same thing two, three, or four times, without employing any additional diaphanous [body]" (37). Here Scheiner has the reader look through a thin plate with two, three, or four holes spaced "smaller than the width of your pupil" (see Fig. 2.6). Closing one eye, if you look through it at stars or other small bright objects at night, you will see them doubled, tripled, or quadrupled, depending on the number of holes. Note, however, that only individuals with ametropia will see the images variously doubled, tripled, etc. Thus, this contrived experience is not universal, contrary to what Scheiner thought.

Experience six shows the reader how to experience the reverse: "Not only can one thing appear multiple, in the way mentioned above, but [it] also may appear as

Fig. 2.6 "Take a round, thin plate (metal is best) with a handle ADC. Pierce holes E & F as shown, smaller than the width of your pupil. Close your other eye, and place it very close to your eye" (Scheiner 1619, 36–37)

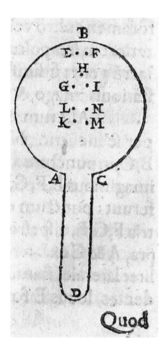

one thing via many apertures" (41). After each of these chapters Scheiner appends another in which he discusses how to perform them properly, what other kinds of phenomena (such as color change seen through the apertures) arise, and so on. At various points he insists that the reader directly experience all the phenomena that will later be used to establish the foundations of optics in books II and III.

Two of the eleven experiences describe individuals with vision defects. Experience four is a report of a man with a peculiar cataract; Scheiner labels this chapter, "The crossing of visible rays into the eye is evident, as nature herself shows."

> There is a man, still among the living, whose left eye is covered with a kind of innate white little cloud, such that access to the pupil is not open to species except for a small space much like the sharp crescent of the new moon... .[21] (36)

If an object (see Fig. 2.7) ILNK is in front of such a person, points I, L, and N will not be seen, but K, via ray HK, will be seen. (K is to the bottom right, but somewhat illegible here owing to a poor impression in the copy I examined.) Again, Scheiner's conclusion here is that "this experience (*experientia*) establishes that things are gazed upon (*aspicere*) through crosswise rays" (36–7). Experience eight likewise discusses general experiences of those with partial *suffusiones* or cataracts, including a notable report copied at length from a text of the physician Ioannes Theodorus. Scheiner thus appeals to direct, first-person contrived experiences, as well as to reliable narratives of the experiences of others in situations where nature seems to act outside of its normal course.

So in addition to (1) first-hand anatomical knowledge, in this first book Scheiner refers to (2) individual experiments, (3) specific, detailed, and contrived first-person experiences, (4) individual, credible narrative reports about extraordinary visual defects, and (5) conclusions derived from general experience in other domains, such as medicine. All methods, for Scheiner, are perfectly valid for developing the broad but well-ordered experience concerning anatomy, physiology, and visual phenomena required to put optics on a secure foundation.

What does Scheiner do with these experiments and experiences? At the end of book I, he writes:

> In the course of this work you will frequently land upon other experiences not touched upon here, which you will also sufficiently elucidate, as I trust, by your worthy reckoning, and you will add many discovered by other performers or by your own *ingenium* or experiments, to which, as if by repeated blows, you will impress the nail of this opinion and doctrine upon your own minds and upon the minds of others, so that you will never undergo any foreign persuasion to tear you from it.[22] (52)

---

[21] "Est homo, etiamnum in vivis, cuius sinsitra oculi pupilla obducta est alba quadam sed nativa nubccula [sic], ut vel in eandem, vel ex eadem pupilla aditus non pateat speciebus nisi spatio tantillo quantum novae lunae falx acutissimo visui largiri dignatur..."

[22] "In alias experientias hic ex instituto non tactas incides frequenter huius totius operis decursus, quas etiam satis elucidatas, uti confiso, calculo tuo approbis, multasque ab aliis actoribus, aut ingenio aut experimento proprio inventas adiunges, quibus velut ictibus repetitis sententiae atque doctrinae huiusce clavum ita tua aliorumque mentibus infinges, ut ab ea. te divelli nulla aliena persuasione ullatenus patiaris."

Fig. 2.7 An account of a
man with a cataract
covering all but the
crescent-shaped area
DEFH, with the object
ILNK in the visual field
(Scheiner 1619, 36–37)

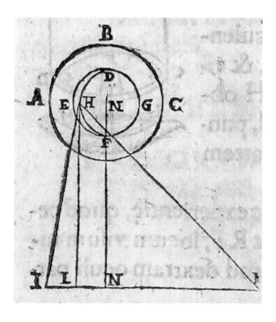

This passage calls to mind Aristotle's *Posterior Analytics* II.19, in which *nous*, or knowledge of first principles which is arrived at via experience, is even more secure than *episteme* or demonstrative knowledge. What in later methodological frameworks might loosely be called "perceptual" experiments are, for Scheiner, pathways for acquiring experience. These experiences are not universal knowledge *per se*, but they are nevertheless more general than any particular experiment, at least in the sense that they do not consist of a single, particular memory but an organized collection of them. A sufficient collection of such experiences allows one to grasp the universal first principles of a science via the process of induction. But again, note that in the quote above Scheiner is not appealing to the psychological state of having experience in order to justify the truth or validity of his first principles. Instead, as we will see, he is building toward a forceful persuasion via argumentation from meticulously curated experience and experiment. However mysterious, according to some modern commentators, this move (discussed both at the end of the *Posterior Analytics* and in *De anima* book III, chapters 4 and 5) from particulars to universals might seem, Scheiner's use of his carefully curated stock of experiences to establish the beginnings of the science of optics is hardly so. He is simply instructing his readers how to gather certain requisite experiences and then drawing out certain conclusions.

Thus, Scheiner begins in book II by eliminating various parts of the eye as contenders for the seat of the visual faculty, arguing that they are incompatible with the results of the experiments and experiences he recorded in the previous book. To do so he cites his list of experiences by number and section. For instance, on the opinion that the cornea is the seat of visual sensation, he writes that this is not possible "because experiences 4, 8, and 9 part 2, chapters 4, 11, and 12, and also 13 and 14, are all alike inconsistent with this opinion" (57). The fact that these experiments and contrived experiences are systematically written down is key—Scheiner is engaged, one might say, in a project of making experience literate.[23] Along the way, in book II, he introduces a number of points about the refraction of light, the relative refractive powers of the various humors (here appealing to both his own anatomical experiments as well as those of others), the nature of rays (or species) of light and color, and so on.

In book III, part 1 he concludes that the retina is the seat of vision, both by process of elimination and owing to the fact that the substance and position of the retina are appropriate for receiving visible rays. In book III, part 2 he reconstructs the visual cone using the retinal theory, thus allowing that theory to accommodate Euclid's axioms. The expanded set of principles found in the later perspectivists, such as Alhazen and Peckham, are not necessarily retained. The crucial benefit of the retinal theory of vision is that, unlike the traditional visual cone model, it combines seamlessly with the burgeoning science of dioptrics—the science for understanding the effects of burning lenses, eyeglasses, and telescopes—particularly as developed by Kepler (1611; Malet 2003). A good portion of book III, part 2 is therefore spent resolving outstanding questions about experiences with eyeglasses, telescopes, vision disorders, and other matters.

2.6   Conclusion

We return to Peter Dear, who asked, "How could 'experience' be established as common property if most people lacked direct access to them?" (1987, 160.) A few pages later he elaborates:

> experimental knowledge was recondite, constituting private rather than public experience, and if it failed to achieve public warrant it could not form part of a science. In order to legitimate experimental statements, therefore, the mathematician had to find ways of extending private experience to his audience through the medium of the mathematical treatise itself. (167)

Finally:

> much of the experiential basis of astronomy and of optics was manufactured by expert practitioners, and could not easily be transformed into the evident experience which would provide adequate principles for a true science. (174)

---

[23] Although Scheiner's work predates Francis Bacon's *Insauratio magna*, I allude to the notion of *experientia literata* found in the latter: "A two-sided activity, *experientia literata* was at the same time concerned with the production of experiments and their presentation in structured, systematic accounts. However, experimentation reached its "literate" stage only if detailed in written reports" (Pastorino 2011, 543). See also Jalobeanu (2016); in addition to Scheiner's carefully written reports, one can arguably find within Scheiner's *Oculus* many of the "patterns of inquiry" that Jalobeanu identifies in Bacon.

I argue that there are several reasons why Dear's suggested obstacles were, for Scheiner, not radical epistemic problems. The first is that Dear, and many other twentieth-century historians and philosophers of science, took Aristotle's *Posterior Analytics* as providing an epistemic justification for grasping first principles. On this account, for an Aristotelian to justify their indemonstrable starting points for a science, they could only appeal to commonly accepted experiences that are the basis for the inductive leap to universal first principles. Against this view, I argue that early modern Aristotelians read the *Posterior Analytics* more in line with the recent scholarly account sketched above. Because the justification of one's axioms or starting points is different for each science, the steps for gathering experience and comprehending the first principles of a science from this experience look different for each science. Such accounts, far more nuanced and elaborate than that given in the *Posterior Analytics*, were given at the beginning of each treatise on that particular science. Indeed Scheiner's *Oculus, hoc est fundamentum opticum* is a treatise on exactly that: the foundations or principles of optics. That is, experiences need not be "common" in the sense that anyone whatsoever would assent to them as a matter of "common sense" (or whatever vague notion of shared experience was thrust on premodern and early modern Aristotelians). Rather, these experiences can be generated in a reader. The author may offer clear directions for contrived experiences and experiments in order to ask that the reader build the experiences up themselves. The author may then ask that the reader accept the reports of reputable observers, like physicians, about both nature in its normal course as well as nature outside its normal course in specific instances. This request would hold provided that the deviations from the norm reveal *different aspects of the normal course of nature*. This last way of generating experience is shown with Scheiner's use of the man with an unusual cataract. The cataract itself is extraordinary, but the small portion of rays that do enter his eye behave as normal, which (he argues) reveals something about the normal action of the eye in refracting incoming rays.

In short, nothing about Aristotelianism prevents these experiences from being complex, from interacting in complicated ways, or from requiring argumentation—or at least ordering—for their consequences to be felt. Furthermore, I argue that Scheiner's work shows that the early modern project of developing methods and techniques for properly establishing first principles or axioms was general—it was not confined to *novatores* such as Bacon or Galileo.

Another issue is that Dear and others take optics to be the science of light, rather than the science of sight. As a result, many authors fail to see the crucial influence of the Galeno–Aristotelian anatomical tradition on optics. This tradition used contrived experiences and experiments and tried to understand final causes and occult "natures" and/or faculties via the dissection of dead bodies (i.e., bodies necessarily *lacking* those natures or faculties). Few in the early modern period doubted the epistemic validity of anatomy, which at the time included what we would call physiology. Thus, to the extent that Scheiner incorporated a similar approach, he was on firm footing. Furthermore, given that the science of optics drew its principles from the physics of the soul, and not merely from the physics of light rays, first-person experience was an ideal starting point for such an investigation.

Again, Scheiner says that he starts not so much from phenomena but from experiments. By "phenomena" he means the standard list of generally accepted visual experiences: that bodies of the same size appear smaller when further away, that square towers in the distance appear round, that objects in flat mirrors appear reversed, and so on. All these phenomena are adequately dealt with via the traditional visual cone, and thus by reconstructing the visual cone within the new retinal theory, Scheiner does not need to worry about addressing the phenomena when starting his inquiry. Scheiner is concerned with sight, understood philosophically (and thus anatomically, physiologically, and psychologically), primarily as it pertains to optics. He wishes to update the physical principles taken for granted at the beginning of a mathematical treatise on optics, an update that became necessary owing to the cracks in perspectivist optics revealed by the works of Paduan anatomists (Fabricius in particular), by Kepler's *Paralipomena,* and by the telescope.

The experimental knowledge that Scheiner directs his readers to obtain is not exactly recondite. I performed most of his experiences successfully, in a few hours, with a few pieces of aluminum foil and a needle. I also found myself largely assenting to his descriptions of these experiences. Such experience is certainly private, but this is precisely what is required to establish the first principles of a mathematical science building on the science of the soul—a science that included the science of sensation and of the sensibles. Private experiences were what the discipline of optics in the second decade of the seventeenth century demanded: they were necessary, but not sufficient. Anatomical knowledge, investigations of light, geometrical knowledge, and other things were also required; moreover, any such contrived experiences and experiments must be repeatable by others. Scheiner does indeed generalize and extend these experiences by means of mathematics—and Dear's description here is correct—but in this there is nothing radically new or strange from the point of view of Aristotelian epistemology or natural philosophy.

Scheiner's *Oculus* belongs only partly to the history of mathematical optics in the modern sense of the science of light and image formation. It belongs more obviously to the long history of perception, as studied by both philosophers and historians of science. This history reaches to Plato and Aristotle, embraces Euclid and Ptolemy, and traces through medieval perspectivists and scholastic philosophers. Moving forward from Scheiner, this history should be seen as eventually feeding into nineteenth-century physiological optics, which was concerned at once with the anatomy and physiology of the visual system together and with issues such as distance, size, and color perception. Helmholtz, for instance, cites Scheiner many times in his 1866 *Handbook of Physiological Optics* (Von Helmholtz 1867).

The larger issue of rigor and control in sciences attempting to understand and explain first-person experiences is worth considering. Scheiner leaned on an Aristotelian psychological model of knowledge, while also drawing from developments in anatomy. Within the latter, final causes helped to structure investigations and offered ways to argue for or against various theories. Appeals to final causes lost epistemic validity as the seventeenth century went on, however, and thus Scheiner's arguments likely lost some of their force over time. We should also note that Scheiner assumes the results of his first-person contrived experiences could be unproblematically generalized. Obviously not all of them can be. For this reason, what has come to be called "Scheiner's experiment" (see Fig. 2.6) is now a test for the visual defect of ametropia: if one looks through a plate or card having two small holes, very closely spaced, with the card placed very close to the eye, that person will see one thing as doubled only if they have a refractive error in their vision whereby distant points do not focus properly on the retina. Scheiner apparently had ametropia. So do I, and when I performed his simple experiment I confirmed what he saw. Thus, Scheiner does not have an explicit method to control for such issues or to resolve them should a controversy arise owing to differences in self-reports. While issues with self-reported perceptual experiments were known and debated in the nineteenth century, what the interim period looks like has not been much explored.

In what way, then, has Scheiner engaged with control practices in the sense relevant to this volume? In attempting to put optics (as the science of sight) on a firm experimental and experiential foundation, Scheiner used experiment and contrived experiences to lay out the possible sites of visual sensitivity in the eye. He then eliminated all but one, often using multiple kinds of evidence; someone gathering the experiences Scheiner sets forth should have no choice but to accept the axioms of optics he presents at the end. However, his experiments and contrived experiences could have been organized differently, from our perspective—for example, in a hypothetico-deductive fashion (Coko, Chap. 8, this volume). While preserving his foundation of rigor and exact anatomy, it would not be hard to rewrite the treatise so that each hypothesis about the manner and seat of vision in the eye would be first presented and then refuted one at a time by using the same experiments he gathered in the second part of book I. In the end, only the retinal hypothesis would remain.

In the *Oculus*, Scheiner does not seem to employ multiple determination in quite the way described by Christopoulou and Arabatzis (Chap. 9, this volume). A sort of second independent determination of the site of sensitivity, however, seems to be used later in Scheiner's *Rosa ursina*. As mentioned above, there he performs the dissection experiment made famous by Descartes 11 years later: the sclera and *uvea* (or choroid tunic) at the back of an eye are removed and the eye is seen to act like a miniature *camera obscura*, with an inverted picture of the world projected on the retina. (In my experience, getting a clear image to appear in this way is not at all straightforward. It is no surprise this was the last method used to establish the retinal theory.) Finally, Kepler's mathematical approach gives us a third independent determination of the site of sensitivity in the eye. He combines a geometrical account of the refraction of rays passing through a sphere with the solution to the problem of pinhole images. To this he adds additional clever mathematical approximations and analogies and then, based on recent anatomical knowledge of the eye, attempts to persuade the reader that the eye casts a picture of the world on the retina. We might label these separate determinations of the *camera obscura* model of the eye the experiential-mathematical-eliminative (Scheiner's *Oculus*), the direct observational (Descartes, Scheiner's *Rosa ursina*), and the mathematical-analogical (Kepler) determinations.

We can also note that, because the private visual experiences of humans can be communicated, challenges to control here seem to be distinct from those involving experiments on non-human animals (Hoffmann, Chap. 11, this volume) and plants (Nickelsen, Chap. 7, this volume; Schürch, Chap. 3, this volume), where only reactions to stimuli can be observed.

The issues of how many experiments are necessary, of replication, and of including parallel trials do not operate in the same way here compared with experiments establishing public facts. For most of Scheiner's self-experiments, one must simply work until one is able to consistently achieve the (private) perceptual experience. Convincing others of the effect demands that they perform the experiment and achieve the experience directly, which again is generally easy to do. The medical reports Scheiner deploys were also apparently understood as reliable and unproblematic, although in any case they are supplementary, something like confirming evidence. It does seem that Scheiner uses something similar to the Baconian strategy of varying the parameters that Coko (Chap. 8, this volume) describes, although certainly in a less sophisticated manner than Coko's actors.

Arguably, any scientific methodology involves control strategies. It seems that premodern and early-modern Aristotelians, however, might easily have been omitted from a history of control. The received account is that experimentation, and therefore the control strategies needed for experimental rigor, developed only in spite of or in opposition to Aristotelian methodological/epistemological precepts. However, Aristotelianism shaped the strategies in Scheiner's *Oculus*, and if anything, his Aristotelianism offered (by the standards of the time) more solutions to experimental issues than it did obstacles. I suggest, but cannot fully argue here, that the decline of Aristotelianism in the seventeenth century was thus not primarily attributable to a fundamental weakness in its methodology or epistemology. To the degree that control strategies are responses to criticism both internal and external to certain disciplines, it seems that one major reason why these strategies were increasingly discussed in the early modern period was the sheer amount of experimental activity and the concomitant explosion in argumentation and criticism. This was a rising tide that lifted all boats.

References

ab Aquapendente, Hieronymus Fabricius. 1600. *De visione, voce, auditu*. Venice: Franciscus Bolzetta.

Aristotle. 1984. In *The Complete Works of Aristotle: The Revised Oxford Translation*, ed. Jonathan Barnes . Princeton, NJ: Princeton University Press.2 vols

Aydede, Murat. 1998. Aristotle on Episteme and Nous: The Posterior Analytics. *The Southern Journal of Philosophy* 36: 15–46.

Bacon, Francis. 2004. In *The Instauratio Magna. Part 2, Novum Organum and Associated Texts*, ed. Graham Rees and Maria Wakely, vol. XI. The Oxford Francis Bacon. Oxford: Oxford University Press.

Baker, Tawrin. 2016. Why All This Jelly? Jacopo Zabarella and Hieronymus Fabricius Ab Aquapendente on the Usefulness of the Vitreous Humor. In *Early Modern Medicine and Natural Philosophy*, ed. Peter Distelzweig, Benjamin Isaac Goldberg, and Evan Ragland, 59–88. New York: Springer.

———. 2019. Dissection, Instruction, and Debate: Visual Theory at the Anatomy Theatre in the Sixteenth Century. In *Perspective as Practice: Renaissance Cultures of Optics*, ed. Sven Dupré, 123–147. Turnhout: Brepols.

———. 2023. The Medical Context of Descartes's *Dioptrique*. In *Descartes and Medicine: Problems, Responses and Survival of a Cartesian Discipline*, ed. Fabrizio Baldassarri, 121–140. Turnhout: Brepols.

Buchwald, Jed Z. 2008. Descartes's Experimental Journey Past the Prism and Through the Invisible World to the Rainbow. *Annals of Science* 65: 1–46.

Cassirer, Ernst. 1906. *Das Erkenntnisproblem in Der Philosophie Und Wissenschaft Der Neueren Zeit. Erster Band*. Berlin: Verlag Bruno Cassirer.

Cunningham, Andrew. 1985. Fabricius and the 'Aristotle Project' in Anatomical Teaching and Research at Padua. In *The Medical Renaissance of the Sixteenth Century*, ed. Andrew Wear, Roger Kenneth French, and Iain M. Lonie, 195–222. Cambridge: Cambridge University Press.

Daxecker, Franz. 2004. *The Physicist and Astronomer Christopher Scheiner: Biography, Letters, Works*, Veröffentlichungen Der Universität Innsbruck 246. Innsbruck: Leopold-Franzens-University of Innsbruck.

de Aguilon, François. 1613. *Francisci Aguilonii e Societate Jesu opticorum libri sex philosophis juxtà ac mathematicis utiles*. Antwerp: Plantiniana.

Dear, Peter. 1987. Jesuit Mathematical Science and the Reconstitution of Experience in the Early Seventeenth Century. *Studies in History and Philosophy of Science Part A* 18: 133–175.

———. 2006. The Meanings of Experience. In *The Cambridge History of Science: Early Modern Period*, ed. Lorraine Daston and Katherine Park, vol. 3, 106–131. Cambridge: Cambridge University Press.

Descartes, René. 1637. *Discours de la methode pour bien conduire sa raison, & chercher la verite dans les sciences. Plus la dioptrique. Les meteores. Et la geometrie. Qui sont des essais de cete methode*. Leiden: de l'imprimerie de Ian Maire.

Distelzweig, Peter. 2013. *Descartes's Teleomechanics in Medical Context: Approaches to Integrating Mechanics and Teleology in Hieronymus Fabricius Ab Aquapendente, William Harvey, and René Descartes*. PhD Dissertation, University of Pittsburgh.

Ekholm, Karin Jori. 2011. *Generation and Its Problems: Harvey, Highmore and Their Contemporaries*. PhD Dissertation, Indiana University.

Euclid. 1557. In *Optica & Catoptrica e Graeco*, ed. Jean Pena. Paris: Andreas Wechelus.

———. 1947. The Optics of Euclid. Trans. Harry E. Burton. *Journal of the Optical Society of America* 35: 357–372.

Galluzzi, Paolo. 1995. La Scienza Davanti Alla Chiesa e al Principe in Una Polemica Universitaria Del Secondo Seicento. In *Studi in Onore Di Arnaldo d'Addario*, ed. Luigi Borgia, 1317–1344. Lecce: Conte editore.

Gillispie, Charles Coulston. 1960. *The Edge of Objectivity*. Princeton: Princeton University Press.

Goldberg, Benjamin Isaac. 2012. *William Harvey, Soul Searcher: Teleology and Philosophical Anatomy*. PhD Dissertation, University of Pittsburgh.

———. 2016. William Harvey on Anatomy and Experience. *Perspectives on Science* 24: 305–323.

Goulding, Robert. 2018. Binocular Vision and Image Location before Kepler. *Archive for the History of the Exact Sciences* 72: 497–546.

Jalobeanu, Dana. 2016. Disciplining Experience: Francis Bacon's Experimental Series and the Art of Experimenting. *Perspectives on Science* 24: 324–342.

Kepler, Johannes. 1604. *Ad Vitellionem paralipomena, quibus astronomiae pars optica traditur*. Frankfurt: Marnius and heirs of Aubrius.

———. 1611. *Dioptrice seu Demonstratio eorum quae visui & visibilibus propter conspicilla non ita pridem inventa accidunt*. Ausburg: Franci.

———. 2000. *Optics: Paralipomena to Witelo & Optical Part of Astronomy*. Edited and translated by William H. Donahue. Santa Fe, New Mexico: Green Lion Press.

Klein, Joel A. 2014. *Chymical Medicine, Corpuscularism, and Controversy: A Study of Daniel Sennert's Works and Letters*. Ph.D. Dissertation, Indiana University.

Lennox, James G. 2006. William Harvey's Experiments and Conceptual Innovation. *Medicina & Storia* 6 (12): 5–26.

———. 2011. Aristotle on Norms of Inquiry. *HOPOS: The Journal of the International Society for the History of Philosophy of Science* 1: 23–46.

———. 2017. William Harvey: Enigmatic Aristotelian of the Seventeenth Century. In *Teleology in the Ancient World: Philosophical and Medical Approaches*, ed. Julius Rocca, 151–168. Cambridge: Cambridge University Press.

———. 2021. *Aristotle on Inquiry: Erotetic Frameworks and Domain-Specific Norms*. Cambridge: Cambridge University Press.

Lindberg, David C. 1986. The Genesis of Kepler's Theory of Light: Light Metaphysics from Plotinus to Kepler. *Osiris* 2: 4–42.

Malet, Antoni. 2003. Kepler and the Telescope. *Annals of Science* 60 (2): 107–136.

———. 2005. Early Conceptualizations of the Telescope as an Optical Instrument. *Early Science and Medicine* 10 (2): 237–262.

Martin, Craig. 2011. *Renaissance Meteorology*. Baltimore: The Johns Hopkins University Press.

Michalik, Jiří. 2019. Johannes Kepler and His Neoplatonic Sources. In *Platonism and Its Legacy: Selected Papers from the Fifteenth Annual Conference of the International Society for Neoplatonic Studies*, ed. John F. Finamore and Tomáš Nejeschleba, 297–318. Lydney: The Prometheus Trust.

Newman, William R. 2005. *Promethean Ambitions: Alchemy and the Quest to Perfect Nature*. Chicago: University of Chicago Press.

———. 2006. *Atoms and Alchemy: Chymistry and the Experimental Origins of the Scientific Revolution*. Chicago: University of Chicago Press.

Newman, Willam R., and Lawrence M. Principe. 1998. Alchemy vs. Chemistry: The Etymological Origins of a Historiographic Mistake. *Early Science and Medicine* 3: 32–65.

Pastorino, Cesare. 2011. Weighing Experience: Experimental Histories and Francis Bacon's Quantitative Program. *Early Science and Medicine* 16 (6): 542–570.

Peckham, John. 1504. *Jo. Archiepiscopi Cantuariensis Perspectiva communis*. Venice: Baptista Sessa.

Plempius, Vopiscus Fortunatus. 1632. *Ophthalmographia, sive Tractatio de oculi fabrica, actione, & usu praeter vulgatas hactenus philosophorum ac medicorum opiniones*. Amsterdam: Henricus Laurentius.

Ragland, Evan R. 2017. 'Making Trials' in Sixteenth- and Early Seventeenth-Century European Academic Medicine. *Isis* 108: 503–528.

Randall, John Herman. 1940. The Development of Scientific Method in the School of Padua. *Journal of the History of Ideas* 1: 177–206.

Raynaud, Dominique. 2020. *Eye Representation and Ocular Terminology from Antiquity to Helmholtz*, Hirschberg History of Ophthalmology: The Monographs 16. Amsterdam: Wayenborgh Publications.

Salmieri, Gregory, David Bronstein, David Charles, and James G. Lennox. 2014. Episteme, Demonstration, and Explanation: A Fresh Look at Aristotle's Posterior Analytics. *Metascience* 23: 1–35.

Scheiner, Christoph. 1619. *Oculus, hoc est, fundamentum opticum*. Oeniponti [Innsbruck]: Agricola.

———. 1626. *Rosa Vrsina sive Sol ex admirando Facularum et Macularum suarum Phoenomeno varius*. Bracciani: Andreas Phaeus.

Shapiro, Barbara J. 2000. *A Culture of Fact: England, 1550–1720*. Ithaca: Cornell University Press.

Shea, William R. 2008. Scheiner, Christoph. In *Complete Dictionary of Scientific Biography*, vol. 12, 151–152. Detroit, MI: Charles Scribner's Sons.

Smith, A. Mark. 1981. Saving the Appearances of the Appearances: The Foundations of Classical Geometrical Optics. *Archive for History of Exact Sciences* 24 (2): 73–99.
———. 2015. *From Sight to Light: The Passage from Ancient to Modern Optics*. Chicago: University of Chicago Press.
Von Helmholtz, Hermann. 1867. Handbuch der physiologischen Optik. In *Allgemeine Encyclopädie der Physik*, ed. G. Karsten, vol. 9. Leipzig: Voss.
Zabarella, Jacopo. 1590. *De Rebus Naturalibus Libri XXX*. Venice: Paulus Meiettus.

Tawrin Baker is an independent scholar, and from 2019 to 2023 was Visiting Assistant Professor in the Program of Liberal Studies and the History and Philosophy of Science at University of Notre Dame. His research focuses on the intersection of anatomy and medicine, natural philosophy, and mathematics in the early modern period.

Chapter 3
One Myrtle Proves Nothing: Repeated Comparative Experiments and the Growing Awareness of the Difficulty of Conducting Conclusive Experiments

Caterina Schürch

3.1 Introduction: From a Proven Truth to a Controversial Physical Problem

By the mid-1780s, physicists across Europe considered the view that electricity accelerated vegetation to be a proven truth (van Troostwyk and Krayenhoff 1788, 134; Rouland 1789, 4). Some 40 years earlier, an Edinburgh teacher had suggested that electricity could be applied "towards the improvement of vegetation" (Demainbray 1747a, 3). His suggestion was soon confirmed by famous electrifying philosophers (Priestley 1767, 140–141), and experiments in the 1770s and early 1780s provided further evidence of the growth-enhancing effect of electricity (D'Ormoy 1791, 29). These seemingly unambiguous results left no doubt as to the correctness of the assumption that electricity promotes vegetation (Senebier 1791, 63).

This certainty disappeared, however, when Jan Ingen-Housz (1786) questioned the validity of the earlier experiments. His criticism convinced many of his contemporaries, who felt compelled to agree that artificial electricity had no influence on vegetation (Senebier 1791, 64). Tiberius Cavallo (1803, 357), for one, concluded in *The Elements of Natural or Experimental Philosophy* that "with respect to vegetation, the most impartial, diversified, and conclusive experiments have shewn, that

C. Schürch (✉)
Institute of History and Philosophy of Science, Technology, and Literature,
Technical University Berlin, Berlin, Germany
e-mail: c.schuerch@tu-berlin.de

© The Author(s) 2024
J. Schickore, W. R. Newman (eds.), *Elusive Phenomena, Unwieldy Things*,
Archimedes 71, https://doi.org/10.1007/978-3-031-52954-2_3

electrization does neither promote nor retard vegetable life."[1] For Alexander von Humboldt (1794, 77–79), on the other hand, the matter was less clear-cut. He suspected that there was no physical problem on which scholars were as divided as about the effect of electricity on vegetation. According to experimental physicists involved in the debate, the case showed how difficult it is to make truly demonstrative experiments (Rouland 1789, 4) and how easy it is to err (Ingen-Housz 1789, 225–226). This chapter takes a closer look at these difficulties. Physicists studying the influence of electricity on plant growth were very well aware that a number of factors could prevent them from drawing correct experimental conclusions. I examine their reports to discover what kinds of errors this group of physicists tried to avoid, and what strategies they used to do so.

The protagonists of this chapter were not attempting to uncover the cause of a puzzling phenomenon, as was the case for the subjects covered by Cristalli and Coko, Chaps. 6 and 8 in this volume, or by Schickore (2023). Rather, they wanted to see whether a well-known factor, electricity, actually affected the growth of plants. To this end, they performed comparative experiments. We will see that they did this as a matter of course, without making their methodology explicit. In line with what Schickore (Chap. 1, this volume) refers to as the "narrower" notion of experimental control, they considered comparative experimental design a prerequisite for drawing safe conclusions. In their experiments they also attended to control practices in the broader sense. They kept the two experimental settings stable to exclude other factors from interfering with the experimental outcomes. They tried to expose their plants to exactly the same conditions, except for electrification, and noted the values for conditions they could not or did not want to control, such as the weather or the temperature. In addition, they detailed their intervention and how they assessed plant growth. Some also measured the intensity of the electricity applied and varied the amount to see if the effect changed accordingly.

While they agreed that comparative tests must support causal conclusions, they disagreed on what exactly would ensure a safe basis for concluding that electricity promotes plant growth. Many felt that a single run of a comparative experiment was sufficient. If the electrified plants grew faster than the non-electrified ones, they took the difference to indicate a vegetation-enhancing electrical effect. Others, however, refused to draw conclusions from single experiments and insisted on many repetitions of the same experiment. Only when they observed the difference consistently were they prepared to assume a causal relationship. The experimenters also disagreed on the number of test objects to be used per experiment. While most physicists compared the growth of a small number of plants, Ingen-Housz monitored thousands of cress seeds. This was intended to compensate for the individual variability of the experimental objects. However, he seems to have been the only one at the time who considered this control measure necessary.

---

[1] Cavallo (1782, 38) had changed his mind on this subject, having previously stated: "By increasing the perspiration of vegetables, Electricity promotes their growth; it having been found, after several experiments, that such plants, which have been often and long electrified, have shewed a more lively and forward appearance, than others of the same kind that were not electrified."

In contrast, many physicists warned against the expectations of experimenters as a source of error. Some even argued that the view of electricity as a growth promotor had only prevailed because it had been reported by famous physicists, and subsequent generations of physicists had not tested it thoroughly enough. Ingen-Housz (1789, 217) therefore urged his colleagues not to rely on authorities, but rather to examine each other's experiments in search of errors.

The chapter is structured as follows. Section 3.2 presents some experiments carried out between 1746 and 1748 comparing the growth of electrified and non-electrified plants. In Sect. 3.3, we will encounter contemporary views on how experimentalists can learn about hidden causes through comparative trials. In addition, we shall devise a list of error sources that physicists associated with experimentation of this type, as well as their suggestions for control practices to avoid them. Section 3.4 examines contributions to the controversy published between 1757 and 1789, in the light of the sources of error and control practices discussed in Sect. 3.2. This includes the elaborate but little-known experiments of Runeberg and Köstlin, as well as the contributions of Ingen-Housz and reactions to them. Finally, Sect. 3.5 summarizes the results of this investigation and suggests: concurrent comparative experimentation was the procedure of choice if the process under study is temporally extended and/or cannot be observed twice on the same object—for example, because it is a directed developmental process.

3.2   Comparing the Growth of Electrified and Non-electrified Plants, 1740s

Stephen Demainbray (1710–1782), a French teacher living in Edinburgh in the 1740s, based his claim that electricity improves vegetation on the following experiment:

> On the 20th of December last I had a Myrtle from Mr. Boutcher's Green-House, which since that time I have electrified seventeen times, and allowed the Shrub half an English Pint of Water each fourth Day, which you'll please to observe was kept in the Room the most frequented of my House and consequently the most exposed to the Injuries of the Air, by the Doors and Windows being oftenest opened.

> This Myrtle hath since by Electrization produced several Shoots, the longest measuring full three Inches; whereas Numbers of the same Kind and Vigour left in the said Green-House have not shewn the least Degree of Increase since that Time.[2]

He compared the growth of several shrubs, one of which he had electrified and the others not. He found that the electrified myrtle, unlike similar non-electrified examples, had produced several new shoots. However, Demainbray was not content with this trial. He set out to perform "a further and more satisfactory Experiment of the

---

[2] Demainbray 1747a, 3. According to advertisements in the *Caledonian Mercury*, e.g. Nr. 3245, of Tuesday 13 January 1741, 3, William Boutcher Jr. was a Nursery-man and Seeds-man at Comely Garden near the Abbey-hill, Edinburgh.

same Nature" and promised to communicate "soon to the Publick some Proofs still more evident of the present Hint."[3]

3.2.1 Demainbray Stabilizes Experimental Conditions to Secure his Discovery

Exactly 1 month after announcing his first experiment, Demainbray made good on his promise and sent another letter to *The Caledonian Mercury*. In this second attempt, he made sure to treat the two plants as similarly as possible, except for electrification:

> On the 17th of January last, Mr. Boutcher favoured me with *two Myrtles* of the greatest Equality of Growth, Vigour, &c. he could chuse; these I placed in the same Room, and allowed them each an equal Quantity of Water.

> On electrifying one of them, it hath produced several Shoots full three Inches. The other Shrub (which I did not electrify) hath not shewn any Alteration since I first had it.[4]

By treating the electrified and non-electrified myrtles as equally as he could, Demainbray had anticipated what an anonymous commentator on the editorial board of *The Gentleman's Magazine* had criticized about his first report. The commentary, following an excerpt from Demainbray's first letter, reads:

> This account is deficient, and, perhaps, no certain inference can be made in favour of the great increase of the plant by electrising only; because it might be occasioned (at least in part) by its having water; which the plants in the greenhouse (by what appears) had not. (Anonymous 1747a, 81)

Both the commentator and Demainbray acknowledged the importance of the experimental actions in drawing conclusions. The commentator observed that the electricity supply was likely not the only difference between what we can call the test-myrtle and the control plants; while the test-myrtle was brought into a house, the others remained in the greenhouse, with unknown amounts of water supplying them. This inequity matters because water promotes plant growth. Because the setup could not rule out other growth factors, the conclusion that the new shoots resulted from electricity does not safely follow from the experiment.

---

[3] Despite the mentioned shortcomings of the experiment, Demainbray's report was reprinted under the titles "An application of electricity towards the improvement of vegetation" in *The Scots Magazine* 9, no. 2: 40; "A remarkable Experiment in Electricity" in *The London Magazine* 16, no. 2: 87; and under "Electricity, effects of, on vegetation" in *The Gentleman's Magazine* 17, no. 2: 80–81.

[4] Demainbray (1747b, 2–3, emphasis in original). This second letter was reprinted as well, e. g. in the *Ipswich Journal* of Friday 21 February 1747: 3, and in *The Scots Magazine* 9, no. 3: 93.

To avoid various sources of experimental error, Demainbray tried to stabilize the conditions. He relied on the resources and expertise of his neighbor, who gave him two myrtles as identical in appearance as possible. Demainbray placed them in the same room and gave them the same amount of water. With these control practices he was confident that electrification was the only difference between the test plant and the control, and he was thus more certain than before that the growth came from electricity. Satisfied with his work, Demainbray concluded his second letter as follows:

> As the Business of my School does not allow me the necessary Time of attending this *now certain* Discovery, I submit it to those whose Leisure will permit them to pursue a Hint which may hereafter be highly beneficial to Society.[5]

Thus, for Demainbray, the certainty of the conclusions depended on the details of the experimental design and its implementation. The commentator of *The Gentleman's Magazine* shared this understanding and applauded the adjustments.[6]

To determine the effect of his intervention on the myrtle, Demainbray counted the number of new shoots and measured their length. On the other hand, the brief report does not mention how he electrified the plant. From his letters of November 1746, we know that Demainbray had become familiar with the latest literature on electricity, in the work of Desaguliers (1742), Hausen (1746), and Bose (1744). His former teacher John Theophilus Desaguliers (1683–1744) had resumed his electrical experiments in the late 1730s, and Demainbray witnessed some of them.[7] After moving to Edinburgh in 1740, he had "all the apparatus that Gravesand [sic] describes made" and reproduced "all the experiments of Hawksbee, and all those which are described in the Brochure by which

---

[5] Emphasis in original. This was Demainbray's last publication on the subject. For his further career, see footnote 37.

[6] A follow-up comment in the same issue reveals that the "same ingenious Scotchman" had repeated his "experiment concerning the effect of Electricity on Vegetation" with "two Myrtles of equal growth, both which are expressly said to have been supplied with an equal quantity of water"; and that the electrified plant "produced several shoots 3 Inches long, and the other remained without alteration" (Anonymous 1747b, 102).

[7] Demainbray told Erasmus King in his letter of 8 November 1746: "I had some general Joys of Electricity before my Arrival in Scotland. By the small Number of Trials I had seen Doct. Desaguliers make of it." Records assembled by the State Paper Office, SP 36/89/1, folios 71–72. National Archives, Kew. For Desaguliers' electrical experiments of the late 1730s, see Desaguliers (1739). Demainbray and his wife came to Edinburgh in 1740 to run "a Boarding-School for Ladies in Bishop's Land." In his letter to Abraham de Moivre, 8 November 1746, Demainbray proudly introduced himself as "a pupil of the late Doct. Desaguliers." Records assembled by the State Paper Office, SP 36/89/1, folios 65–66. National Archives, Kew. Desaguliers instructed Demainbray in mathematics and natural philosophy in the 1720s (Anonymous 1795, 317–318). At the age of seventeen, Demainbray allegedly left London to study in Leyden. Macray (1888, 330), however, noted that Demainbray's name cannot be found in the Leyden *Album Studiosorum*.

M. Desaguliers won the prize in Bordeaux." He further claimed to "have by the means of Hawksbee's Globe set fire to spirits of Wine."[8]

   None of the many authors who later mentioned Demainbray's experiments criticized their design or his lack of detail about how he conducted them. This, however, tells us less about contemporary standards for good experiments than it does about the reception of Demainbray's original reports: hardly anyone seems to have read them (see footnote 15). The only author who explicitly responded to them was Stephen Hales (1677–1761). *The Gentleman's Magazine* for April 1747 states that "[n]otwithstanding what has been inserted of the efficacy of electricity on plants" in an earlier issue, "the Rev. Dr Hales finds his suspicion, that electricity will not promote vegetation, confirmed by several experiments made by Mr King, at his experiment room, near the king's Meuse, London, and by Mr Yeoman at Northampton." We know even less about these trials than we do about Demainbray's. The experiments of Erasmus King, a lecturer in natural and experimental philosophy, and Thomas Yeoman, an engineer, did not become part of the accepted body of studies on vegetation.[9] In contrast, news (though not the exact wording) of Demainbray's experiments reached Jean Jallabert in Geneva and Jean-Antoine Nollet, the dean of French electricians of the time (Heilbron 1979, 254), in Paris.

3.2.2   Nollet's, Jallabert's, and Menon's Comparative Experiments

Two months after Demainbray's second letter appeared in print, another experimental philosopher began a similar study. Jean Jallabert (1712–1768), then professor of experimental philosophy and mathematics as well as curator of the Geneva public library, spent "a part of the month of April and the whole month of May in regularly electrifying for 1 or 2 h, each day, various plants; among others, a yellow wallflower placed in a box full of earth."[10] Like Demainbray, he attended to how the electrified

---

[8] Demainbray to de Moivre, 8 November 1746, loc. cit. In his letter to King, 8 November 1747, loc cit., Demainbray specified that he "made a Wheel with a Treadle to whirl a Globe" and thereby "fired Spirits of Wine etc." In vain, he had "attempted Beatification," an experiment described by Bose (1744). Demainbray saw "the wavering Fire round the Feet but no more." For a sketch by Père Chabrol of the electrical machine used by Demainbray at Bordeaux in 1753, see Fig. 2 in Morton and Wess (1995, 174). The device is similar to the electric globe machine (object nr. 1927-1186 Pt1) in the King George III collection at the Science Museum. The object consists of a glass sphere that can be rotated around a vertical axis by means of a gear inside the brass casing.

[9] The trials were most likely conducted after Hales wrote to the Rev. Mr. Westly Hall on February 23, 1747, see Hales (1748). This letter, which was read before the Royal Society 16 months later, contains no information about the two experiments mentioned in the *Gentleman's Magazine* (Anonymous 1747c, 200). For King and his cooperation with Hales, see Appleby (1990).

[10] Jallabert (1748, 80). The French original reads: "Une partie du mois d'Avril, & tout le mois de Mai, furent emploiés à électriser régulièrement une ou deux heures, chaque jour, diverses plantes; entr' autres, un giroflier jaune ou violier placé dans une caisse pleine de terre."

and non-electrified plants grew. He observed that "[a]ll these plants increased considerably in stem & branches; & in particular the wallflower made beautiful sprays & flowered."[11] But in contrast to Demainbray, Jallabert found the difference between the electrified and non-electrified plants too small to take as an effect of electricity:

> [T]he progress of these electrified plants, compared with that of other plants of the same age, raw in vases full of the same soil &c., did not seem to me to be sufficiently considerable to dare to conclude that the material of electricity was capable of accelerating vegetation.[12]

According to Jallabert, his first experiments gave no evidence that electricity accelerated vegetation. Minor differences did not warrant such a conclusion.[13]

Just as Jallabert (1748, 81) was about to repeat these experiments in the fall of 1747, he heard that myrtles electrified in Edinburgh had grown sprays three inches long within a few days. This growth occurred in a season when other myrtles had not yet budded. Shortly afterwards, his friend Jean-Antoine Nollet (1700–1770) informed him of "some very curious experiments" he had made with mustard seeds. Nollet had also heard about the electrified plants:

> I learned that in England, plants and shrubs had been electrified and felt in such a way as to make people believe that electric virtue promotes or hastens vegetation; but as no details of these experiments have come down to us, I was unable to draw any advantage from them, other than to embolden myself in my intention to devote myself to these tests.[14]

Nollet reported hearing "that two myrtles had been electrified, and that they had grown a few buds; but I don't know what was done to be entitled to attribute this effect to electric virtue."[15] Thus, Nollet, like Demainbray and his commentator,

---

[11] Jallabert (1748, 81): "Toutes ces plantes augmentèrent considérablement en tige & en branches; & en particulier le giroflier fit de très beaux jets & fleurit."

[12] Jallabert (1748, 81): "Cependant les progrès de ces plantes électrisées, comparés à ceux d'autres plantes de même âge, crues dans des vases pleins de la même terre &c. ne me parurent pas assés considérables pour oser en conclure que la matière de l'électricité étoit capable d'accélérer la végétation."

[13] Jallabert seems to have had a quantitative expectation of how big the difference between the test plants and the controls would have to be to warrant attributing the effect to electricity. But he did not specify what would have been a sufficient difference.

[14] Nollet (1749, 356): "[…] j'appris qu'en Angleterre on avoit électrisé des plantes & des arbustes, qui s'en étoient ressenti de maniéré à faire croire que la vertu électrique favorise ou hâte la végétation; mais comme il ne nous est venu aucun détail de ces experiences, je n'ai pu en tirer d'autre avantage, que celui de m'enhardir dans le dessein où j'étois de me livrer à ces épreuves."

[15] Nollet (1752, 172): "J'ai oui dire depuis, qu'on avait électrisé deux myrthes, & qu'ils avoient poussé quelques boutons; mais j'ignore ce que l'on a fait pour être en droit d'attribuer cet effet à la vertu électrique." In a footnote he added that he had since—that is, since reading the Mémoire before the Académie Royale in April 1748—learned that this experiment was made "in Edinburgh by Mr. Mambray, that two myrtles having been electrified during the whole month of October 1746, grew at the end small branches & buds; which similar non-electrified shrubs did not." This footnote probably served as the template for most of the later references to Demainbray's work. It looks as though few of the naturalists who referred to Demainbray's experiments read his original reports. Soon the erroneous view crept in that the experiment (singular) was carried out in *October*; that *two* myrtles were electrified at the same time, and that they began to *bloom*. Priestley (1767, 135), for instance, wrote: "Mr. Maimbray at Edinburgh electrified two myrtle trees, during the

distinguished several dimensions of experimentation. The physical activities for conducting the experiment were one, and the mental activities for interpreting the results were another. Demainbray's former teacher Desaguliers (1727, 264) referred to this distinction when he noted that "a Mechanical Hand, and a Mathematical Head are the necessary Qualifications of an Experimental Philosopher: The first alone may enable a Man to make a great many Experiments, but not to judge of them."

To judge whether growth-promoting effects could be attributed to electricity, Nollet (1748, 189), like Demainbray and Jallabert, compared the growth of electrified and non-electrified plants. He otherwise treated them as equally as possible:

> I took two Garden-Pots, filled with the same Earth, and sowed with the same Seeds; I kept them constantly in the same Place, and took the same Care of them, except that one of the two was electrified for fifteen Days running, for two or three, and sometimes four Hours a Day. This Pot always shewed its Seeds raised two or three Days sooner than the other, a greater Number of Shoots, and those longer, in a given Time.

This result made him "believe, that the electrical Virtue helps to open and display the Germs, and facilitates the Growth of Plants." However, Nollet said that he advanced this "only as a Conjecture, which deserves further Confirmation" (189–190). His letter was read at a meeting of the Royal Society of London in February 1748. Two months later, a more detailed account was read at the Académie Royale des Sciences at Paris. From this mémoire, we learn that on Monday, October 9, 1747, he took two similar tin bowls filled with the same soil, and sowed in each an equal quantity of mustard seed taken from the same packet.[16] After leaving them in the same place for two days, Nollet and his collaborators electrified one of the bowls (Fig. 3.1):

> I placed one of the bowls marked with the letter A in the tin cage where it was electrified for ten hours, namely, in the morning from seven o'clock until noon, and in the evening from three o'clock until eight. During this time, the other bowl was kept apart, but in the same room, where the temperature was quite uniformly 15 ½ degrees according to M. de Reaumur's thermometer.[17]

The following day the bowls were exposed to the sun and watered equally. When they entered the house in the evening, Nollet still did not see anything. But on October 13 at nine o'clock in the morning, he "saw three seeds in the electrified bowl, whose stems were three lines above the ground" while "the non-electrified bowl had none." Nollet and his assistants continued the experiment: "We took the same care of one & the other as the previous day, & in the evening electrified the one

---

[16] Nollet (1752, 173): "Le 9 Octobre 1747, je fis remplir de la même terre deux petites jattes d'étain toutes semblables; je semai dans chacune une égale quantité de graine de moutarde prise au même paquet; je les laissai deux jours dans le même lieu, sans y faire autre chose que les arroser & les exposer aux rayons du soleil, depuis environ dix heures du matin jusques à trois heures après midi."

[17] Nollet (1752, 173): "[…] je plaçais une des jattes marquette de la lettre A dans la cage de tôle où elle fut électrisée pendant dix heures, savoir, le matin depuis sept heures jusqu'à midi, & le soir depuis trois heures jusqu'à huit. Pendant tout ce temps-là, l'autre jatte était à l'écart, mais dans la même chambre, où la température était assez uniformément de 15 degrés $^1/_2$ au thermomètre de M. de Reaumur."

Fig. 3.1 Left: Plate 7 in Nollet (1752, 200). Nollet placed objects to be electrified (e.g. the mustard seeds in bowl A) in a cage made of three large sheets of cloth, held at the four corners by iron mounts. The cage was suspended by two metal rings on a large silk cord stretched horizontally. An iron chain conducted the electricity, generated by rubbing a glass globe, to the cage. Two "strong men," replaced by two others from time to time, turned this ball while a third person rubbed it with their hands (316–317). Right, top: Fig. 4 from plate 2 in Nollet (1749, after 162), showing an electrometer. Right, middle and bottom: Plate 1 from Nollet (1765, after 24) showing the glass globe and the apparatus by which it is turned

whole month of October 1746; when they put forth small branches and blossoms sooner than other shrubs of the same kind, which had not been electrified." Anonymous (1752, 75–76) reads: "This acceleration in plants was tested in Edinburgh, by Mr. Mambrai. Two myrtles having been electrified during the whole month of October 1746, grew at the end small branches & buds, which similar non-electrified shrubs did not." According to Sigaud de Fonds (1771, 372), the conclusion that electricity "must hasten the effects of vegetation […] was confirmed by a number of observations made with care by several famous Physicists. Doctor Mimbray was one of the first who applied himself to this research. As early as October of the year 1746, he found that two electrified myrtles grew small branches and buds, which similar non-electrified shrubs did not." De la Cepède (1781, 175) wrote: "Mr. Mambrai having electrified two myrtles in Edinburgh during the month of October 1746, saw them grow small branches and put on buds, which was not the case with other myrtles to which no attempt was made to give a new quantity of fluid." Bertholon (1783, 152) wrote: "Dr. Mainbrai electrified two myrtles in Edinburgh, during the whole month of October

intended for this test for three hours."[18] The next morning, the difference between the plants in the two bowls was even more marked: "the electrified bowl had nine stalks out of ground, each of which was seven to eight lines long, & the other one had still absolutely nothing raised; but in the evening, I saw one in this one starting to show."[19] That afternoon the first bowl was electrified again for 5 h. Nollet summarized the rest of his experiment as follows:

> Until the 19th of October, I cultivated these two small portions of seeded land similarly, continuing to electrify one, & always the same one, for several hours every day; […] after eight days of experiments, the electrified seeds were all raised & had stems from fifteen to sixteen lines high, while there were barely two or three of the others out of the ground with stems of three or four lines at most.[20]

With this result, Nollet initially feared that he had not adequately controlled for all relevant experimental conditions: "This difference was so marked, that I was tempted to attribute it to some accident that escaped my knowledge." He first imagined the accident to be a factor inhibiting plant growth in the non-electrified bowl. But a few days later, when he noticed that all the seeds in the control bowl had sprouted, he began to "believe with some confidence, that electricity had truly accelerated the vegetation & growth of the others."[21] Nollet emphasized that he "only came to this conclusion after several repeated tests on different seeds, with

---

1746; they grew small branches and buds at the end, which similar non-electrified shrubs did not. The shoots they gave on this occasion were even three inches long, which is astonishing in a season when the other trees were not yet budding." See also Runeberg (1757, 15), Gardini (1784, 15), Duvarnier (1786, 94), and Rozieres (1791a, b, 351). Miller (1803, 23–24) wrote: "The application of electricity to growing vegetables was first made by Mr. Maimbray, of Edinburgh, who found that, in certain cases, it expedited the progress of vegetation." As the great exception, Anonymous (1795, 317) cited Demainbray's first letter published in *The Caledonian Mercury*.

[18] Nollet (1752, 173–174): "Le 12, ces deux jattes furent exposées ensemble au soleil & arrosées également; on les rentra de bonne heure le soir, & je n'y aperçus encore rien de levé. Le 13 à neuf heures du matin, je vis dans la jatte électrisée, les trois graines levées dont les tiges étaient de trois lignes hors de terre; la jatte non électrisée n'en avait aucune. On eut de l'une & de l'autre le même soin que le jour précédent, & l'on électrisa le soir pendant trois heures celle qui était destinée à cette épreuve."

[19] Nollet (1752, 174): "Le 14 au matin, la jatte électrisée avait neuf tiges hors de terre, dont chacune était longe de sept à huit lignes, & l'autre n'avait encore absolument rien de levé; mais le soir, j'en aperçus une dans celle-ci qui commençoit à se montrer: la première fut encore électrisée ce jour-là pendant cinq heures de l'après-midi."

[20] Nollet (1752, 174): "Jusqu'au 19 d'Octobre, je cultivai également ces deux petites portions de terre ensemencées, en continuant d'en électriser toûjours une, & toûjours la même, pendant plusieurs heures tous les jours; & qu'au bout de ce terme, c'est-à-dire, après huit jours d'expériences, les graines électrisées étaient toutes levées & avoient des tiges de quinze à seize lignes de hauteur, tandis qu'il y en avait à peine deux ou trois des autres hors de terre avec des tiges de trois ou quatre lignes au plus."

[21] Nollet (1752, 174): "Cette différence était si marquée, que je fus tenté de l'attribuer à quelque accident qui aurait échappé à ma connoissance: mais au retour d'un petit voyage que je fus obligé de faire, je trouvai toutes les graines levées dans la jatte qui n'avait pas été électrisée; & je commence à croire avec quelque confiance, que l'electrité avait accéléré véritablement la végétation & l'accroissement des autres."

more or less similar results: I almost always saw a considerable difference between electrified seeds and those that were not; the former sprouted more quickly, in greater numbers in a given time, and grew more rapidly."[22]

Motivated by the results of Demainbray and Nollet, Jallabert returned to his experiments on the influence of electricity on vegetation in the last days of 1747. He put several daffodil, hyacinth, and narcissus bulbs on water-filled carafes. Most of the plants had already sprouted roots and leaves, and some even had advanced flower buds. Jallabert measured the length of their roots, stems, and leaves, and then placed the carafes on resin cakes. This last measure was a "precaution or preparation" to ensure that the carafes were ready to receive electricity.[23] To electrify the plants he used archal wires that, starting from an electrified bar, plunged into the water of each carafe. Thus, from December 18 to 30, except for December 24 and 25, he electrified the bulbs for 8 to 9 h a day. Like Nollet, Jallabert had a thermometer from Mr. de Réaumur in his cabinet, which throughout the tests stayed between the 8th and 10th degree above freezing.[24] This time the difference between the test and control plants was marked:

> The difference in the progress of the electrified onions, compared to that of other onions of the same species equally advanced & situated & treated the same except for electrification, was very noticeable. The electrified onions increased more in leaf, & in stem; their leaves spread more; & their flowers bloomed more promptly.[25]

According to Jallabert, this experiment suggested that electricity hastens plant growth. From another trial he argued that electricity also promotes transpiration: electrified daffodil bulbs lost more weight than non-electrified ones.[26] In a third experiment, Jallabert put cress and mustard seeds on the outer surface of a vessel. He reported that at "the end of the second day of 8 to 9 h of electricity each day,

---

[22] Nollet (1752, 174–175): "Quoique cela parût assez clairement indiqué par l'expérience que je viens de citer, je ne me suis rendu à cette conséquence qu'après plusieurs épreuves réitérées sur différentes graines, & suivies de résultats à peu près semblables: j'ai presque toûjours vû une différence assez considérable entre les semences électrisées & celles qui ne l'étaient pas; les premières ont levé plus promptement, en plus grand nombre dans un temps donné, & leur accroissement s'est fait plus vîte."

[23] Nollet (1752, 67–68): "Il est essentiel d'indiquer quelques précautions ou préparations nécessaires pour les [corps] mettre en état de recevoir la vertu électrique. Us doivent être isolés de tout autre corps non électrique. On les en sépare, soit en les suspendant à des cordons de soye exempts de toute humidité; ou, en les posant sur des gateaux de résine, sur des caisses pleines de poix, sur des guéridons de verre séchés exactement." Nollet used this second form of insulation by hanging the electrified cage from silk cords.

[24] Jallabert (1748, 82–83): "Depuis le 18 jusqu'au 30 Décembre, excepté le 24 & le 25, j'électrisai de cette maniéré plusieurs oignons 8 à 9 heures chaque jour; & pendant toute cette opération, un thermomètre de Mr. de Reaumur fut, dans mon cabinet, entré le 8me & 10me degré au-dessus de la congélation."

[25] Jallabert (1748, 83): "La différence du progrès des oignons électrisés, comparé à celui d'autres oignons de même espèce également avancés & situés & traités de même à l'électrisation près, a été très sensible. Les oignons électrisés ont plus augmenté en feuilles, & en tige; leurs feuilles se sont étendues davantage; & leurs fleurs se sont épanouïes plus promptement."

[26] See Jallabert (1748, 83–85).

several mustard sprouts had grown. And, without electricity, by the 4th day, only a few had sprouted. The stems of the electrified sprouts rose, and their first two small leaves opened much more rapidly."[27] Later in his monograph, Jallabert argued that the results compelled a certain conclusion: electricity increases the speed of moving fluids. "Experience demonstrates it, and this is enough to account for the rapid vegetation of electrified plants."[28]

The basic design of the experiments of Demainbray, Nollet, and Jallabert was the same: they compared the growth of electrified plants with that of non-electrified ones. They intended to attribute differences in plant growth to the action of electricity. At the same time, another naturalist conducting analogous experiments was the priest and Doctor of Theology at the University of Angers, François Menon (?–1749). On December 2, 1747, Menon wrote René Antoine Ferchault de Réaumur, Nollet's former teacher and collaborator:

> I sowed and electrified lettuce seeds that I watered before electrifying them. They sprouted three days earlier than those I had sown at the same time and which I watered with the same quantity of water and at the same times. I put in the ground some ranunculus and I tried to plant them equally. Eight days after I electrified them for an hour each day, they are as advanced as the ones that are a month old, if they continue to prosper, I'll have flowers in January at the latest.[29]

Giovanni Battista Beccaria (1753, 125), professor of physics at the University of Turin, summarized the situation as follows: at about the same time, Demainbray in Edinburgh, Nollet in Paris, Menon in Angers, Bose in Wittenberg, and Jallabert in Geneva had been experimenting on the same subject. In Beccaria's view, these researchers, "with their different experiences, had known the same truth."[30] In fact, we have seen that their studies differed in many ways. Not only did they electrify the plants differently, but they also

---

[27] Jallabert (1748, 85): "De la semence de cresson, & de moutarde, appliquée le 26 Décembre à la surface extérieure de ce vase de terre poreuse […] a germé plus promptement sur ce vase électrisé, que lors qu'il ne l'est pas. A la fin du 2^d jour d'une électricité de 8 à 9 heures chaque jour, plusieurs germes de moutarde avoîent poussé. Et, sans électricité, à peine le 4me jour en parutil quelques-uns. Les tiges des germes électrisés s'élevèrent, & leurs deux prémiéres petites feuilles s'épanouïrent aussi beaucoup plus promptement."

[28] Jallabert (1748, 196): "[…] [on est] forcé de convenir que l'électricité augmente la vitesse des fluides qui se meuvent déja. […] l'expérience le démontre; & cela suffit pour rendre raison de la promte végétation des plantes electrisées."

[29] Menon to Réaumur, 2 December 1747, Fonds Réaumur 69 J, 67/24, Académie des Sciences, Paris. The French original reads: "J'ay semé et electrisé des graines de Laitue que y'arrois avant de les electriser. Elles ont levé trois jours plutôt que celles que j'avoit semé en même temps et que j'arrosois avec la même quantité d'eau et aux mêmes heures. J'ay mis en terre des renoncules et j'ay essayé de les rendres égales. Celles qui y etoient depuis plus d'une mois et qui etoient desja bien avancées il y a huit jours que j'ay les electrisé une heure chaque jour, et je suis prêt de les voir aussi avancées que Celles d'un mois, si elles continuent a prosper j'auray des fleurs dans le mois de janvier au plûstard." For the last 18 months of his life, Menon worked as demonstrator of Réaumur's cabinet (Birembaut 1958, 167).

[30] In January 1748, Nollet received a letter from Georg Matthias Bose (1710–1761) informing him that he had electrified several species of plants and shrubs, and that the vegetation appeared to him to be constantly accelerating (Nollet 1749, 356–357).

used different types of plants at different stages of development and focused on different aspects of growth.[31] In one respect they even came to different conclusions: unlike Jallabert, Nollet did not find thicker stems in the electrified plants than in the non-electrified ones. Rather, it seemed to him "that the grains whose germination and growth had been accelerated by electricity had grown smaller and softer stems than those that had sprouted on their own." But he was cautious and "would not dare to say for sure", as he had "not had a large enough number of experiments to be sure."[32]

Finally, the experimenters also had different motivations for working on electricity and plants. Demainbray realized that electricity was the "modish Topick of all Europe" and "the subject of conversation for all the Savants, half-Savants and Ignorants." He therefore "endeavoured to strike out some few Things on this Subject" himself.[33] Menon started electrifying plants because he learned that Mr. Bose (1747) had written a treatise entitled *Tentamina Electrica Tandem Aliqvando Hydravlicae Chymiae Et Vegetabilibvs Vtilia.*[34] Jallabert's and Nollet's experiments, in turn, grew out of their extensive study of electrical phenomena. Both had studied electrical effects in inanimate bodies and wondered whether electricity affected organized, living beings.[35] After noticing that electricity spreads easily in plants, Jallabert (1748, 80) wanted to investigate whether it helped or hindered development. Nollet (1749, 363), on the other hand, found that electricity accelerated the flow of liquids through narrow channels, so he suspected that it had some effect on plant sap. Knowing about this possible influence seemed useful, "especially now that so many people have electrified themselves, & that anyone can easily do so."[36]

---

[31] Demainbray worked with growing myrtle shrubs, Nollet and Menon (and Jallabert, in his last experiment) worked with mustard, cress, and lettuce seeds, while Jallabert used bulbous plants. Some of those had already formed roots, stems, and leaves. Demainbray reported the number of newly formed shoots and their length, while Jallabert initially assessed whether the plants increased in stem, branches, and twigs, and whether they flowered. Later, he also recorded how extensively the leaves spread and how quickly they flowered, how much the plants weighed, how many of them sprouted in a given time, and how quickly they emerged and opened their first leaves. Nollet again measured the number and height of seedlings rising from the ground after a certain period of time. He and Menon paid attention to how quickly the plants germinated and sprouted.

[32] Nollet (1752, 175, footnote): "Il m'a semblé que les grains don't l'électricité avoit accéléré la germination & l'accroissement, avoient poussé des tiges plus menues & plus foibles que celles qui avoient levé d'elles-mêmes; mais je n'oserois l'assurer, n'ayant point eu un assez grand nombre d'éxpériences pour m'en rendre bien certain."

[33] Demainbray's letters to Erasmus King and Abraham de Moivre, both written on November 8, 1746. Records assembled by the State Paper Office, SP 36/89/1, folios 71–72 and 65–66. National Archives, Kew. Heilbron (1979, 261) agrees with Demainbray's assessment that in the second half of the 1740s, nothing was more fashionable than electricity.

[34] Menon to Réaumur, 2 December 1747, Fonds Réaumur 69J, 67/24, Académie des Sciences, Paris.

[35] Nollet had been introduced to this field by his teacher Charles François de Cisternay du Fay (Benguigui 1984, 11).

[36] Nollet (1752, 172): "[…] l'électricité entraîne les liquides qui sont obligés de passer par des canaux fort étroits, je commençai à croire que cette vertu […] pourroit avoir quelque effet sur la sève des végétaux […]. Soit qu'on en dût craindre de mauvaises suites, soit qu'on en dût attendre

3.3   The Purpose of Comparative Experiments and the Need for Control

We have seen that many natural philosophers from across Europe conducted comparative experiments to determine whether electricity promotes plant growth. This section explores their broader attitudes toward how nature should be investigated. Demainbray, Nollet, and Jallabert were among those offering lecture series on Experimental Philosophy, and they advocated the approach of investigating nature by experiment.[37] In what follows, we see that physicists associated experimentation with finding causes. They also used several strategies to avoid drawing erroneous conclusions from comparative experiments.

3.3.1   Discovering Causes from their Effects

In his inaugural lecture as professor of experimental philosophy and mathematics, delivered at the University of Geneva in 1739, Jallabert placed himself in a tradition with Nollet and Demainbray's teacher Desaguliers. A central idea of this tradition, he said, is that good experiments can discover the works of nature:

---

de bonnes, il me paroissoit également utile de le savoir, présentement sur-tout, que beaucoup de personnes se sont électriser, & que tout le monde le peut aisément." Jallabert (1748, 236) followed this reasoning: "The acceleration of the course of the water […] through the capillary pipes by the action of the electric matter & the phenomena which electrified plants give are a strong prejudice that the electric fluid increases the movement of the liquors which the plants contain & that it consequently contributes to pushing & introducing into their extremities the juices necessary to develop them, extend them & increase them."

[37] Jallabert taught as professor of experimental philosophy and mathematics at the University of Geneva. Nollet published his *Leçons de la philosophie expérimentale* in six volumes and read lessons in experimental philosophy to the Duke of Savoy, the Duke of Penthievre, and the Duke of Chartres (Nollet 1743, xiii–xiv). He also reported that several Colleges and Oratories as well as the University of Rheims adopted his plan of introducing experimental proofs into their public exercises. In 1744 and 1745, he taught physics to Prince Louis, son of King Louis XV, and his wife Marie-Thérèse at Versailles. Demainbray started his career as an itinerant lecturer in natural philosophy in the winter of 1748/9. In *The Caledonian Mercury* of June 28, 1748, no. 4324, he "proposeth to give a Course of Experimental Philosophy, consisting of 51 Lectures, and to begin on Monday 7th of November next." Besides all the experiments of his former teacher John Theophilus Desaguliers, he promised "some additional in Mechanicks, Hydrostaticks, Pneumaticks, Opticks, and Astronomy: The Properties of Magnetism will be examined, and the Doctrine of Fire, with the Nature of Electricity attempted." Gentlemen and ladies were asked to pay two guineas for the whole season, or one shilling for a single lecture. After lecturing in Edinburgh, Demainbray travelled through the north of England before moving on to Ireland and France (Morton 1990). Late in 1754 he returned to London, where he began to lecture in 1755. During that year, he gave a course of lectures to the Prince of Wales, the future King George III, and Prince Edward (Morton 1990, 420). By the end of the 1750s he gave up lecturing and became an official of the Excise, and was later superintendent of the observatory built for King George III at Richmond. He was also librarian for Queen Charlotte and instructed her in experimental philosophy and natural history (Rigaud 1882, 281).

You will see there [i.e. in the most famous lyceums of Europe] how successfully the famous Desaguliers's Gravesande, Muschenbroek, and the expert Nollet, who is closely associated with me, teach physics, while whatever they bring forward to the medium, they confirm with their own experiments; [...] [their disciples] want all the Works of Nature to be discovered by certain experiment.[38]

Nine years later, he spoke of what it takes to get a glimpse of "the mechanism by which Nature operates." According to Jallabert, this could only be achieved "by gathering a great number of facts, and by considering them in all their circumstances."[39] Thus, in his *Expériences sur l'électricité*, he set out to "describe the main electrical phenomena accurately, & to arrange them in an order that would facilitate the deduction of the resulting consequences."[40] From the consequences, he hoped to identify the causes and then develop a theory: "For such is, & especially in Physics, the slow but necessary gradation of our knowledge; it is only by the consequences that we can go back to the causes, & arrive insensibly at a theory."[41]

For Demainbray's teacher, too, discovering causes was also a central goal of scientific activity. Desaguliers (1745, iii) wrote that the "business of science" was to "contemplate the Works of GOD, to discover Causes from their Effects, and make Art and Nature subservient to the Necessities of Life, by a Skill in joining proper Causes to produce the most useful Effects." Nollet, for his part, warned against confusing effect with cause, adding that it is easier to recognize the former than the latter.[42]

---

[38] Jallabert (1740, 8): "Perlustra, quaeso, celebriora Europae Lyceae: Videbis & ibi quam feliciter Physicam doceant Celeberrimi *Desaguliers, 's Gravesande, Muschenbroek*, mihique conjunctissimus peritissimus *Nollet*, dum quaecunque in medium proferunt, suis experimentis confirmant; videbis & ibi, quantos faciant profectus in perscrutatione rerum naturalium beati tantorum virorum discipuli, dum Magistrorum exemplo omnia Naturae Opera certo experimento comperta esse volunt." In 1734, Nollet visited Desaguliers in England as well as 's Gravesande and Musschenbroek in the Netherlands (Anstey and Vanzo 2016). Later, Nollet (1770, xiii) referred to the textbooks of both 's Gravesande and Desaguliers. Morton (1990, 413) and Schofield (1970, 81) describe Desaguliers as the doyen of lecturers in natural philosophy in England.

[39] Jallabert (1740, v): "Ce n'est qu'en rassemblant un grand nombre de faits, & en les considérant dans toutes leurs circonstances, qu'on peut entrevoir le mécanisme par lequel la Nature opére." Home (1756, 7): "The operations of bodies are to be accounted for only from their known qualities ascertained by experiment. Reasoning on any other plan, can never certainly lead to truth."

[40] In collecting, organizing, and reflecting on experiments and observations, Jallabert (who described the work of "Nature Historians") thus claimed to be doing what, according to Anstey and Vanzo (2016), corresponded to central activities of the Baconian method of natural history.

[41] Jallabert (1748, iii–iv): "Je ne me suis proposé que de décrire avec exactitude les principaux phénomènes électriques, & de les ranger dans un ordre qui facilitât la déduction des conséquences qui en résultent. Car telle est, & surtout en Physique, la lente mais nécessaire gradation de nos connoissances; ce n'est que par les conséquences que nous pouvons remonter aux causes, & arriver insensiblement à une théorie." He thus endeavored "to describe the main electrical phenomena accurately, and to arrange them in an order that facilitates the deduction of the resulting consequences."

[42] According to Nollet (1743, xxxv), the effect could be known by the least educated peasant, while the cause would not be known by the most learned philosopher.

Now the goal of the experiments described in Sect. 3.1 was not so much to discover unknown causes for mysterious phenomena. Rather, the physicists wanted to decide whether any extraordinary or diminished plant growth could be "attributed" to the action of electricity. To decide this question, they conducted comparative experiments: they performed sets of simultaneous parallel experiments, where one set showed the unperturbed course of nature, or the plant's normal growth. The other set was used to determine how changing one variable, electricity, affected the growth outcome.

The use of comparative experiments to test assumptions about causal relations seems to have been a widespread practice: neither Demainbray nor the anonymous commentator felt the need to introduce or defend the procedure in detail. Neither did Nollet, Jallabert, or any of the other experimentalists discussed here. The only exception is M. d'Ormoy (1789), who sowed an equal number of electrified and non-electrified seeds and assured that everything else "was completely equal, as should always be observed in experiments of this nature."[43] The others did not comment on the design of their trials, perhaps because, like William Marshall (1745–1818), they thought that "the Mode of making Experiments—requires little explanation" (1779, introduction).[44] To make a comparative experiment, according to Marshall, one needs to observe "an identity of *place, time, element*, and *process*, [...] in every particular, excepting only the *intended difference* which constitutes the Experiment" (1779, introduction, emphasis in original). Half a century later, the botanist Augustin-Pyrame de Candolle (1778–1841) stated that to perform an experiment with any certainty means to do so in a comparative manner (1832, 1130). He explained that

> a test proves nothing, as long as another comparative test is not placed next to it [...]. We must place the beings we want to study comparatively in all similar circumstances, except for one, which we will establish as positive in one case, negative in the other. Then we can conclude on one point.[45]

Marshall and de Candolle agreed that the processes to be compared should occur at the same time and place, and should differ in only one factor. The experimentalists mentioned earlier tried to follow this rule. Apart from electrifying the test plants, they treated their tests and controls as equally as possible. In line with Schickore's (Chap. 1, this volume) narrower notion of experimental control, they considered the

---

[43] D'Ormoy (1789, 162): "[E]n un mot, tout a été entièrement égal, ce qu'on doit toujours observer dans les expériences de cette nature."

[44] Schickore (2017), Bertoloni Meli (2009), and Boring (1954) have identified a few experimentalists who applied this strategy in the seventeenth century.

[45] De Candolle (1832, 1535): "[U]n essai ne prouve rien, tant qu'on ne place pas à côté de lui un autre essai comparatif; je m'explique: une expérience ne peut donner qu'un seul résultat. On doit placer les êtres qu'on veut étudier comparativement dans toutes les circonstances semblables, sauf une seule, qu'on établira positive dans l'un des cas, négative dans l'autre. Alors on pourra conclure sur un point." See also de Candolle (1832, 1130): "[J]e demanderdai, [...] si l'expérience a jamais été faite avec quelque degré de certitude, c'est-à-dire d'une manière comparative."

comparative experimental design a prerequisite for making "certain" inferences or discoveries, or for concluding "safely" that electricity promotes vegetation.

3.3.2   Potential Errors and Strategies to Avoid them

In their comparative experiments, the physicists studying electricity and vegetation used what Schickore (Chap. 1, this volume) calls control practices in the broader sense: they kept the two experimental settings stable to rule out as much as possible that the differences in growth were due to factors other than electrification. Nollet and Jallabert not only implemented these strategies in their work, but also discussed them in writing. Nollet (1749, 104) reminded his readers that "we can be fooled by a fact, because we will have changed the circumstances without knowing it, or without paying attention to it." He thus urged that "we must have great regard for these circumstances" known to influence the result, "since they can be an occasion of error, for anyone who neglects to pay attention to them" (127). Jallabert (1748, ix) explained that electrical experiments were particularly susceptible to minute changes in setup—their outcomes can vary infinitely due to slight differences in performance or external circumstances. Nollet's and Jallabert's comments are consistent with Marshall's (1779) more explicit methodological discussion of comparative experiments. He stressed that experimenters have to act prudently and accurately, guarding against any dissimilarity of factors that should be kept constant such as the soil or the seeds.[46]

In the following, we focus on individual elements of experimentation and how experimenters tried to control them. In this way we can identify differences between the methodological views of individual authors and thus have a template for later discussion of further developments.

3.3.2.1   Stabilize (and Monitor) Experimental Conditions

The physicists kept their experimental plants under similar conditions: they moved them together from room to room, to expose them to the same air and equal amounts of water. In addition to ensuring that the control and test plants were at similar temperatures, Nollet and Jallabert also reported the results of their temperature measurements. Jallabert was particularly interested in temperature because of his understanding of the mechanism of plant growth. One of his experiments suggested

---

[46] Marshall (1779, introduction). Home (1756, 3) blew the same horn when he warned: "What a disagreement from a small difference in one of these circumstances!" Another author of an agronomical textbook, Thaer (1809, 9), similarly emphasized that one must prevent as far as possible the interference of anything foreign or unknown that might influence the outcome. For more on Thaer's and Marshall's conception of comparative experiments and their implementation in the agronomical sciences, see Schickore (2021).

that electricity increases transpiration in plants. This increase initially leads to a loss of substance, he thought, but ultimately to growth because the loss is "repaired by food." According to Jallabert, "nourishing juices make plants and animals grow" and electricity "accelerates the flow of fluids and the movement of plant juices."[47] Plant sap and its movement thus played a central role in Jallabert's understanding of how electricity promotes plant growth. Aware that "the sap seems to be in total inactivity in winter," Jallabert admitted that the "experiment made in England on myrtles" seemed to "combat his conjectures." He therefore regretted that he had not been able to determine the temperatures at which Demainbray's myrtles thrived:

> It would have been desirable that in publishing these curious observations the degree of the thermometer in the place where they were made would have been marked. However diligent I may have been to find out about this fact, I was unable to do so, and I do not know if this precaution was not neglected.[48]

Jallabert thought it likely that the sap in Demainbray's electrified myrtles was not not entirely without movement, both because it is warmer indoors than out and because "perhaps the myrtles that the electric virtue had caused to bud were handled before the experiments & then surrounded by spectators." This would have further warmed the room. Jallabert was referring to plant-specific knowledge when he added that "it is certain that the myrtle does not need as much heat to grow as most of the plants that are removed during the winter in greenhouses." He quoted Hales (1727, 62) that pineapples thrive at 29 degrees on John Fowler's thermometer, aloes at 19, Indian figs at 16, orange trees at 12, and myrtles at 9.[49] This episode shows nicely that the decision about which experimental conditions are relevant and worth measuring depended on how the experimenters conceptualized the process under investigation. In Sect. 3.4 we shall meet a physicist who decided to record the weather because he believed that it affected the strength of electricity.

---

[47] Jallabert (1748, 266–267): "[...] les sucs nourriciers qui les font croitre, la dissipation de leur substance causée par la transpiration & que la nourriture répare [...]." And: "On ne doit donc pas trouver étrange que l'électricité qui accéléré le cours des fluides & le mouvement des sucs des plantes, exerce encore son action sur les êtres animes." On pp. 236–237, Jallabert made sense of his and Nollet's observations: "[A]s the nourishing juice flows more easily & more abundantly in the tender organs of a young plant than in those of a plant already strong by the ease it finds in passing through vessels which yield & expand easily, it is doubtless the cause of the rapidity with which the seeds germinated in the ground by the Abbé Nollet and those with which I covered the vase of porous earth of which I spoke. It is apparently by the same mechanism that electricity noticeably hastens the blossoming of the flowers which make all the parts of the plant the most delicate & those where the juices are carried most easily & in greatest abundance. The leaves and the petals that electrification seemed to revive seem to lend a new force to these conjectures since the juice made more abundant in their fibers must, by swelling them, shorten them and consequently straighten them."

[48] Jallabert (1748, 239): "Il auroit été à souhaiter qu'en publiant ces curieuses observations on eût marqué le degré du thermomêtre dans le lieu où elles ont été faites. Quelque diligence que j'aye apporté à m instruire de ce fait je n ai pû y réussir & j'ignore si cette précaution n'a point été négligée."

[49] Jallabert (1748, 240). Jallabert added that this 9th degree at Fowler's thermometer does not quite correspond to the 5th above zero of de Reaumur's thermometer.

3.3.2.2   Stabilize Experimental Objects

Demainbray and Nollet attempted to stabilize their experimental plants by picking myrtles of similar shapes and by using seeds from the same package. None of the experimenters, however, explained why they had chosen myrtles, wallflowers, or cress seeds in the first place. Nor did they specify the exact number of experimental plants (except for Demainbray, who compared one myrtle to another). Jallabert and Menon performed the same experiments on different plant species, and Nollet emphasized that he had repeated his tests on different seeds. We will see that, four decades later, Ingen-Housz called for better control of individual variability in experimental plants.

3.3.2.3   Control Intervention and Detection

Nollet (1749, 103–104) warned about a fundamental source of error when he explained that "in Electricity, as in all other matters of Physics, it is on the report of our senses that we judge things." Because our senses could deceive us, he advised suspending judgment "until we have sufficiently verified the fidelity of their testimony." As a control strategy, he committed to the maxim of making an observation "several times & in the same circumstances" and having "other eyes agree" with his: "Why not hear all the witnesses who can testify to a fact, if the unity of their voices should give more certainty to our knowledge?"

Nollet called for more than just multiple observers to witness an experimental result. He also wanted the same result to occur multiple times. This goal sets him apart from the other physicists. Demainbray was confident that comparing one electrified myrtle with one non-electrified myrtle was sufficient to draw a firm conclusion. Jallabert and Menon made several experiments with different plants, but they did not repeat the same experiment. Nollet, on the other hand, emphasized the need for many experiments to consistently show the same effect. After making repeated tests with mustard seeds, he still felt that more experiments were needed to determine with greater certainty the effect of electricity on vegetation. In experiments preceding his trials with organized bodies, Nollet (1752, 168–169) claimed to have repeated each experiment at least three or four times, adding that the results were identical or differed only slightly. He therefore felt that he could draw safe conclusions. Hales (1727, vii), for his part, maintained that to pry on the operations of Nature, physicists must take the "pains of analysing Nature, by numerous and regular series of Experiments."[50]

---

[50] Desaguliers (1727, 266) picked up on this rhetoric and praised Hales for following in Newton's footsteps, "averting nothing but what is evidently deduc'd from those Experiments, which he has carefully made, and faithfully related; giving an exact Account of the Weights, Measures, Powers and Velocities, and other Circumstances of the Things he observ'd; with so plain a Description of his apparatus, and manner of making every Experiment and Observation, that as his Consequences are justly and easily drawn, so his Premises or Facts may be judg'd of by any Body that will be at the Pains to make the Experiments, which are most of them very easy and simple."

One must avoid errors in the determination of experimental results, for these results are the basis for causal conclusions (Desaguliers 1745, i). If physicists do not measure the effects of their interventions accurately, then they risk drawing false conclusions.[51] For the same reason, we might expect Demainbray and the others to have tried to control their intervention, perhaps even to varying the amount of electricity, to see if the effect varied accordingly. But apparently this was not a concern for most of the physicists introduced so far. Nollet and Jallabert gave more detailed information about how and for how long they electrified their plants. But they did not further quantify the electricity applied.[52] They decided to compare two scenarios—the growth of electrified and non-electrified plants—rather than considering a third group, which they might have electrified twice as long, for example. The one exception was Erasmus King. He conducted what today we would call a sensitivity analysis. He "electrified 12 new laid eggs, three thrice, three 5 times, three 15, and the other three 20 times." However, the experiment was inconclusive.[53]

In contrast to Marshall (1779) and de Candolle (1832), other authors writing about comparative experiments required that the factors vary in order to examine them for their effects. Albrecht Daniel Thaer (1809), for example, maintained that "in order to investigate the effect of a thing under our control," we must

> add and omit, quantitatively and qualitatively change, only this *single* thing in various experiments, set up at the same time and next to each other, but keep everything else as constant as possible. The success will then tell us what part the *single* altered circumstance played in it.[54]

Later, we will meet physicists who tried to measure and vary the electricity supplied to their plants. But only one, Köstlin (1775), made a quantitative argument by

---

[51] Desaguliers (1719, 2) warned that "we must not go about to define a Cause, unless we know its Effects" and advised experimental philosophers to "measure the Quantity of the Effects produc'd, compare them with, and distinguish them from each other" in order "to find out the adequate Cause of each single Effect, and what must be the Result of their Action." In his review of Hales' *Vegetable staticks*, Desaguliers (1727, 264) warned that "without being able to observe, compare, and calculate the exact Quantity of Weight, Force, Velocity, Motion, or any other Change to be taken notice of in making Experiments; Effects may be attributed to Causes which are not adequate to them, and sometimes expected to be produc'd even without a Cause."

[52] Nollet (1749, 157–158) did describe a procedure for measuring the decreases or increases of electricity. His electrometer consisted of a linen thread on an iron rod suspended horizontally, the two ends of which hung parallel to each other (see Figure 3.1). He explained that "as long as the two ends of the wire diverge from each other, it is certain that the body from which they hang is electric, and the angle they form, moving away from each other, is a kind of compass that marks more or less electricity."

[53] Anonymous (1747c, 200). The report continues: "One of these latter eggs produced a chick, and in all there were but 7 chickens hatched, six being addled eggs, among which was one unelectrified egg; so that nothing can be inferred from the experiment." It is unclear whether the one egg remained unelectrified by design or by accident.

[54] Thaer (1809, 10, emphasis in original). The German original reads: "[...] so müssen wir, um die Wirkung eines in unserer Gewalt stehenden Dinges zu erforschen, nur dieses *einzige* in verschiedenen zugleich und neben einander angestellten Versuchen zusetzen und weglassen, quantitativisch und qualitavisch verändern, alles übrige aber möglichst gleich erhalten. Der Erfolg wird uns dann über den Antheil, den der *einzige* veränderte Umstand darauf hatte, belehren [...]."

comparing the amount of electricity with the amount of effect (i.e. the speed of germination).

3.3.2.4 Neutralize Expectations and Report Accurately

A third potential source of error was the experimenter's expectations about test outcomes. Nollet (1748, 191–192) presented his experimental outcomes as confirmation of his theoretical reasoning. He wrote that his conception of plants as hydraulic machines led him to perform the experiments and made him "foresee their Success." This claim is remarkable in light of the fact that authors such as John Keill—"the first who publicly taught Natural Philosophy by Experiments in a mathematical Manner" (Desaguliers 1745, v)—saw this same point as a weakness of the experimental method. According to Keill (1700, 3), experimenters had "too often distorted their Experiments and Observations, in order to favour some darling Theories they had espoused." He therefore urged that comparisons of ratios with the phenomena of nature be made with great caution:

> [W]e are well apprised how fond these Gentlemen are of their Theories, how willing they are that they should be true, and how easily they deceive both others and themselves, in trying their Experiments. Such therefore as are produced by all, and which succeed upon every trial, we receive as undoubted Principles or Axioms: as likewise we ought sooner to give credit to those Experiments that are more simple and easy to be shewn, than to those that are more compounded, and difficult to be performed. (Keill 1700, 7)

In order to proceed with "greater safety, and, as much as possible, avoid all Errors," Keill advocated presupposing only those definitions necessary to arrive at the knowledge of things, and concentrating on one problem, ignoringall irrelevant aspects. He also urged others to start with the simplest cases (7–9).[55] However, neither Nollet nor Jallabert claimed to have considered particularly simple cases of plant growth. In contrast to Keill, Nollet emphasized the advantages of a hypothesis-driven approach. In his view, it was "useful […] in Physics to form a point of view early on, & to establish on first discoveries a system of explanations that one is nevertheless always ready to abandon, as soon as it is contradicted by sufficient reasons […]."[56] Ultimately, Nollet made it clear that physicists should submit only to facts and not to opinions: "When it comes to physics, one shouldn't

---

[55] Keill (1700, 2–3) described the philosophers who proceed upon experiments as one of four most eminent "sects of Philosophers that have wrote on physical subjects" (the others being the Pythagoreans and Platonists, the Peripatetics, and the mechanical philosophers). The Experimenters, according to Keill, "make it their sole business, that the Properties and Actions of all Bodies may be manifested to us, be the means of our Senses." When describing his own "Manner of Proceeding, in the investigating the Causes of Natural Things," Keill proposed to combine useful elements of all four "ways of Philosophizing."

[56] Nollet (1752, 166): "On jugera par les expériences que je vais rapporter, […] combien il est utile en Physique de se former de bonne heure un point de vûe, & d'établir sur les premières découvertes un système d'explications que l'on soit cependant toûjours prêt d'abandonner, dès qu'il sera démenti par des raisons suffisantes […]."

be a slave to authority; one should be even less of a slave to one's own prejudices […]."[57]

Accordingly, physicists were careful to report unexpected results. Jallabert, for
example, initially hoped that a certain experiment would "serve to show more
clearly the way in which the electric fluid accelerates vegetation." But when it
seemed to prove the opposite of his suspicion, Jallabert did "not conceal the fact":
"although [the experiment] did not yield what I expected of it, I must nevertheless
relate it so as not to omit any fact that has some influence on the discovery of the
cause of such interesting phenomena."[58] He (1748, viii) reminded his readers that
"honesty and accuracy in the detail of observations should be the main characteristics of the Nature Historian." Hales (1727, vi) had also emphasized the importance
of accurate description when he asserted: "I have been careful in making, and faith-
ful in relating the result of these Experiments, and wish I could be as happy in drawing the proper inferences from them."

3.3.2.5   Conclude Safely

According to Nollet, natural philosophers question nature by experiment, study
nature's secret by assiduous and well-considered observations, and allow as knowledge of only that which appears to be obviously true.[59] According to Nollet (1749,
189–190; 1752, 168–169), Desaguliers (1727, 264), and Hales (1727, vii), this
method is only reliable when physicists draw conclusions about causal relations
from numerous, well-confirmed comparative experiments with consistent results.

3.4   Comparing the Growth of Electrified and Non-electrified Plants, 1750s–80s

Almost 40 years after Demainbray wrote to *The Caledonian Mercury*, Ingen-Housz
caused a stir by denying that electricity affected plant growth at all. His contemporaries believed that overwhelming evidence for this view had accumulated over the
previous four decades. But a closer look at the experiments from the 1750s to the

---

[57] Nollet (1743, xxi): "[E]n matière d'Physique, on ne doit point être esclave de l'autorité; on dev-
roit l'être encore moins-de ses propres préjugés, reconnoître la vérité; partout où elle se montre, &
ne point affecter d'être Newtonien à Paris, & Cartésien à Londres."

[58] Jallabert (1748, 237–238). Ultimately, Jallabert suggested that "from the fact that the electric
fluid could not in this experiment overcome the resistance occasioned by the gravity of the water
& the friction of the walls of the tube, it should not be concluded that in still narrower pipes such
as those of plants the electric fluid cannot lift and set in motion the liquors they contain."

[59] Nollet (1743, ix): "[O]n prit le parti de l'interroger par l'expérience, d'étudier son secret par des
observations assidues & bien méditées, & l'on se fit une loi de n'admettre au rang es connoissances, que ce qui paroîtroit évidemment vrai."

early 1780s reveals that not all naturalists were convinced that electricity promoted plant growth. The Swedish naturalist Edvard Fredrik Runeberg (1757, 15), for example, said that "far too much" had been concluded from Demainbray's experiments, "for although two electrified myrtle twigs [sic!] have grown more rapidly than unelectrified ones, one cannot be sure that electricity causes the same rapid growth, or that it has the same effect on all twigs of the same kind."

This section examines contributions to the controversy published from the 1750s to the late 1780s in light of the errors and control measures highlighted in Sect. 3.3.2.

3.4.1   No Causal Inference on the Basis of a Single Experiment

Runeberg (1721–1802) was not convinced that electricity caused extra growth in Demainbray's test-myrtle, nor that electricity promoted plant growth in general. In his own experiments, he emphasized details of the intervention and a condition, the weather, that had not previously been considered. Perhaps most remarkably, he refused to draw any causal conclusions.

On July 4, 1754, at 9 o'clock in the morning, Runeberg distributed 22 almonds in four containers of equal depth. He placed eight almonds in each of two nearly identical wooden boxes and three almonds in each of two unglazed stone pots. Of these, one wooden box and one stone pot were electrified, while the other wooden box and the stone pot stood next to each other without being electrified.[60] He stored the electricity generated by rubbing a glass ball in a Leyden jar and transferred it to iron bolts near the almonds (Fig. 3.2). Of his detection regime he wrote: "At 12 o'clock every day, both the electrified and non-electrified plants are measured with a yardstick set up for that purpose and divided into decimal inches and lines" (17). In a table (Fig. 3.3), he noted the weekly growth of the plants, with the electrified plants marked with Latin letters and the non-electrified with Greek letters.

Runeberg decided to run his experiment longer than those of his predecessors, and to electrify the test plants more extensively.[61] He monitored growth over 16 weeks, during which time he electrified the eleven test plants between five and seventy-four times per week. He also built an electrometer to "take into account the relative strength of the electricity" (16), and for each of the 16 weeks he checked the strength (strong, medium, weak) of the electricity applied. In addition to the intensity and frequency of the intervention, he also registered the weekly weather, regretting that he had to do without a barometer and thermometer. Based on his experimental results, Runeberg concluded,

---

[60] Runeberg (1757, 17–18). In the same year that he published his experiments, Runeberg, who was the inspector of weights and measures in Stockholm, became a member of the Royal Swedish Academy of Sciences in Stockholm.

[61] Runeberg (1757, 15).

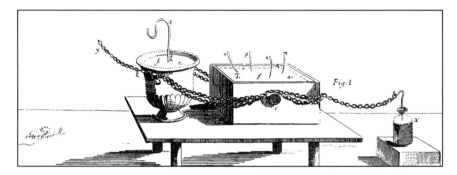

Fig. 3.2 Plate from Runeberg (1757, after 78), depicting the setup to electrify the iron bolts (m, n, o, p, and q) in the stone pot and the wooden box. The bolts are connected to C, the head of an iron bolt of 4 lines in diameter. This bolt receives electricity from the iron chain connected to a Leyden bottle. The position of the almonds is given by the letters a, b, k, d, g, e, f, l

Tàbell, fom utvifar plantornas långder i linier, för hvar åttonde dag.

	$L.$	$F.$	$A.$	$K.$	$G.$	$D.$	$\lambda.$	$\gamma.$	$\delta.$
1 Aug.	6								
8	29								
15	62	44	61	42½	1		6	7	
22	87	67	80	54	22		16	21	
29	104	82	90	60	50		22	41	
5 Sept.	118	86	98	62	71		25	70	
12	126	88	103	65	79		25	83	
19	128	88	107	67	80		25½	90	
26	128	88	111	67	83		27	100	
3 Oct.	128	88	113	67	84		29	101	
10	128	88	115	67	89		30	105	
17	129	88	115	67	90		30	105	0
24	129	88	115	67	90		30	108	4
31	129	88	115	67	90		30	110	7
7 Nov.	129	88	116	67	91	0	30	111	10
14	129	88	116	67	91	4	30	111	20

Fig. 3.3 Table from Runeberg (1757, 28) showing the lengths of the plants in lines, for every 8 days. The almonds L, F, A, K, G, and D were electrified while α, γ, and δ were not. The seven almonds that did not germinate at all are not included in this table nor in Runeberg's calculated averages of growth

(1) that the electrified plants came up first, and probably fastest, but if the electricity caused that speed, several rounds must show. (2) That more of the electrified than of the unelectrified almonds came up. (3) That none of the unelectrified almonds made as many shoots per day as the plant L, namely 8 lines. (4) That none of the unelectrified plants have reached the height of two of the electrified ones. (5) That since the cold began and the strength of the electricity was reduced, the growth of the electrified plants was slowed down. (6) That the electrified plants did not lose in coarseness and steadiness against the unelectrified ones, even as the plant L, which grew strongest of all, shot even from its root a shoot, which was trimmed by violence, and was on the 2nd of September 40 lines long, and as thick as any of the unelectrified ones of the same length. (7) That the average of the growth of the electrified plants was 82 ½, while the average of the unelectrified plants was only 53 $^2/_3$, at the same time as the former were electrified 491 times.[62]

Runeberg distinguished finely between various effects of electricity: (1) the rate of upcoming, (2) the proportion of upcoming plants among those planted, (3) the maximum growth length of a shoot within a day, (4) the height attained, (5) the growth rate of electrified plants as a function of weather, (6) the coarseness and firmness of the plants, and (7) their average height. Although he described marked differences between electrified and non-electrified almonds, Runeberg did not draw any causal conclusions from the differences. His comment suggests that he would not do so until he had repeated the experiment several times with the same result.

3.4.2 Inferring Causes from Constant Effects Rather than Single Observations

Runeberg was cautious and avoided drawing conclusions from a single comparative experiment. At the opposite extreme, we find authors who considered mere observations to indicate the influence of atmospheric electricity on plant growth. Carl Heinrich Köstlin (1755–1783), a student of medicine at the University of Tübingen, stated that we know the positive influence of storms on vegetation "by common experience, especially if with rain the [electric] material of the lightning melts away"; "We see that after such storms the plants, which were previously weak, recover new strength, and the next generation grow happier plants" (1775, 34). He also observed that "the regions subjected to more storms are known as the most

---

[62] Runeberg (1757, 26–27): "Håraf finner man 1), at de electricerade plantorne vål kommo fõrst up, och skõto måstadels fortast, men om electriciteten fõrorsakat den skyndsamheten, måfte flera rõn utvisa. 2) At flera af de oelectricerade, ån af de oeletricerade mandlarna upkommit. 3) At ingen af de oelectricerade mandlarna gjordt så stora skott om dygnet, som plantan L, nåmligen 8 linier. 4) At ingendera af de oelectricerade plantorna hunnit til den hõgd, som tvånne af de electricerade. 5) At sedan kõlden tog til och electricitetens styrka af, saktades de electricerade plantornas tilvåxt. 6) At de electricerade plantorna ej fõrlorat i grofhet och stadighet emot de oelecticerade; hållt som planten L, hvilken våxte starkast af alla, skõt åfven ifrån sin rot en telning, som trifdes vål, och var. den 2 Sept. 40 linier lång, och rått så tjoek som någondera af de oelectricerade af samma långd. 7) At medium af de electricerade plantornas våxt år 81½, då medium af de oelektricerade år allenast 53 $^2/_3$, linea, på lika tid, då de fõrre blifvit electricerade 491 gångor."

fertile regions." Louis-Hyacinthe d'Everlange-Witry (1719–1791) similarly referred to the "observation confirmed by Gardeners, that natural rain, being more or less impregnated with a certain portion of electric fire, is better suited to plants than watering made with other water: You will judge by this the effect of the appreciably electrified rain that is observed at all times" (1777, 18). Based on this observation, he was willing to ascribe a growth-promoting effect to natural electricity.[63]

Jan Ingen-Housz (1730—1799) rejected this argument. He commented that one "cannot doubt the fact" that stormy rains revive vegetation, "but one could doubt whether these rains would not produce the same effect, if they were not electric" (1789, 202). Another four decades later, de Candolle (1832, 1090–1091) noticed similarly that such observations "always leave a little doubt, because it is difficult to isolate by thought the effect of electricity from that of heat and humidity, which more often than not combine with it." Moreover, these events are difficult to observe, "because we are surprised by thunderstorms, and seldom have the presence of mind to measure them exactly." Nevertheless, de Candolle believed that these observations "tend to prove, at least vaguely, the influence of atmospheric electricity on vegetation." The naturalists therefore differed in their view of the value of observations in assessing causal relationships.

Another oft-quoted observation concerned the growth of wild jasmines in the garden of Senator Quirini. Pierre Bertholon de Saint Lazare (1741–1800), who taught physics in Montpellier, quoted from a letter he had received from the priest, physicist, and professor of astronomy at the University of Padua, Giuseppe Toaldo (1719–1797):

> Two of these jasmines which are contiguous to the chain of the conductor […] rose to an extraordinary height, and after two years one saw them surpass the roof of the house, at thirty feet in height; while the other jasmines which are cultivated with the same care, have hardly four feet in height. These two shrubs which are twisted to the mast & to the chain of the conductor, are of a triple size of the others & give flowers before them & in much greater quantity; they still continue to give some several days & several weeks after the others.[64]

Toaldo wrote to Bertholon that "this confirms what you say in your book [*De l'électricité des végétaux*] that the plants grow better and are more vigorous around the lightning conductors, when there are some of them." Indeed, Bertholon

---

[63] We find similar reasoning in Gardini (1784, 25), who stated that "[o]bservations of natural atmospheric electricity further show that the greatest influx into vegetation originates from the same. For the plants begin to develop, grow and flourish, while in spring many stormy clouds begin to appear scattered everywhere, which possess and give the greatest amount of electricity to the air, but this development and vigor of the plants continues until in the autumn such clouds cease; in fact, at the end of the summer the storm clouds decrease in frequency and number."

[64] Bertholon (1787, 371–372): "Deux de ces jasmins qui se trouvent contigus à la chaîne du conducteur dans l'endroit où il s'enfonce en terre se sont élevés à une hauteur extraordinaire, & au bout de deux ans on les a vus surpasser le toit de la maison, à trente pieds de hauteur tandis que les autres jasmins qui sont cultivés avec le même soin, ont à peine quatre pieds de hauteur. Ces deux arbrisseaux qui se sont entortillés au mât & à la chaîne du conducteur sont d'une grosseur triple des autres & donnent des fleurs avant eux & en beaucoup plus grande quantité ils continuent encore à en donner plusieurs jours & plusieurs semaines après les autres."

considered observations to be decisive even when they did not come from comparative experiments. In his view, there was "nothing more decisive than this beautiful observation."[65] Ingen-Housz disagreed. For him, the debate provided a moment to write about the effects of chance and how to control them:

> In order to decide on the existence of a law of nature of this kind [*i. e.* that atmospheric electricity accelerates vegetation], it is necessary that a large number of direct and comparative facts demonstrate its reality by a uniform result. Now, the fact in question is an isolated one, and consequently does not decide anything as such. Pure chance could have produced it among the jasmines, as chance sometimes produces a giant among men.[66]

Ingen-Housz (1789, 330) reminded his readers that "from a particular case of this nature, we cannot legitimately deduce a general consequence." Quirini's experiment "would not decide the question, as long as other similar experiments repeated and observed with care have not had the same effect constantly and obviously" (328). Ingen-Housz suggested that rigorous comparative trials would more accurately identify the cause of the jasmine's extraordinary growth. If "a similar shrub had been planted near a pole which was not topped by a conductor," one would probably have observed a similar effect. According to Ingen-Housz, Quirini would have had a basis for causal conclusions only if he had repeatedly compared the growth of jasmines near lightning rods with that of jasmines near ordinary poles (225–226).

Ingen-Housz asked his fellow physicists "that they have the goodness not to allege single, isolated facts, or such as they hold from hearsay, second or third hand." In his view, the idea of electric force as an accelerator for vegetation had "already served as a basis for endless works & theories, & for costly practices (see Fig. 3.4), which could find themselves quite uninstructed, if, unfortunately, the foundation of the system itself were found to be lacking" (217). He argued that "the public, in a matter of such superior importance", must be able to base decisions on "well-detailed and carefully observed reports of experiments, made by those who present them."[67]

---

[65] Bertholon (1787, 372).

[66] Ingen-Housz (1789, 222): "Pour décider de l existence d une loi de la nature de ce genre, il faut qu'un grand nombre de faits directs & comparatifs en démontrent la réalité par un résultat uniforme. Or le fait dont il s'agit est un fait isolé, & qui par conséquent ne décide rien comme tel. Un pur hasard auroit pu le produire parmi les jasmins, comme un hasard produit quelquefois un géant parmi les hommes." Ingen-Housz (1789, 220–221) met a scholar who had visited Quirini's garden in 1786 and corrected Toaldo's account as follows: "There are not two jasmines, [...] but only one, which is contiguous to the mast surmounted by a lightning rod." The scholar moreover assured Ingen-Housz that this jasmine is "at least three times as big as all the others, [...] leaned against the mast" and that "Senator Quirini, & all those who witnessed this fact, attribute the extraordinary height of this jasmine to the fact that the conductor supplied him with an extraordinary quantity of electric fluid."

[67] Ingen-Housz (1789, 219). Regarding the letter of Toaldo to Bertholon, for example, Ingen-Housz (1789, 219) complained that "some articles of this letter seem to me to lack clarity, which probably comes from the little care that the one who copied or translated it from the original employed there."

Fig. 3.4 Plate II, Fig. 1, from Bertholon (1783, after 468), depicting a means to spread electric rain on trees, in order to increase their vegetation. Bertholon (1783, 406–407) expected that the electric rain would carry to all plants "a principle of fecundity, a particular virtue which has the greatest influence on all the vegetable economy." Contact between a man and an electric machine set in motion is established by means of a chain E, attached to the conductor (D). Standing on a large insulating stool, the man waters the tree (G) by pushing the piston (C). In order to communicate electricity to the tray (B) filled with water, he places one foot on a small plate of tin F

3.4.3 Sensitivity Analyses and Varying the Amount of Electricity Applied

Although many physicists of the 1770s also cited observations to support the view that electricity promoted plant growth, they primarily conducted comparative experiments. Among these efforts, the experiments of Köstlin (1775) stand out as particularly elaborate. Köstlin, one of the few who had read Runeberg's article (in its German translation), studied how electricity affected the development of chicken and butterfly eggs, as well as the growth of certain plants. He listed the "precautions which are to be observed in experiments, if I wished to elicit anything certain from them." These precautions all concern the fact that the test and control plants should be treated as equally as possible.[68]

To get a sense of Köstlin's elaborate experiments, let us look at one that lasted more than two weeks. Köstlin sowed seeds of *Cheiranthus cheiri* into twenty

[68] Koestlin (1775, 7) used equal vessels, and, as far as possible, filled them with the same kind of earth in the same quantity. Each vessel was irrigated at the same time, in the same way, and with the same quantity. The seeds were placed deep in the ground and were, as far as possible, equal seeds. The same number were placed in the electrified and non-electrified vessels. The pots placed outside the windows were positioned next to each other so that they were exposed in the same way to temperature, sunlight, and other factors. Moreover, those not electrified were kept in the same room for the same time while the others were electrified.

vessels. The vessels themselves were of different materials and filled with different substances, and those that were electrified received the electricity in different ways.[69] The "system of electrification" from June 8 to June 16 was as follows:

> On the 8th of Jun. in the morning immediately after sowing, at noon and in the evening, they were electrified, Nr. 2, 3, 5, 9, 10, 12, 14, 16, & 18. for 30 min. Nr. 4. but for 45. min. From Nr. 7. five times in the morning, at noon, and in the evening, as many simple sparks were fired as they could fire after 50 revolutions of the wheel. And to Nr. 8. were applied 5. spark concussions three times a day, so that the individual concussions succeeded each other after 50. rotations of the ball. Simple sparks and concussions were evoked from the surface of the ground and usually in one and the same place.[70]

Köstlin recorded the effects of these treatments by noting the order of the vessels in which the seedlings (not single seeds, but many) germinated:

> June 10 evening in nr. 4.
> D. 11. — morning in nr. 12. at noon in nr. 3. and in the evening in nr. 14.
> D. 12. — morning in nr. 5. at noon in nr. 13. & 16.
> D. 13. — morning in nr. 18. & 15. & evening in nr. 2.
> D. 14. — morning in nr. 5. & 10. noon in nr. 1. & evening in 17.
> D. 15. — morning in nr. 19. & 20. evening in nr. 9.
> D. 16. — morning in nr. 11.
> Note 1.) in nr. 10 & 11. not all the seeds germinated, but more did germinate in nr. 10. than in 11. Furthermore, the seedlings in both pots were very weak.
> Note 2.) in nr. 7. & 8. No seedlings sprouted in those places, from which simple electric sparks and concussions were elicited. Versus the walls of the vessels sprouted indeed some, but they seemed to be weak, and burned immediately if sparks were drawn from them.
> The seedlings grew in the ratio of germination, so that the difference was always noticeable until the 22nd, for on that day the vessels were emptied.[71]

---

[69] From 15 plate vessels, 9 were filled with humus. Vessels Nr. 1. and Nr. 6 were treated in the ordinary way, i.e., they were not electrified. Electrical material was passed through vessel Nr. 2. Vessels Nr. 3. and Nr. 5. were exposed to simple electrification (positive, without sparks). Nr. 4. likewise received simple electrification, but the electrical material was communicated over a longer period of time. The ground of vessels 5 and 6, in addition, was manured with cow dung. Vessel Nr. 7 was treated with simple electric sparks, Nr. 8 with electric shocks, and Nr. 9 was exposed to negative electrification. Plate vessels Nr. 10. and 11. contained clay, 12. and 13. river sand, and 14. and 15. groves of wood. Two glass vessels (16. and 17.) and three shell vessels (18., 19., and 20.) were filled with earth. Vessels 10., 12., 14., 16., and 18. received simple electrification, while vessels 11., 13., 15., 17., 19., and 20. were treated in the ordinary way.

[70] Koestlin (1775, 21–22): "Die 8 Jun mane statim post sationem meridie & vesperi electrisabantur. nro 2, 3, 5, 9, 10, 12, 14, 16, & 18. per 30 min. nro. 4. autem per 45. min. Ex nro. 7. eliciebantur quinquies mane, meridie & vesperi tot scintillæ simplices, quot post 50. gyrationes rotæ elici potuerant. Et nro. 8. adplicabantur 5. scintilla concussoria ter quotidie ita, ut singulæ concussiones post 50. gyrationes globi se invicem succederent. Evocabantur scintillæ simplices & concussoriæ ex superficie humi & plerumque in uno eodemque loco. Continabatur hæc electrisandi ratio ab 8. Jun.—16. Jun."

[71] Koestlin (1775, 22): "Die 10. Jun. vesperi in nro. 4. D. II. — mane in nro. 12. meridie in nro. 3. & vefperi in nro. 14. D. 12. — mane in nro. 5. meridie in nro. 13. & 16. D. 13. — mane in nro. 18. & 15. & vesperi in nro. 2. D. 14. — mane in nro. 5. & 10. meridie in nro. 1. & vesperi in 17. D. 15. — mane in nro. 19. & 20. vesperi in nro. 9. D. 16. — mane in nro. 11. Not. 1.) in nro. 10. & 11. non omnia semina progerminabant, progerminabant vero plura in nro. 10. quam in 11. Plantulæ porro in utroque vase erant valde debiles. Not. 2.) in nro. 7. & 8. plantulæ nullæ in illis locis pro

Köstlin repeated these experiments three more times—twice with *Cheiranthus cheiri*, and once with *Cheiranthus incano*. Seeds of the latter were sown into six equal-sized vessels filled with the same earth, and again they were electrified in different ways.[72] Vessels Nr. 1. and 3., 5, were electrified at the same hours over the same intervals. Köstlin found that the seedlings in Nr. 1. and 3. sprouted after two and a half days, and the seedlings in Nr. 5. sprouted 1 day earlier than those in Nr. 2., 4., and 6 (23).

From these and other experiments, Köstlin concluded that electric matter influences the development of certain plants in the germination of freshly sown seeds. He was convinced that germination was accelerated *by* the passage of electrical material. In contrast to everyone else considered so far, Köstlin was able to argue that germination was accelerated in proportion to the quantity of electrical matter applied. Experiment §. 28., according to Koestlin, shows that electricity is able

> (α.) to accelerate germination *in several types of soil* […]; (β.) with respect to *vessels* that contain earth; a.) mostly in plate vessels, b.) less in glass vessels, c.) not at all in shell vessels. (γ.) germination seems to be more accelerated by electric material than by the *fermentation of cow dung*. (δ.) […] *in free air and in a closed room* […] (ε) even if the *soil*, including the seeds, *is not watered*. (ζ.) to hasten germination if the earth containing the seeds is irrigated by means of *water*, to which *previously the electric charge* has been communicated.[73]

Köstlin further concluded that "negative electrification retards germination," and that seedlings do not grow in places where electric sparks are applied. His experiment §. 29 showed that the electrical material also affects vegetation in plants already germinated.[74] Overall, Köstlin was pleased that some of his results agreed with the "notable" experiments of Runeberg, Jallabert, and Nollet. In his opinion,

---

germinabant e quibus eliciebantur scintillæ electricæ simplices & concussoriæ. Versus vasorum parietes progerminabant quidem aliquot sed debiles esse videbantur & torrebantur statim, si ex illis ipsis scintillæ eliciebantur. Crescebant plantulæ in ratione progerminationis, ita, ut differentia usque ad diem 22. semper effet notabilis, nam illo die vasa evacuabantur."

[72] Koestlin subjected pot nr. 1. to "simple electricity" and kept it "in closed air." Pot nr. 2., was kept in the open air and treated "in a common manner." Nr. 3. was subjected to simple electricity. Nr. 4 was treated in a common way. Vessels nr. 3. & 4. received no water or any other fluid. He irrigated the soil in vessel nr. 5., with water, to which he had previously communicated simple electricity. Finally, he irrigated nr. 6. at the same times at nr. 5, but with non-electrified water.

[73] Koestlin (1775, 29): "α.) accelerare progerminationem *in pluribus terræ speciebus* 1.) in humo, 2.) argilla, 3.) arena, & 4.) scobe lignorum. β.) respectu *vasorum*, quæ terram continent & quidem a.) in vasis bracteatis maxime, b.) in vasis vitreis minus, c.) in vasis testaceis minime. γ.) progerminationem videri magis accelerari a materia electrica, quam *stercoratione simi bubuli*. δ.) accelerare progerminationem *in aëre libero & in clauso*, scilicet in conclavi, cui liber aditus aëris non patet. ε) accelerare progerminationem, quanquam *terra*, quæ semina includit non irrigetur. ζ.) accelerari progerminationem, si modo *aqua*, cui *antea simplex electricitas* est communicata terra semina continens irrigetur" (emphasis in original).

[74] In annual plants, Koestlin (1775, 30) found vegetation to be accelerated by means of electrical stimulation "in such a way that out of two equal plants, if the size of one is increased by electrification, the other becomes equal again to the former; when the former's electrification ceases, the latter is electrified."

the experiments together all "show in a similar manner the earlier germination of the seeds of certain fresh crops, and the increased vegetation of certain plants by the aid of electrification." However, the question of *how* electric matter produces these effects remained open.[75]

One of the few physicists to respond to Köstlin's work was Francesco Giuseppe Gardini (1740–1816), a former student of Beccaria practicing medicine. In 1782 Gardini was appointed professor of philosophy at the college in Alba.[76] Gardini (1784) repeated Köstlin's experiments with cotton and observed "that the seeds of cotton impregnated with electricity germinated more quickly than others, which I kept for comparison under the same circumstances" (18). Like Köstlin, Gardini studied the influence of electricity on annual plants, and "used and changed electricity in many and various ways and observed its influence" on vegetation. He claimed to have obtained the same results as Köstlin, except that he "could not observe a notable difference between positive and negative electrification" (19). To Gardini, from his experiments and those of other authors, it seemed "sufficiently proven" that "artificial electricity influences the life of plants, and that this influence is different in different circumstances and promotes their vegetation" (25).

3.4.4   Less Intricate Comparative Experiments

Other experimental reports were far less intricate. D'Everlange-Witry (1777, 18), for one, merely opined rather than reported an experiment.[77] Bernard Germain Etienne de La Ville-sur-Illon, comte de la Cepède (1756–1825) claimed that the electric fluid's action on vegetation had been "proven by incontestable experiments." But he remained vague in describing his own experiments.[78] Jean-Paul Marat (1743–1793) compared the growth of electrified and non-electrified lettuce seeds:

---

[75] Köstlin (1775, 30–31) wrote that "this is a question to which fate can never give a definite answer, since so much about the very nature of electricity is still hidden from us. For the labors of the natural researchers, as most of their writings sufficiently testify, have certainly made little progress in this matter so far. It must therefore be emphasized in the probable explanation of such phenomena by the other effects of electricity known to us."

[76] Bertholon (1783, 154) outlined an experiment of a certain Édouard-François Nuneberg (sic!), reported by the physical and economic society of Stuttgart. He must have read the strongly abbreviated and erroneous second-hand report (Anonymous 1777), but not Runeberg's (1757) original account or its German translation, Runeberg (1759).

[77] D'Everlange-Witry, a noble canon of Tournai, was the superintendent of the cabinets of rarities at the Court of Brussels of Prince Charles Alexander of Lorraine. In 1773, he became a member of the Imperial and Royal Academy of Sciences and Belles Lettres of Brussels.

[78] De la Cepède (1781, 175–176). He maintained: "Whenever I have electrified a plant, I have also seen it grow and rise more strongly than usual, and I have always succeeded perfectly in hastening the vegetation of plants whose onions are made to germinate and grow in vases full of water." On de la Cepède, see (Schmitt 2010).

On December 3, 1780, I filled six fayance pots with moist soil. I sowed lettuce seed picked on the same stem, and I maintained the fresh soil by watering it. Three of these jars were placed at the bottom of a very large jar on an insulator with a high glass column & in the middle of a chamber, where the thermometer was at two degrees above freezing. The other three were placed on an insulator in the middle of an adjacent room, equally exposed, equally without fire, and where the air was at the same temperature. For fifteen consecutive days, I kept the jar constantly loaded for seventeen hours out of twenty-four, and all this time, the thermometer only varied by one degree.

Now, from the seventh day, we could see the beginning of vegetation in the first ones: it continued to grow little by little; and at the end of the fortnight, the little plants were as advanced as those of another pot which had been sown at the same time and kept in a room where the thermometer was constantly nine degrees above zero. But in the last three pots, there was no appearance of vegetation.[79]

Marat worked with the same experimental plants as Menon and said that the seeds came from the same stem. As with other experiments considered so far, it is unclear how many seeds were sown and how many germinated.

Finally, Bertholon (1783, 166) sowed poppy seeds in two identical vases and electrified one of them "from time to time." He observed "an acceleration in the germination and growth of the parts of the plant, […] and also a multiplication of small branches, leaves, flowers, capsules and seeds, which the poppies in the non-electrified vase did not show, although the cultivation and everything connected with it were the same on both sides."[80]

Bertholon repeated these experiments on tobacco plants with equal success. He found that "the ratios varied, but the plant multiplication in the electrified individuals was always constant" (167). He considered his experiments to be "decisive":

Having electrified some plants for a certain time, & having observed […] that their branches, twigs, and leaves were considerably multiplied, by comparing them with plants of the same

---

[79] Marat (1782, 359–360): "Le 3 Décembre 1780, je remplis de terreau humide six pots de fayance. J'y semai de la graine de laitues cueillie sur la même tige, & j'entretins la terre fraîche en l'arrosant. Trois de ces pots furent placés au fond d'une fort grande jarre sur un isoloir à haute colonne de verre & au milieu d'une chambre, où le thermomètre étoit à deux degrés au dessus de la congélation. Les trois autres furent posés sur un isoloir au milieu d une chambre voisine également exposée également sans feu, & où l'air étoit à la même température. Pendant quinze jours consecutifs je tins la jarre constamment chargée dixsept heures sur vingt-quatre; & tout ce tems thermomètre ne varia que d'un degré. Or, dès le septième jour, on appercevoit un commencement de végétation dans les premiers: elle continua à se faire peu à peu; & au bout de la quinzaine, les petites plantes étoient aussi avancées que celles d'un autre pot qui avoit été ensemencé en même tems & tenu dans une chambre où le thermomètre se soutint constamment neuf degrés au dessus de zero. Mais on n'apperçut dans les trois derniers aucune apparence de végétation." At the time, Marat was still serving as a physician in the household of the Comte d'Artois, but already focused on his career as an experimental physicist (Conner 2012, 23).

[80] Bertholon calculated "average numbers," suggesting to him that the "ratios of multiplication," or the "differences in excesses," were "for the branches of eight more; for the leaves, of thirty; for the flowers & fruits, of six; for the seeds contained in the capsules, of ten."

species in the same circumstances, I always noticed that the roots of the electrified plants were larger, more abundant, better supplied with radicles & hair.[81]

Bertholon reported that, when he examined this object carefully, he found that the ratios of multiplication of the roots and hairs were about the same as that of the branches and leaves, namely 8 to 30. It is striking how precisely he measured the effects of his interventions, while saying little about what the intervention was.

In contrast, Franz Karl Achard (1753–1821), director of the physical classes of the Royal Academy of Sciences in Berlin, distinguished between the application of positive and negative electricity, just as Köstlin and Gardini did. He "filled three Leyden bottles to the half with moistened garden soil, & after having equalized it, I covered it with wet flannel, on which I put cress seed: one of these bottles was not electrified, the other was positively electrified, and the third negatively." However, he did not vary the electricity applied to his plants and did not report the duration of the experiment. He only stated that "every hour, [he] gave back to the bottles their electricity charge, and observed":

1. That the cress seed in the two electrified Leyden bottles germinated more than the one in the non-electrified bottle;
2. That the growth of the germ took place in the two electrified bottles with the same speed.
3. That the plants increased more in height in these two bottles than in the non-electrified bottle.[82]

Note that Achard's goal was not to evaluate the influence of electricity on plant growth; rather, he wanted to compare the effects of positive and negative electricity. He found that the value of the charge did not change the rate of growth.

Experiments in the 1770s and early 1780s by Marat, Achard, Bertholon, and Gardini seemed to confirm the view that electricity stimulates plant growth. And these experiments were not considered questionable at all; on the contrary. Bertholon and Gardini were awarded the *prix de physique* by the Académie des Sciences, Belles-Lettres & Arts de Lyon in 1782.[83] Moreover, the Société Royale des Sciences de Montpellier concluded that Bertholon's monograph (1783) deserved the praise of

---

[81] Bertholon (1783, 167–168): "Ayant électrisé quelques plantes pendant un certain tems, & ayant observé, comme je l'ai dit, que leurs branches, leurs rameaux, leurs feuilles, &c. étoient considérablement multipliées, en les comparant à des plantes de même espece dans les mêmes circonstances, j'ai toujours remarqué que les racines des plantes électrifées étoient plus grandes, plus abondantes, mieux fournies de radicules & de chevelus."

[82] Achard (1784, 432): "1°. Que. la semence de cresson, dans les deuæ bouteilles de Leyde électrisées, germa plutôt que celle qui étoit dans la bouteille non électrisée; 2°. Que. l'accroissement du germe se fit dans les deux bouteilles électrisées avec la même vîtesse. 3°. Que. les plantes augmentèrent plus en hauteur dans ces deux bouteilles que dans la bouteille non électrisée."

[83] In 1782, the Académie des Sciences, Belles-Lettres & Arts de Lyon had formulated the following questions for the *prix de physique*: Does the electricity of the atmosphere have any influence on the plants? What are the effects of this influence? And if it is harmful, what are the means to remedy it?

the public and the approval of the Société because it contained "a large number of interesting, ingenious & decisive observations & experiments."[84]

3.4.5   No Difference and no Reason for Inferring Causal Relevance

In the early 1780s, someone else was studying electricity and plant growth: Jan Ingen-Housz, the court physician to the Austrian empress Maria Theresa. After reviewing other experiments and finding them ill-judged, he sought to test the assumption that electricity promotes vegetation "very carefully by repeated facts."[85] This decision was welcomed by the businessman and chemist Adriaan Paets van Troostwyk (1752–1837) and by the physician Cornelis Rudolphus Theodorus Krayenhoff (1758–1840). These two believed that, all in all, only a few physicists had been concerned with the subject. They also believed they knew why: the experiments that had already been done "seemed to be sufficient in the eyes of the majority of physicists." Instead of repeating the experiments, which seemed useless, physicists simply accepted them—because "the names of Nollet, Jallabert, Menon, Achard, and a few others were authoritative enough to place the acceleration of vegetation by electricity among the best established principles."[86]

After eight years of study, Ingen-Housz (1789) concluded that "the experiments that have so far been offered to show that the electric force accelerates vegetation are not decisive." Crucially, he did not criticize the design of the earlier experiments. His trials were also comparative, but he found no consistent differences between electrified and non-electrified plants. Like the others, Ingen-Housz saw experiments as a means of substantiating causal hypotheses. He described the goal of physics as "the contemplation, in detail, of the intermediate causes & phenomena whose examination is within its reach, or which it produces, by combining different agents" (197). He illustrated this abstract account with the following example: "Rains prodigiously speed up the vegetation." For Ingen-Housz, there was little doubt about this causal relationship: "We see the obvious effects of this. We imitate them by artificial watering which produces the same effect, without ever missing it." But the situation is rather different with the electric fluid, he argued:

> Its influence on plants [...] does not yet seem to me to be specially demonstrated; and I believe that from my experiments I shall be able to conclude that by artificially sprinkling

[84] Extrait des Registres de la Société Royale des Sciences de Montpellier du 1er Juin, 1783, at the end of Bertholon de Saint-Lazare (1783).

[85] Ingen-Housz (1789, 183). Ingen-Housz criticized the work on *mimosa* for attributing to electricity an effect whose occurrence is in fact independent of the presence of electricity (see Ingen-Housz 1786, 92).

[86] Van Troostwyk and Krayenhoff (1788, 134).

plants with this fluid [...], an effect has been attributed to the electric force which, in reality, was produced by the faintness of the light.[87]

Ingen-Housz did not doubt that other physicists had found growth differences, but he did not believe that they were due to electricity. He offered an alternative explanation, attributing the differences to the different light intensities to which the test and control plants were exposed—those less exposed to light grew faster.

But let us start from the beginning. In the spring of 1781, Ingen-Housz (1789, 183–184) placed some daffodils and hyacinths on an insulator, electrified them continously during the day, and placed other similar plants some distance away but did not electrify them. He found no difference in growth. These preliminary trials, Ingen-Housz recalled, showed him that the effect of electricity on vegetation "was not so evident" as he had believed, "according to the writings of the physicists who had established or confirmed this system." Over the next two years, he repeated the experiments but never obtained the same results as the other physicists.

3.4.6 Compensating Individual Variability with Many Experimental Objects

Determined to judge the matter more carefully, Ingen-Housz decided not to work with daffodils or hyacinths. He considered these bulbous plants to be unsuitable test objects "because of the great difference which one often observes in the progress of their vegetation; in such a way that one rarely finds three in a row which grow in a uniform way" (184–185). From this we can deduce that, for Ingen-Housz, suitable test objects should exhibit the target behavior consistently. In other words, the experimental plants should grow uniformly under similar conditions. After all, any differences in growth would provide the basis for causal inferences. Instead of bulbous plants, Ingen-Housz used seeds of mustard and cress, which are plants that grow much more uniformly.

He sprinkled 60 to 100 seeds on a "floating island" made of slices of cork wrapped with pieces of fog paper or linen. He then used different methods to electrify the seeds. At "the same time, in a place far from all electricity," he performed "an equal number of *experiments of comparison*, exactly uniform to those

---

[87] Ingen-Housz (1789, 197): "Son influence sur les végétaux, dont on ne sauroit douter ne me paroît pas encore spécialement démontrée; & je crois que d'après mes expériences je pourrai conclure qu'en arrosant s'il est permis de'm exprimer ainsi, artificiellement les plantes de ce fluide, on a attribué à la force électrique un effet qui, en réalité, étoit produit par la foiblesse de la lumière." Ingen-Housz (1789, 188) explained in more detail: "I have observed that sunlight, so beneficial to adult plants, is very harmful to the development of seeds, and to the growth of very young plants. This is why the seeds of mustard, cress, and probably any other plant, develop better when placed at the bottom of a room, than when they are placed near the windows; and it is probably for lack of this attention, that we have made an erroneous judgement (if it is an erroneous judgement) on the cause of the sudden growth of electrified plants."

mentioned above" (186, emphasis added). "The constant result" was that the electrified plants, "placed in exactly the same circumstances as the others," did not grow faster. Ingen-Housz assured his readers:

> [I]n all these experiments, varied in every way I could imagine, it was evident that the electric force had no effect in advancing vegetation; it was evidently from the greater or lesser degree of light, and in no way from the electric force, that the difference in vegetation acceleration depended. Also no difference could be found between electrified & non-electrified plants, when both were placed at exactly the same distance from the windows.[88]

But he was still not satisfied, and so proceeded to make "infinitely more conclusive" experiments by sowing mustard and cress seeds on the largest fayence dishes he had. This experiment was supposed to be more conclusive because it involved more plants. Each dish contained more than 1000 seeds. Although Ingen-Housz "kept the dishes electrified night and day," the vegetation "was always more or less precocious, […] and the electricity did not contribute in any way to make them grow more rapidly." He thus summarized:

> Seeing that vegetation was always at least as good in the non-electrified jar as in the one that was constantly electrified, it seemed quite clear to me that it was the weakness of the light and not at all the electric force that was the cause of the early growth of the seeds placed in these electrified jars.[89]

3.4.7   The Need for Perfectly Equal Conditions

Van Troostwyk and Krayenhoff (1788) supported Ingen-Housz's findings. They also found no consistent difference between electrified and non-electrified plants. During their study, they observed nothing "that could provide the slightest reason to defend the influence of electricity on vegetation" (140). In the summer and fall of 1786, they experimented with Turkish beans, cress, and horseradish. Unlike Ingen-Housz, however, they compared the growth of individual seeds. On August 3, for example, they chose from many Turkish beans "four beans which appeared to the eye to be exactly alike" and treated them equally with the exception of electrifying

---

[88] Ingen-Housz (1788, 324): "En un mot, dans toutes ces expériences variées de toutes les manières que je pouvois imaginer, il étoit évident, que la force électrique n'avoit aucun effet pour avancer la végétation; c'étoit évidemment du degré de lumières, & nullement de la force électrique, dont la différence dans l'accélération de la végétation dépendoir. Aussi ne pouvoit-on trouver aucune différence entre les plantes électrisées & non-électrisées, lorsque les unes & les autres étoient placées exactement à là même distance des fenêtres."

[89] Ingen-Housz (1786, 92): "En voyant que la végétation se faisoit toujours au moins aussi-bien dans la jarre non électrisée que dans celle qui l'étoit constamment, il me paroissoit assez décidé, que c'étoit la foiblesse de la lumière & nullement la force électrique, qui étoit cause de l'accroissement précoce des semences placées dans ces jarres électrisées."

two of them.[90] The two electrified beans germinated first and continued to grow faster than the non-electrified ones:

> [O]n the 26th of August, when we finished this experiment, one of the electrified plants had a height of 16 inches and a quarter: the other of 21 inches and a quarter: while one of the non-electrified plants was only 8 inches and a quarter, and the other of 10. Since the beginning, 455 hours of electricity had been used.

> [...] although the two electrified plants surpassed the other two in height, they did not appear to be more advanced in other respects, nor more vigorous: for they grew their second and third stems at about the same time as the other two; and all four resembled each other in this respect.[91]

While this first experiment suggested that electricity positively affects vegetation, other attempts yielded different results. On September 1, van Troostwyk and Krayenhoff took three small bean plants, left one "in its natural state," and electrified the others for 76 h. Twelve days later the three plants were "perfectly in the same state, which continued without the slightest difference until the 20th." On the same day they started an experiment with three vases and five beans in each—one vase was not electrified, with the other two positively and negatively electrified respectively. This time the non-electrified plants grew best. Further experiments with cress seeds sown on pieces of wool also showed no difference between electrified and non-electrified plants: "vegetation was equal in all directions." After van Troostwyk and Krayenhoff cut these stems to the same height, "the vegetation started again with an equal vigor without being able to notice the least difference." Repeating this experiment with negative instead of positive electricity gave the same result: "Expansion, germination, growth, and the production of new stems, after the first ones had been cut off: everything, in a word, happened on one of the two pieces of wool as on the other, without us being able to notice the slightest difference."

The two were puzzled that their experiments were "so diametrically opposed to those that were made before" them, since those experiments had been done by physicists whose names will be "forever celebrated in the history of electricity and will always have much authority" (141). They did not doubt the "good faith" of those

---

[90] Van Troostwyk and Krayenhoff (136–137) reported that they "placed each [bean] in a glazed earthenware pot, filled with an equal quantity of the same earth, provided with a hole in the lower part, & placed on a saucer which contained the same quantity of water. We suspended two of them (with their saucers) from metal wires: & two others from silk cords: in order to raise them by a bottle of Leide [...]."

[91] Van Troostwyk and Krayenhoff (137): "[...] le 26 d'Août, que nous terminâmes cette expérience, une des plantes électrifées avoit une hauteur de 16 pouces & un quart: l'autre de 21 pouces & un quart: tandis qu'une des plantes non électrisées n'étoit que de 8 pouces & un quart, & l'autre de 10. On avoit employé depuis le commencement 455 heures d'électricité. Nous croyons devoir ajouter, que, quoique les deux sèves électrifées surpassassent les deux autres en hauteur, elles ne paroissoient cependant pas plus avancées à d'autres égards, ni plus vigoureuses: car elles poussoient leurs secondes & troisiémes tiges à peu près dans le même temps que les deux autres: & toutes quatre se ressembloient à cet égard."

figures, who were "endowed with all the talents necessary to observe Nature well, & who so often gave proofs of their genius & of their exactitude." Van Troostwyk and Krayenhoff maintained that it was difficult to explain their results, "since there are a great number of circumstances which can accelerate or retard the vegetation of plants," and, hence, there are many potential confounders. Nevertheless, they had a suspicion: "[It] seems likely to us that not enough care and precaution was taken in the first experiments on this subject to make all the circumstances of the plants that were electrified and those that were not electrified perfectly equally" (141). They followed Ingen-Housz in suspecting that "perhaps care was not taken to provide the same degree of light to these two types of plants: a circumstance which nevertheless has the greatest influence on vegetation."

3.4.8   No Other Authorities beside Comparative Experiments

Ingen-Housz (1788, 337) did not conclude that the electric fluid has no influence on plants. But he claimed that the experiments hitherto thought to establish electricity's growth-promoting effect "do not have all the authenticity that has been attributed to them." Still, Ingen-Housz hoped that his experiments would motivate other physicists "to imitate them, or to imagine new ones, in order to be able to judge whether, and to what extent, I have been mistaken in my observations." He urged his peers to examine his work and said that nothing would give him more pleasure than to "see my experiments invalidated by others more conclusive." He looked forward to embracing the growth-promoting effect again as soon as a physicist presents "to the court of the public an exact detail of experiments analogous to" his own, or others which would have had a "constant success" opposite to his findings.

Ingen-Housz (1789, 191) demanded that objections be based on observation and not, like the criticism of M. Duvarnier (1786), on "the respectable authority of all the nations and of the most famous physicists they have produced." But Duvarnier was not alone in his position. Thomas Nicolas Jean de Rozieres (1791a, 352), too, considered it legitimate to decide the question "according to recommendable & respectable authorities." Rozieres was concerned that many people "were put off by the numerous contradictions of the scholars, in their writings, which make that after having read a lot, one is often not more informed," and thus wished for more unity among scholars (354, fn 2).

Compelled to defend the call for rigor, Ingen-Housz (1789, 225–226) assured that he was "by no means guided by the spirit of contradiction or criticism […] but by a sincere desire to discover the light in the middle of darkness; by a desire to lift the veil under which nature often likes to hide herself." He emphasized how difficult it is to discover nature's secrets, and how easy it is to err. This was precisely the lesson that Jean-Claude de la Métherie drew from the controversy.[92] The editor of the *Journal de Physique*, where both Ingen-Housz's and Duvarnier's articles appeared, suggested that the experiments "must still be repeated to know finally on which side the truth lies." Van Troostwyk and Krayenhoff (1788, 142), for their part, wished physicists to follow Francis Bacon's lesson in not imagining or supposing, but *discovering*, what nature does or may be made to do.[93] This lesson requires that one avoid the influence of previous studies when experimenting—a demand for which Nollet (1743) had already advocated.

According to Ingen-Housz (1789, 182–183), the problem with the work on electricity and plant growth in the 1770s was the physicists' expectations. He assumed that they were convinced that electricity accelerated vegetation, and so wanted to see this influence confirmed by outdoor experiments. Four decades later, de Candolle revisited the problem. According to de Candolle (1832, 1535), "most of those who make experiments like to see them succeed." As a result, experimenters "always tend, by a very forgivable inclination of the mind, to exaggerate the favorable results of their trials, and to conceal the contrary results." But if naturalists fail to report experiments that find no effect, while exaggerating effects when they do find them, then they distort the facts. Hence one sees a multitude of procedures praised by authors and unchecked by newspapers—a situation that in reality "cannot be sustained in practice, nor enlighten the theory."[94]

De Candolle proposed two countermeasures. First, experimenters should take systematic notes to prevent their expectations from inflating their observations. "Without precise notes, without rigorous labels, without exhibits," de Candolle (1832, 1536) believed, "the most exact minds are prone to strange illusions about long-lasting phenomena." Second, de Candolle criticized learned societies and journal editors, because they were supposed to be the institutional bodies of control. They should act as gatekeepers, publishing only reports that meet certain standards. He deemed it "desirable that this mass of agricultural and horticultural societies which cover Europe today, accept in principle to give some attention only to those experiences which are really comparative and expressed by formal figures" (1535).

---

[92] The editor's note on Duvarnier (1786, 94) reads: "Such opposite results in experiments made on the one hand by Physicists as famous as those quoted here by Mr. Duvarnier, and on the other by Physicists no less famous, Messrs. Ingen-Housz, Schwankhardt, etc., must surprise, and show all the difficulties that the art of experiments presents."

[93] After Bacon (1620, Liber Secundus, Aphorismus X).: "Primo enim paranda est historia naturalis et experimentalis, sufficiens et bona; quod fundamentum rei est: neque enim fingendum, aut excogitandum, sed inveniendum, quid natura faciat aut ferat."

[94] De Candolle (1832, 1534–1535) complained: "Every day one reads in the books on cultivation, and hears in conversation, the use of such and such a process, and proclaims it good or bad, without an exact term of comparison. The product is related to an approximate average that each one has thought of the product of his fields; and when one comes to a more careful examination, one recognizes that this average is almost arbitrary within large limits; that, consequently, the vague assertion that a process has succeeded well or not so well is very often due to the personal character of the observer."

3.5 Comparative Experimentation in the Eighteenth Century and beyond

Psychologist Edwin Boring (1954, 589) suggested understanding the methodological status of control as check or comparison with reference to John Stuart Mill's (1806–1873) method of difference. Mill (1843, 459) praised this method as the only way to "arrive with certainty at causes" through direct experience. The method shares the essential features of a strategy we have seen in the writings of various naturalists—two settings are kept perfectly equal except for one intended difference (Marshall 1779); a single circumstance is altered (Thaer 1809); or one is established as positive (de Candolle 1832). One then evaluates how the two settings compare. Other authors who discussed comparative trials were Matthias Jacob Schleiden (1804–1881) (see Nickelsen, Chap. 7, this volume) and Claude Bernard (1813–1878). According to Bernard (1865, 224), comparative experimentation allows physiologists to isolate a phenomenon to be studied from all the complications surrounding it. It does so by adding to a comparison organism all the experimental modifications except one, which is the one they wish to identify.[95] In Mill's (1843, 455) words, experimentalists strive to bring about two instances—one where the phenomenon occurs, and one where it does not—that have all circumstances in common except one. The circumstance "occurring only in the former; the circumstance in which alone the two instances differ" is the "effect, or cause, or a necessary part of the cause, of the phenomenon."

This section reinforces Boring's proposal by summarizing the main findings about eighteenth-century control practices that emerge from the controversy on electricity and vegetation. We shall further discuss the connection between comparative experimentation and the study of biological phenomena.

3.5.1 Comparative Experimentation and Strategies of Control

We have seen that the concept of comparative experimentation was not only mentioned in methodological discussions, but also guided physicists in the design of their experiments. Between the 1740s and 1780s, physicists compared the growth of electrified and non-electrified plants in the kingdoms of Great Britain, France, Savoy-Piedmont, Sweden, Prussia, the Netherlands, the Republic of Geneva, the

---

[95] Bernard (1865, 224): "[I]l nous suffira de bien isoler le seul phénomène sur lequel doit porter notre examen en le séparant, à l'aide de l'expérimentation comparative, de toutes les complications qui peuvent l'environner. Or, l'expérimentation comparative atteint ce but en ajoutant dans un organisme semblable, qui doit servir de comparaison, toutes les modifications expérimentales, moins une, qui est celle que l'on veut dégager." For the differences between Mill's method of difference and Bernard's comparative experimentation, see Schickore (2017, chapter 7).

Archduchy of Austria and the Holy Roman Empire. Many of the authors earned their living as itinerant lecturers or university professors of experimental or natural philosophy, or of physics (Demainbray, Nollet, Jallabert, Beccaria, Gardini, Bertholon). Others funded their research through employment as (court) physicians (Marat, Ingen-Housz, Krayenhoff), and a third group belonged to the clergy (Menon, d'Everlange-Witry). Given this diverse group of experimentalists, it is remarkable that they all conducted comparative experiments: they agreed that the method was essential to draw conclusions about cause-effect relationships. The practice seemed so familiar to physicists (and to journal editors) that they rarely defended it explicitly.[96]

As the following summary shows, it is instructive to consider the control strategies of these physicists in light of their goal of inferring causes from differences.

3.5.1.1 Stabilize (and Monitor) Experimental Conditions

A key control strategy was to treat the test and control plants as equally as possible except for the intervention. The physicists spent many words testifying that they had indeed maintained equal treatment except for electrification. If the test and control plants were consistently different, they felt justified in identifying electricity as the cause of extraordinary plant growth. But if there were no consistent differences, they concluded nothing about the causal role of electricity.

While everyone seemed to agree on this, they differed on how many experimental runs they needed to draw conclusions. Some were content to draw far-reaching conclusions based on single comparative trials. Nollet (1749) and Runeberg (1757), on the other hand, insisted that multiple replications were necessary and that the test and controls should be consistently different before any conclusions could be drawn. After Ingen-Housz's unexpected findings, the problem of control took on new urgency. Ingen-Housz (1789) and van Troostwyk and Krayenhoff (1788) again emphasized the need for many rounds of comparative experiments with consistent results in order to draw reliable causal inferences. Otherwise, one runs the risk of attributing an effect to the intervention when in fact it occurred by chance. In contrast to what Schickore (2011, 516–520; 2017) found in her analysis of seventeenth- and eighteenth-century snake venom experiments, the strategy of many repetitions of the same experiment was less firmly anchored in the minds of the physicists considered here.

---

[96] The practice was by no means limited to the circle of "electrifying philosophers." Schickore (2021, 487) found the same for practitioner-authors in agricultural science for the same period.

3.5.1.2   Stabilize Experimental Objects

There was no extensive debate about the choice experimental plants. Runeberg (1757) tested whether electricity would also benefit nut growth, after it had been shown to do so for shrubs, onion plants, and seeds. Köstlin (1775) and Gardini (1784) were careful to study both annual and perennial plants. Ingen-Housz (1789) was the only author to discuss the growth characteristics of different plants, and he argued that cress was more suitable for growth studies than other species because it grew uniformly under the same conditions. Other physicists chose seeds from the same stem, or took shrubs or beans that looked as identical as possible on the outside. Their goal was to minimize the risk that different results were caused by individual variability in the plants.

The task of stabilizing the experimental objects has challenged physiologists ever since. Bernard (1865, 225–226) noted that "no animal is ever absolutely comparable to another." In his view, experimentalists can therefore only assume that the "two animals being compared are sufficiently similar" so that the "difference observed in them as a result of the experiment cannot be attributed to a difference in their organism." Decades later, botanist F. A. F. C. Went (1863–1935) admitted that "the material being experimented with, the living plant, cannot be kept completely constant" (Went 1931, 173). This was a problem because, according to Went (1933, 137), in order to examine the "influence of any factor on a life process," one "needs to keep all other factors constant, let only one change and then wait for the result." Went's own son Frits Went (1903–1990) struggled with the same problem. When his experimental plants showed different responses in reaction to a given intervention, he suspected that they were "not all equal." Went (1928, 27–28) determined the reaction of a "larger number of reaction plants" and was thus able to "arrange the obtained numbers in the form of a binomial curve." This approach, without the statistical model, was exactly what Ingen-Housz (1788) had chosen 140 years earlier. Instead of comparing a few plants like other physicists, he followed the growth of thousands of cress seeds. His contemporaries did not follow suit. Van Troostwyk and Krayenhoff (1788), for example, compared five electrified beans with five non-electrified beans. Bernard spoke of comparing *two* animals.

3.5.1.3   Control Intervention and Detection

Some physicists described how they supplied electricity to their test plants, although we learn little about whether and how they tried to control the intervention. Only one, Köstlin (1775), intentionally varied the amount of electric force in order to evaluate whether the putative effect also varied. Nor did the physicists make fine distinctions in conceptualizing the problem. They took their trials all as contributions to one and the same problem, even though they were investigating different aspects of vegetation such as the formation of new branches (Demainbray), opening of seeds in a given time (Nollet), formation of additional leaves (Jallabert), amount of growth in length during a certain interval (Runeberg), order of germination (Köstlin), strength of growth (de la Cepède), or the recovery of weak plants (Bertholon).

According to the historian of biology Brigitte Hoppe (2010, 107), the plants were electrified without measuring the amount of static electricity in most of the experiments. This fact points to the role the experiments played for the experimenters. In Hoppe's view, they demonstrated the wonders of nature in an entertaining way. In contrast, she credited Ingen-Housz with a genuine interest in plant physiological mechanisms, which would explain his more careful experimentation.

On closer inspection this explanation is not valid. For one, Ingen-Housz's methodological ideas coincided quite closely with those of Nollet (apart from his using thousands of plants). For another, Nollet and Jallabert were actually interested in how electricity promotes plant growth (see Sect. 3.3.2). What did change between the 1740s and 1780s, however, were the conceptions of plant growth and of nutrition. In the 1770s, the "simple" view of plant growth, on which plant material was no more than transmuted water, was undermined (Nash 1957, 344). New studies emphasized the role of light and the atmosphere on vegetation (350). Under these circumstances it is hardly surprising that, in the 1770s, a new generation of physicists attempted to prove that atmospheric electricity promotes vegetation. It is equally understandable that Ingen-Housz, after his studies on the influence of light on plants (Ingen-Housz 1779), was prepared to give a central role in plant development to light.

3.5.1.4 Neutralize Expectations, Report Accurately, and Conclude Safely

For the protagonists considered in this paper, the details of experimental procedure were crucial for assessing the safety of the experimental conclusions. At the same time, though, they often did not have access to those details. In all likelihood, none except for Hales had read Demainbray's reports. Jallabert's experiments were probably known to many through the writings of Nollet, and the same is definitely true for the experiments of Bose and Menon.[97] Another contribution rarely read in the original was that of Runeberg (1757). Nevertheless, van Troostwyk and Krayenhoff (1788) felt free to criticize it.[98]

Similarly, several authors worried about how physicists' expectations affected their work. Keill (1700) warned that experimental results are often distorted by physicists wishing to confirm their favorite theories. Van Troostwyk and Krayenhoff (1788, 141)

---

[97] In any case, no one mentioned that Jallabert at first could not find any clear effect.

[98] Despite the fact that Runeberg was careful not to draw any conclusions at all, van Troostwyk and Krayenhoff (1788, 141) were "astonished that some Physicians & especially MM. Achard & Nunebert [sic!], have dared to decide a question of such importance on the basis of so few facts." Most probably, they were referring to Anonymous (1777, 436), an erroneous secondhand report on Runeberg's trial.

suggested that knowing others' conclusions about the same issue might lead physicists to interpret experimental results too hastily, such that they confirm earlier findings. Since famous and capable experimenters had found that electricity positively affected plant growth, their successors were well advised to find the same. Because the physicists believed the matter to be "sufficiently decided by the experiments of their predecessors," they were "satisfied with a single experiment which by chance succeeded in confirming them in the feeling for which they were so strongly advised." The two admitted that "the same thing could have happened to us, if we had wanted to be satisfied with a small number of observations: since our first experiments seemed to confirm the doctrine of electricity in plants." To counteract this dynamic, Ingen-Housz asked his colleagues for a rigorous review of his experiments. Van Troostwyk and Krayenhoff suggested, referring to Bacon, that physicists should not expect experimental outcomes in a way informed by earlier experimental findings. Rather, they should investigate without bias what nature does or can be made to do. In the 1830s, de Candolle advised naturalists to systematize their note-taking. He suggested that journal editors accept only those contributions that met certain methodological standards. We can understand these suggestions as attempts to discipline the community of experimentalists and to ensure the quality of their experiments.

3.5.2 Controlling Complex Systems

De Candolle (1832, 1534–1535) claimed that the "logical method" of "rigorously comparative experiments" was "well known in all the other sciences." Marshall (1779, 17) similarly considered "comparative Experiments" to be the hallmark of science, and necessary for the acquisition of knowledge. In contrast, Albrecht Daniel Thaer (1752–1828), another author of an agronomic textbook who characterized comparative experiments, considered the comparative method appropriate for many, but not all, empirical sciences. According to Thaer (1809, 9–10), comparative experiments are the strategy of choice when experimenters do not have full control over all conditions—for example, when they cannot introduce or remove conditions at will, or even measure and weigh them. In contrast, in an isolated room such as the chemist's laboratory, Thaer thought it possible to perform completely perfect and pure experiments.[99] Bernard explained that it would never be possible, on the other hand, "to experiment with any degree of rigor on living animals" because physiological phenomena are so complex.[100] Comparative experimentation,

---

[99] Thaer (1809, 9–10). In the chemical laboratory, naturalists "allow known and measured substances and potencies to interact, cut off the influence of other substances and potencies, and note the success of the experiment."

[100] Bernard (1865, 223): "Les phénomènes physiologiques sont tellement complexes, qu'il ne serait jamais possible d'expérimenter avec quelque rigueur sur les animaux vivants, s'il fallait nécessairement déterminer toutes les modifications que l'on peut apporter dans l'organisme sur lequel on opère."

however, can reduce this complexity and "eliminate en bloc all known or unknown causes of error." In other words, the great advantage of comparative experimentation is that experimenters do not need to have control nor stabilize many potentially relevant conditions at all. Conditions such as the weather or temperature in the plant experiments, can vary as long as they do so in the same way for test and control objects. Thus, the variation poses no threat to the validity of the experiment.

This technique greatly facilitates the study of animals and plants, but it was also used in studying less complex and inanimate systems. For example, Nollet (1749, 140–141) used the comparative approach to study the process of cooling liquids and the influence of electricity on it.[101] Presumably he thought it would be less work to study the process in two separate vessels simultaneously than to control the room temperature precisely during the two successive cooling processes. In another case, Nollet compared the velocity of electrified and non-electrified water streams. In this case, the experiments to be compared did not run parallel, but one after the other. After measuring the flow of electrified water, Nollet used "the same water and the same vase" when he repeated the experiment without electrification. He noted the duration of this flow "for comparison with that of the first" (346). Since little time passed between the test and control instances, Nollet could assume that the environmental conditions had not changed much. We can thus conclude that simultaneous, comparative experimentation is the procedure of choice in two situations: when the process takes a long time, and/or when the process cannot be observed twice on the same object (as in the case of a directed developmental process, such as plant growth).

3.6   Conclusion: The Need for Rigor

This chapter has examined physicists from across Europe who, between the mid-1740s and the mid-1780s, investigated whether electricity promoted plant growth. Reports of their experiments were presented at the meetings of illustrious societies like the Académie Royale in Paris, the Royal Society in London, and the Royal Swedish Academy of Sciences in Stockholm. The controversy attracted attention even beyond the circle of practicing experimental philosophers.[102] Ingen-Housz's experiments by no means settled the question. De Candolle (1832, 1097),

---

[101] Nollet described this experiment as follows: "I filled two cylindrical glass vases of the same height and capacity with water; I plunged the ball of a very sensitive thermometer into one and the other, so that it did not reach the bottom of the vessel; I put the whole thing in a hot water bath until the liquor of the two thermometers had risen to 40 degrees; then I placed one of the two vessels on the metal cage to be electrified and I put the other one on a table a little apart, but in the same place. I observed the two thermometers, whose constant reading on both sides taught me that electricity neither delayed nor accelerated the cooling."

[102] Theologian Samuel Miller (1803, 27), for example, wrote that "the correction of former errors, with respect to the influence of electricity on vegetables, by Dr. Ingenhouz, may be considered among the most interesting of recent improvements" in the study of electricity.

who thought it probable that electricity stimulated plant life, suggested that the subject "must be elucidated by precise experiments under the direction of a physicist familiar with the phenomena of plant life." However, he warned that "such comparative experiments are difficult to rid of all causes of error" (1094).

This impression was shared by Ingen-Housz and his contemporaries. The controversy reminded them how error-prone experimental work is and demonstrated the difficulty of systematically investigating causes and effects (Schickore 2021, 502; Schickore 2023). Some authors used the opportunity to call for stricter methodological standards, hoping that increased rigor would help to uncover the secrets of nature more efficiently.[103] The example illustrates that practicing experimentalists have given a lot of thought to sources of error. They incorporated these considerations into their study designs and into the organization of their scientific communities. These are compelling reasons for further study of historical practices of experimentation to improve our understanding of how these discussions and practices have evolved.

Acknowledgments I would like to thank Jutta Schickore and the Mellon-Sawyer seminarians for the warm welcome in spring 2022 and the stimulating discussions on experimental control and beyond. Further, I thank Jutta Schickore, Claudia Cristalli, Cesare Pastorino, as well as the other participants of the October 2023 chapter workshop for their many helpful comments on a first draft of this chapter. Finally, I would like to thank Marc Ratcliff for his valuable support in the early stages of this project.

References

Achard, Franz Karl. 1784. Mémoire renfermant le récit de plusieurs expériences électriques faites dans différentes vues. *Observations sur la physique, sur l'histoire naturelle et sur les arts* 25 (2): 429–436.

Anonymous. 1747a. Comment on Electricity, effects of, on vegetation. *The Gentleman's Magazine* 17 (2): 81.

———. 1747b. Comment on Electricity, effects of, on vegetation. *The Gentleman's Magazine* 17 (2): 102.

———. 1747c. N.B. *The Gentleman's Magazine* 17 (4): 200.

———. 1752. *Histoire générale et particuliere de l'électricité 1*. Paris: Rollin.

———. 1777. Effet de l'éléctricité sur la végétation. In *Introduction aux observations sur la physique, sur l'histoire naturelle et sur les arts*, vol. 1, 436. Chez Le Jay.

———. 1795. Dr. Demainbray. *The Environs of London* 3: 317–318.

Anstey, Peter R., and Alberto Vanzo. 2016. Early Modern Experimental Philosophy. In *A Companion to Experimental Philosophy*, ed. Justin Sytsma and Wesley Buckwalter, 87–102. Malden: Blackwell.

Appleby, John H. 1990. Erasmus King: Eighteenth-century Experimental Philosopher. *Annals of Science* 47 (4): 375–392.

Bacon, Francis. 1620. *Instauratio magna* [Novum organum]. London: John Bill.

---

[103] According to Senebier (1772, 40), it is necessary for observers to shorten their labours by prescribing a rigorous method for the study of their subjects.

Beccaria, Giambattista. 1753. *Dell'elettricismo artificiale e naturale 2*. Torino: Filippo Antonio Campana.

Benguigui, Isaac. 1984. *Théories électriques du XVIIIe siècle*. Geneva: Georg et cie.

Bernard, Claude. 1865. *Introduction à l'étude de la medicine expérimentale*. Paris: J. B. Baillière et fils.

Bertholon de Saint-Lazare, Pierre-Nicolas. 1783. *De l'électricité des végétaux*. Paris: P. F. Didot Jeune.

———. 1787. *De l'Électricité des Météores 2*. Lyon: Bernuset.

Bertoloni Meli, Domenico. 2009. A Lofty Mountain, Putrefying Flesh, Styptic Water, and Germinating Seeds. In *The Accademia del Cimento and its European Context*, ed. Marco Beretta, Antonio Clericuzio, and Larry Principe, 121–134. Sagamore Beach: Science History Publications.

Birembaut, Arthur. 1958. Les liens de famille entre Réaumur et Brisson, son dernier élève. *Revue d'histoire des sciences et de leurs applications* 11 (2): 167–169.

Boring, Edwin Garrigues. 1954. The Nature and History of Experimental Control. *American Journal of Psychology* 67: 573–589.

Bose, Georg Matthias. 1744. *Commentatio de electricitate inflammante et beatificante*. Wittenberg.

———. 1747. *Tentamina electrica, tandem aliquando Hydraulicae, Chymiae et vegetabilitas utilia 2*. Wittenberg: Johann Joachim Ahlfeldt.

Cavallo, Tiberius. 1782. *A Complete Treatise of Electricity in Theory and Practice with Original Experiments*. London: Edward and Charles Dilly.

———. 1803. *The Elements of Natural or Experimental Philosophy 3*. London: T. Cadell and W. Davies.

Conner, Clifford D. 2012. *Jean Paul Marat: Tribune of the French Revolution*. London: Pluto Press.

d'Everlange-Witry, Louis-Hyacinthe. 1777. Mémoire sur l'Electricité, relativement à sa qualité de fluide moteur dans les végétaux & dans le corps humain. *Mémoire de l'Académie impériale et royale des sciences et belles-lettres de Bruxelles* 1: 181–192.

D'Ormoy, M. 1789. De l'influence de l'électricité sur la végétation, prouvée par de nouvelles Expériences. *Observation sur la physique, sur l'histoire naturelle et sur les arts* 35 (2): 161–176.

———. 1791. Experiments on the Influence of Electricity on Vegetation. *Annals of Agriculture and other Useful Arts* 15: 28–60.

de Candolle, Auguste-Pyrame. 1832. *Physiologie végétale, ou, Exposition des forces et des fonctions vitales des végétaux pour servir de suite a l'organographie végétale, et d'introduction a la botanique géographique et agricole 3*. Paris: Béchet Jeune.

de la Cepède, Bernard Germain Etienne de La Ville-sur-Illon, comte de. 1781. *Essai sur L'Électricité Naturelle et Artificielle 2*. Paris: Imprimerie de Monsieur.

de Rozieres, Thomas Nicolas Jean. 1791a. Essai sur cette question: Quelle est l'influence de l'Electricité sur la Germination & la Végétation des Plantes. *Observations sur la physique, sur l'histoire et sur les arts* 38 (1): 351–365.

———. 1791b. Suite de l'Essai sur cette question: Quelle est l'influence de l'Electricité sur la Germination & la Végétation des Plantes. *Observations sur la physique, sur l'histoire et sur les arts* 38 (1): 427–446.

Demainbray, Stephen. 1747a. "Letter to the publisher of *The Caledonian Mercury*, 19 January 1747." *The Caledonian Mercury* of Tuesday 20 January 1747, no. 4100: 3.

———. 1747b. "Letter to the publisher of *The Caledonian Mercury*, 19 February 1747." *The Caledonian Mercury* of Friday 20 February 1747, no. 4113: 2–3.

Desaguliers, John Theophilus. 1719. *A System of Experimental Philosophy, Prov'd by Mechanicks*. London: B. Creake, J. Sackfield, W. Mears.

———. 1727. An account of a book entitul'd vegetable staticks [...]. *Philosophical Transactions* 34(398): 264–291.

———. 1739. Some Thoughts and Experiments concerning Electricity. *Philosophical Transactions* 31 (454): 186–194.

———. 1742. *A Dissertation concerning Electricity*. London: W. Innys and T. Longman.

———. 1745. *A Course of Experimental Philosophy*. 2nd ed. London: W. Innys, T. Longman and T. Shewell.

Duvarnier, M. 1786. Observation relative à la Lettre de M. Schwankhardt, au sujet de l'influence de l'electricité sur la vegetation. *Observations sur la physique, sur l'histoire naturelle et sur les arts* 28 (1): 93–94.

Gardini, Francesco Giuseppe. 1784. *De influxu Electricitatis Atmosphericae in Vegetantia. Dissertatio ab Academia Lugdunensi praemio donata an MDCCLXXXII*. Torino: Gianmichele Briolo.

Hales, Stephen. 1727. *Vegetable Staticks: or, an account of some statical experiments on the Sap in Vegetables*. London: W. & J. Innys.

———. 1748. VII. Extract of a Letter from the Rev. Dr. Stephen Hales F. R. S. to the Rev. Mr. Westly Hall, Concerning Some Electrical Experiments. *Philosophical Transactions* 45: 409–411.

Hausen, Christian August. 1746. *Novi Profectus in Historia Electricitatis*. Leipzig: Theodor Schwan.

Heilbron, John L. 1979. *Electricity in the 17th and 18th centuries a study of early modern physics*. Berkeley, Los Angeles and London: University of California Press.

Home, Francis. 1756. *The Principles of Agriculture and Vegetation*. Edinburgh: Kincaid and Donaldson.

Hoppe, Brigitte. 2010. Experimentation in the Early Electrophysiology of Plants in the Frame of an Experiment-oriented Mechanical Natural Philosophy. *Annals of the History and Philosophy of Biology* 15: 101–117.

Humboldt, Friedrich Alexander. 1794. *Aphorismen aus der chemischen Physiologie der Pflanzen*. Leipzig: Voss & Compagnie.

Ingen-Housz, Jan. 1779. *Experiments Upon Vegetables, Discovering Their Great Power of Purifying the Common Air in the Sun-shine, and of Injuring it in the Shade and at Night*. London: P. Elmsly and H. Payne.

———. 1786. Lettre de M. Ingen-Housz à M. N. C. Molitor, au sujet de l'effet particulier qu'ont sur la germination des semences et sur l'accroissement des plantes formées, les différenes espèces d'air, les différens degrés de lumière et de chaleur, et l'électricité. *Observations sur la physique, sur l'histoire naturelle et sur les arts* 28 (1): 81–94.

———. 1788. Lettre de M. Ingen-Housz à M. Molitor, au sujet de l'influence d'Electricité athmosphérique sur les Végétaux. *Observations sur la physique, sur l'histoire naturelle et sur les arts* 32 (1): 321–342.

———. 1789. *Nouvelles Expériences et Observations sur divers Objets de Physique, 2*. Paris: Théophile Barrois.

Jallabert, Johannis. 1740. *De philosophiae experimentalis utilitate, illiusque et matheseos Concordia. Oratio inauguralis*. Genevae: Barrillot et Filii.

Jallabert, Jean. 1748. *Experiences sur l'electricité avec quelques conjectures sur la cause de ses effets*. Geneva: Barrillot et Fils.

Keill, John. 1700. *An Introduction to Natural Philosophy: or, Philosophical Lectures Read in the University of Oxford Anno Dom. 1700*. To which are Added, The Demonstrations of Monsieur Huygens's Theorems concerning the Centrifugal Force and Circular Motion. London: William and John Innys, and John Osborn.

Köstlin, Carl Heinrich. 1775. *Dissertatio physica experimentalis de effectibus electricitatis in quaedam corpora organica*. Tübingen: Litteris Sigmundianis.

Macray, W.D. 1888. Demainbray, Stephen Charles Triboudet (1710–1782). In *Dictionary of National Biography 14*, ed. Leslie Stephen, 330–331. London: Smith, Elder, & Co.

Marat, Jean Paul. 1782. *Recherches physiques sur l'électricité*. Paris: Clousier.

Marshall, William. 1779. Experiments and observations concerning agriculture and the weather. London: J. Dodsley.

Mill, John Stuart. 1843. *A System of Logic, Ratiocinative and Inductive 1*. London: John W. Parker.

Miller, Samuel. 1803. *A Brief Retrospect of the Eighteenth Century*. New York: T. and J. Swords.

Morton, Alan Q. 1990. Lectures on Natural Philosophy in London, 1750–1765: S. C. T. Demainbray and the 'Inattention' of his Countrymen. *British Journal for the History of Science* 23 (4): 411–434.

Morton, Alan Q., and Jane A. Wess. 1995. The Historical Context of the Models of Stephen Demainbray in the King George III Collection. *Journal of the History of Collections* 7 (2): 171–178.

Nash, Leonard K. 1957. Plants and the Atmosphere. In *Harvard Case Histories in Experimental Science 2*, ed. James Bryant Conant and Leonard K. Nash, 323–436. Cambridge, MA: Harvard University Press.

Nollet, Jean-Antoine. 1743. *Leçons de physique expérimentale 1*. Paris: Frères Guerin.

———. 1748. Part of a Letter from Abbé Nollet, of the Royal Academy of Sciences at Paris, and F.R.S. to Martin Folkes Esq; President of the same, concerning Electricity. *Philosophical Transactions of the Royal Society of London* 45 (486): 187–194.

———. 1749. *Recherches sur les causes particulières des phénoménes électriques*. Paris: Frères Guerin.

———. 1752. Éclaircissemens sur plusieurs faits concernant l'Électricité. Quatrième Mémoire. Des effets de la vertu électrique sur les corps organisés. *Histoire de l'Academie Royale des Sciences, Année* 1748: 164–199.

———. 1765. *Essai sur l'électricité des corps*. 2nd ed. Paris: Frères Guerin.

———. 1770. *L'art des expériences, 1*. Paris: P. E. G. Durand.

Priestley, Joseph. 1767. *The History and Present State of Electricity*. London: J. Dodsley, J. Johnson, B. Davenport, T. Cadell.

Rigaud, Gibbes. 1882. Dr. Demainbray and the King's Observatory at Kew. *The Observatory* 66: 279–285.

Rouland, Urbain-François. 1789. Lettre de M. Rouland à M. de la Méthrie: Sur l'Electricité appliquée aux Végétaux. *Observations et Mémoires sur la Physique, sur L'Histoire Naturelle et sur les Arts et Métiers* 35 (2): 3–8.

Runeberg, Edward Frederik. 1757. Försök, At med Electricitetens tilhjålp drisva Våxter, gjorde I Stockholm, år 1754. In *Kongl. Vetenskaps Academiens Handlingar for År 1757*, vol. 18, 14–28.

———. 1759. Versuche, mit Beyhülfe der Electricität Gewächse zu treiben. In *Der Königl. Schwedischen Akademie der Wissenschaften neue Abhandlungen aus der Naturlehre, Haushaltungskunst und Mechanik, auf das Jahr 1757*, vol. 19, 15–26.

Schickore, Jutta. 2011. What Does History Matter to Philosophy of Science? The Concept of Replication and the Methodology of Experiments. *Journal of the Philosophy of History* 5: 513–532.

———. 2017. *About Method Experimenters, Snake Venom, and the History of Writing Scientifically*. Chicago and London: University of Chicago Press.

———. 2021. The Place and Significance of Comparative Trials in German Agricultural Writings around 1800. *Annals of Science* 78 (4): 484–503.

———. 2023. Peculiar Blue Spots: Evidence and Causes Around 1800. In *Evidence: The Use and Misuse of Data*, ed. Robert Mason Hauser and Adrianna Link, 31–55. Philadelphia: American Philosophical Society Press.

Schmitt, Stephane. 2010. Lacepède's Syncretic Contribution to the Debates on Natural History in France Around 1800. *Journal of the History of Biology* 43 (3): 429–457.

Schofield, Robert E. 1970. *Mechanism and materialism: British natural philosophy in an age of reason*. Princeton: Princeton University Press.

Senebier, Jean. 1772. Réponse a la question, proposée par la Societé de Harlem: Qu'est ce qui est requis dans l'Art d'Observer? Et jusques où cet Art contribue-t-il à perfectionner l'Entendement? *Verhandelingen uitgegeeven door de Hollandsche Maatschappye der Weetenschappen te Haarlem* 13 (2): 98–127.

———. 1791. Physiologie végétale. In *Encyclopédie Méthodique, Forêts et Bois; Arbres et Arbustes; Physiologie végétale*, ed. Louis-Marie Blanquart de Septfontaines and Jean Senebier. Paris: Panckoucke.

Sigaud de la Fond, Joseph-Aignan. 1771. *Traité de l'Électricité. Dans lequel on expose, & on démontre par expérience, toutes les découvertes électriques y saites jusqu à ce jour*. Paris: Laporte.

Thaer, Albrecht Daniel. 1809. *Grundsätze der rationellen Landwirthschaft 1*. Berlin: Realschulbuchhandlung.

van Troostwyk, Adriaan Paets, and Cornelis Rudolphus Theodorus Krayenhoff. 1788. *De l'Application de l'électricité à la physique et à la médecine*. Amsterdam: D. J. Changuion.

Went, Frits W. 1928. Wuchsstoff und Wachstum. *Recueil des travaux botaniques néerlandais* 25: 1–116.

Went, F. A. F. C. 1931. Wachstum. In *Lehrbuch der Pflanzenphysiologie 2*, ed. Sergej Pavlovic Kostytschew. Berlin: Julius Springer.

———. 1933. *Lehrbuch der Allgemeinen Botanik*. Jena: Gustav Fischer.

Caterina Schürch is Junior Professor of History of Science at the Institute of History and Philosophy of Science, Technology, and Literature at the Technical University Berlin. Her research focuses on the history of biophysics, biochemistry, and physiology in the early twentieth century, collaborative and interdisciplinary research, as well as on history and philosophy of experimentation in the life sciences.

Chapter 4
Controlling Induction: Practices and Reflections in David Brewster's Optical Studies

Friedrich Steinle

4.1 Introduction

The term "induction" has many meanings, although modern philosophical discussions often understand it as enumerative induction. The early nineteenth century, in contrast, had a wider understanding. When speaking of "inductive science," philosopher-scientists such as John Herschel or William Whewell or philosophers such as John Stuart Mill had in mind sciences based on empirical input—as opposed to, for example, mathematics, logic, or metaphysics. Although they had different ideas on how the inductive procedure should work, they shared that general understanding. This is what I mean by induction in this article's title.

My interest has long been to understand different types of learning from experiments or observation, which includes induction in the broad sense. In this chapter, I am interested in how this process of learning has been both conceived and practiced as more or less rigorous and strictly controlled in its various steps. Rigor and control might appear on many levels, such as conceiving and performing the experiment or drawing conclusions from its outcomes. They secure or enhance the reliability of the inductive process and its results.

Here, I shall begin with a specific example of eighteenth-century optical research, and from there shall develop wider considerations. The historical case will serve as illustration for three theses. First, and not surprisingly, experimental control in the physical sciences has different dimensions. These are connected to different experimental traditions. Second, the way experimental control was practiced and reflected in historical cases stems from certain specific epistemic goals. Third and last, in nineteenth-century experimental optics, at least two different traditions of experimental control and rigor intertwined, which gave rise to the most remarkable optical achievements.

F. Steinle (✉)
Institute of History and Philosophy of Science, Technology, and Literature,
Technical University Berlin, Berlin, Germany
e-mail: friedrich.steinle@tu-berlin.de

© The Author(s) 2024   105
J. Schickore, W. R. Newman (eds.), *Elusive Phenomena, Unwieldy Things*,
Archimedes 71. https://doi.org/10.1007/978-3-031-52954-2_4

4.2   Optical Research in the Early Nineteenth Century

In the early nineteenth century, optical theory went through a turbulent if not dramatic phase—the debate over the corpuscular and wave theories of light, with the latter's final 'victory.' This process has sometimes been called a scientific revolution.[1] In its first phase, the debate took place mainly in Paris, with prominent researchers such as Jean-Baptiste Biot, François Arago, Augustin Fresnel, Étienne Louis Malus, and Siméon Denis Poisson involved. It involved fierce debates and complicated frontlines that were shaped by personal and institutional relationships.

On the empirical side, one key event was E. L. Malus' analysis of double refraction and his discovery of light's polarization by reflection in 1809. Other arguments cited carefully conducted experiments, often with new findings. As a rule, however, these experiments could support specific positions, but did not rule out others. Hence the debate could find no easy resolution.

Wave theory was also discussed in Britain, of course, with London polymath Thomas Young a key figure.[2] Others pursued the experimental side and took up the question of polarization by reflection, including an unexpected researcher in Edinburgh: David Brewster. Originally a clergyman with scientific interests, he later became professor of physics and an important academic in Britain. He had been working in optics since the century's first decade and began his most intense experimental studies in response to Malus' findings. Over a period of approximately 40 years he published many papers, often in the *Philosophical Transactions,* plus several books on optical topics. He made many optical discoveries and developed new instruments for both scientific and public use, such as the kaleidoscope. He was also active in reorganizing British science. He became Fellow of many learned societies and academies in Europe and received prestigious prizes, such as the Royal Society's Copley Medal (1815), the Annual Prize of the Paris Institut de France/Academy of Sciences (1816), the Royal Society's Rumford Medal (1818) and its Royal Medal (1830).[3] His research had a specific profile, however: he is usually taken as a central figure in optical *experimentation*, while being thought weak with respect to *theory*. Commenting on a statement of B. Airy, Whewell would later call him "Father of Modern Experimental Optics" (Whewell 1859, 133). The wording was deliberate: while highlighting Brewster's outstanding achievements in experimentation, Whewell—a dedicated wave-theory promoter—remained politely silent about the theoretical side.

From the outset, Brewster's approach differed from what he saw in Paris, and deliberately so. His distancing from the Paris approach is illuminating: he emphasized that he was not interested in theory debates, at least for the time being, but rather in finding and establishing laws. To be sure, his stance on theoretical matters

---

[1] For profound studies, see Cantor (1983), Buchwald (1989), and Darrigol (2012), among others.

[2] The British discussion has specifically been treated in Cantor (1975), James (1984), and Buchwald (1992).

[3] There is no up-to-date biography of Brewster; for a starting point, see Morse (1973).

was far from neutral: he was convinced of the corpuscular theory (or, more generally, the "selectionist" theory, as Jed Buchwald has appropriately called it). At the same time, however, Brewster was also convinced that the debate was premature, and could be fruitfully conducted only after further advancements: with laws based on empirical research and more 'rigor,' as we would say, than he saw in others' research. This he set out to achieve in his own work.

Given that background, it is instructive to look more closely at Brewster's experimentation and its outcomes. After all, his prominence depended not only on discovering new effects and instruments, but even more on his formulations for numerous optical laws. The hallmark of his research was reasoning from experiments, in order to formulate new laws. His case is thus instructive for the question of how to control induction in order to support those laws, and my purpose for the next sections is to analyze whether and how Brewster realized that ambition.

One remark on terminology before we continue. Brewster did not discuss methodological questions at length, and as far as I can see never used "induction" in his work for procedures that his contemporaries would have described using that term. So when I speak of induction here there is a certain degree of anachronism. But given my remarks above about the meaning of induction as reasoning from experience to general statements, there should be no obstacle.

My analysis of the historical case has three steps. I first ask about the nature of Brewster's claims, and then about how his experimental approach led him to them. I then ask how he himself reflected, mostly in passing, on that approach.

4.3   Brewster's Epistemic Goal

It was characteristic for Brewster that he started his optical publications with a book, in 1813, on new optical instruments. Here, he focused on the tools that he had invented or improved so far, such as micrometers, goniometers, telescopes, microscopes, and instruments for measuring distances, dispersive, and refractive powers, among other things. He described them in detail with ample illustrations. These instruments he used in all his further research. His tool-based focus underscores a point that he often stressed: to make all experiments as secure as possible, and take seriously all irregularities. Central to this point were close attention to the experimental apparatus and measuring instruments, and rigorous analysis of their workings.

In the book Brewster also described the many findings he had achieved with his instruments so far. At the beginning he gives an overview, and a brief look here will be instructive:

4. All doubly refracting crystals possess a double dispersive power, the greatest refraction being accompanied with the highest power of dispersion.
11. Light is partially polarized when reflected from polished metallic surfaces.
12. The light reflected from the clouds, the blue light of the sky, and the light which forms the rainbow, are all polarized. (Brewster 1813a, x–xii)

These are all empirical statements, some more specific and some more general. Some have the logical form of a conditional: if certain conditions obtain, a certain effect will occur. The statements are bold, moreover, often claiming that *all* things of a certain kind behave in the way described. Later I shall discuss a case illustrating the reasons for such confidence. But statements such as these would characterize his results for decades.

In other texts, Brewster reflected explicitly on the goal of his research, expressing it in various ways:

> ...to discover several new properties of light, and to establish the laws which regulate the most remarkable of the phenomena. (1814b, 397)

> ...my next object was to ascertain the law of the phenomena in relation to the number of plates and the angle of incidence at which the polarization was effected. (1814a, 220–221)

Elsewhere, following optical experiments on double refraction with other substances (carbonate of strontites, carbonate of lead, and chromate of lead), he felt justified to reach a conclusion and to

> establish the general law, that each refraction of crystals which give double images is accompanied with a separate dispersive power. (1813b, 108)

We also find these later formulations:

> ... and we obtain the important law, *That when two polarised pencils reflected from the surfaces of a thin plate lying on a reflecting surface of a different refractive power interfere, half an undulation is not lost, and WHITE-centred rings are produced, provided the mutual inclination of their planes of polarisation is greater than 90°; and that when this inclination is less than 90°, half an undulation is lost, and BLACK -centred rings are produced; when the inclination is exactly 90°, the pencils do not interfere, and no rings are produced.* (1841, 50, emphasis in original)

These examples illustrate the character of the claims he was aiming for. Sometimes he also used more methodological language: "[Philosophy] ... can *reduce* to a satisfactory generalization the anomalous and capricious phenomena" (1813a, 314, my emphasis). He also said he was able "to *reduce* the results obtained from *glass* under the same principle" (1815, 126, my emphasis).

These statements show that Brewster had a clearly defined epistemic goal through all his research: to find and establish laws, and to "reduce" individual cases to those. He did not explicate what he meant by "reduce," but others often talked of "reducing" particulars to general laws in his time. I shall return to this meaning, because it differs from later ones. In all of this, of course, the core idea was to make experiments the sole foundation for those laws. We must have that goal in mind when I reflect later on the specificities of his inductive procedure.

And, indeed, Brewster was successful in formulating some laws, with the one we still call "Brewster's law" the most prominent example (with which I will deal in detail below). Whewell probably had this in mind when he characterized Brewster as the "Father of Experimental Optics" (Whewell 1859, 133). Whewell had previously compared Brewster's achievements in optics to those of Kepler in astronomy (Whewell 1837, 462). That analogy is significant, because Whewell regarded Kepler as the one who established the laws of planetary motion—not from theoretical considerations, but from empirical data. He saw and acknowledged the same achievement in Brewster.

4.4 Brewster's Experimental Approach: From Measurements to Laws

How did Brewster arrive at those laws, and what did his experimental procedures look like? To answer these questions, I shall analyze one of his more prominent papers in detail.

In 1815 Brewster announced something entirely new. He had begun studying in detail a phenomenon discovered by Malus: when light impinged obliquely on the surface of a transparent body, the reflected part came out fully polarized at a specific angle of incidence. That angle came to be called the polarizing angle. The challenge, of course, was to determine that angle in many substances, and to find a law connecting to the substances' other optical properties. Malus had set out to do exactly this. He had figured the angle for glass and water and had attempted to correlate the results with optical properties before coming to a negative conclusion:

> The polarising angle neither follows the order of the refractive powers, nor that of the dispersive forces. It is a property of bodies independent of the other modes of action which they exercise upon light." (quoted from Brewster 1815, 125/6).[4]

Brewster had been skeptical about that conclusion and wished to check it with more experiments. He expanded the number and variety of materials, going beyond water and glass to precious stones and other objects. He also used great precaution in his instruments and measurements to achieve high-precision results. In his paper he did not describe his experimental procedure in detail, but some passages indicate that he had always tried to determine "the angle at which the intensity of the evanescent pencil is a minimum" (129). He had already established that in this minimum setup, the "pencil" (his name for a beam of light) was fully polarized, but he explained no further. From many experiments, he was led to suggest—contra Malus—that the angle was in fact correlated with the optical properties of the materials. In addition, he formulated a specific law in mathematical terms.

There was one major obstacle, however: the experiments with glass did not fit the law, even though glass was the most important optical material. This was a difficult epistemic situation indeed.

Like Malus had done, Brewster at first gave up. After a year, though, he returned and found that another precious stone followed the law. He then focused on glass again and saw that the experiments were irregular: he got different results on different surfaces of one and the same piece of glass (126). This puzzling result made him consider the possibility of an unknown factor or source of error.

---

[4] In this section, all page numbers refer to Brewster (1815), unless noted otherwise.

He set to analyze the surface of the glass plates, because polarization by reflection is a phenomenon of the surface. In a series of original experiments performed with careful scrutiny, he established that some surfaces had undergone chemical changes across long contact with air. He was able to reproduce those changes experimentally and found ways to avoid them. In other words, he had gained control over the changes—i.e., over a previously unknown factor in the experiment that had caused irregular deviations. Now it could be controlled.

In returning to the original question and performing experiments with unchanged glass surfaces, he was finally successful and found that glass followed the same law he had found for all other materials. He was therefore able to triumphantly formulate the law in all generality:

> Having thus ascertained the cause of the anomalies presented by glass, I compared the various angles which I had measured, and found that they were all represented by the following simple law. *The index of refraction is the tangent of the angle of polarisation.* (127, emphasis in original)

Elsewhere he said he was now able "to reduce the results obtained from *glass* under the same principle" (126, emphasis in original).

This episode illustrates, among things, what he meant by "reducing": he had shown that the individual phenomenon was just a special case of the general law. This understanding of "reduction" under a law or principle means to demonstrate that a specific phenomenon is consistently covered by the law or principle. The understanding also fits well with the earlier quotations. And although the terminology might seem strange for us, as reduction has different connotations to us, this sense was not uncommon in Brewster's time. I have found the term with that specific meaning in Ampère's and Faraday's research on electromagnetism, for example (Steinle 2016), but also as early as the eighteenth century with Dufay and d'Alembert. I discuss more details in a forthcoming paper (Steinle forthcoming), but a broader historical picture remains to be completed.

The episode also illustrates how Brewster dealt with "anomalies," or irregular outcomes that gave different measurements even with the same piece of matter. Such an anomaly could occur only, or so he was convinced, when the experimenter had overlooked some experimental factor. The events thus point to a specific aspect of experimental control: ensuring that the experimenter has a complete view of all the experimental conditions with an effect on the outcome. As the above quotation indicates, Brewster regarded these experimental conditions as the "causes" of the result, which suggests an understanding of causes that resonates with what Mill would describe in his 1843 *System of Logic*. Later, I shall return to that aspect of control.

The relation between polarizing angle and refractive index is what we today call "Brewster's law," and the specific angle "Brewster's angle." For this result, Brewster achieved considerable recognition: he was immediately made Fellow of the Royal Society of London and received its prestigious Copley Medal. A short time later, the Paris Institut de France/Académie des Sciences honored the result and its author.

It awarded to him half the annual prize, carrying a significant monetary award, for the most important scientific discovery in the physical sciences.

To support the generalization, Brewster's publication presented his results in a table showing the strikingly varied materials he had used: from water and various sorts of glass to diamond, crystals, and precious stones, as well as mother of pearl and birdlime. It was a vast collection indeed and must have been costly, even though he had no institution funding his experiments (Table 4.1).

In the columns he gave the material (column 1), the polarizing angle as calculated from the refraction index with the tangent law (column 2), the same angle as measured with his instruments (column 3), and the difference between the two numbers (column 4). Column 5 presented the calculated angles for the material's second surface (e.g., the lower surface of a glass plate with two parallel surfaces), which he discussed later in the paper (in section II of his paper, from p. 134 onwards), but this information was not relevant to formulating the law. Giving an argument with tables was characteristic for him, and he often used the strategy in later writings and with other cases to support general claims from a mathematical formula. The table was the central means to support the inductive claim, and it did so in two ways. First, it made obvious that the measured values had a "very remarkable" (128) coincidence with those calculated from the law. Second, it suggested that, because the law held for so many different substances, it could be generalized to all materials without

Table 4.1 Brewster's table of polarizing angles for various materials (Brewster 1815, 128)

Table containing the calculated and observed polarising angles for various bodies.

Names of the Bodies..	Calculated polarising angles for the *first* surface.			Observed polarising angles for the *first* surface.			Difference between the calculated and observed angles.			Calculated polarising angles for the *second* surface.		
	°	′	″	°		°		°		°	′	″
Air - - - -	45	0	32	45	or	47				44	59	28
Water - - - -	53	11	0	52°	45′		0°	26′	—	36°	49′	
Fluor spar - - -	55	9	0	54	50		0	19	—	34	51	
Obsidian - - -	56	6	0	56	3		0	3	—	33	54	
Birdlime - - - -	56	40	0	56	46		0	6	+	33	20	
Sulphate of lime -	56	45	0	56	28		0	17	—	33	15	
Rock crystal - -	56	58	0	57	22		0	24	+	33	2	
Opal coloured glass	58	33	· 0	58	1		0	32	—	31	27	
Topaz - - - -	58	34	0	58	40		0	6	+	31	26	
Mother of pearl -	58	50	0	58	47		0	3	—	31	10	
Iceland spar - -	58	51	0	58	23		0	28	—	31	9	
Orange coloured glass	59	28	0	59	12		0	16	—	30	32	
Spinelle ruby - -	60	25	0	60	16		0	9	—	29	35	
Zircon - - - -	63	0	0	63	8		0	8	+	27	0	
Glass of antimony	64	30	0	64	45		0	15	+	25	30	
Sulphur - - - -	63	45	0	64	10		0	25	+	26	15	
Diamond - - -	68	1	0	68	2		0	1	+	21	59	
Chromate of Lead	68	3	0	67	42		0	21	—	21	56	

much risk. Hence it justified the bold inductive step. That step was also supported by another aspect of the specific case: even the material that had appeared at first to contradict the law could, under careful scrutiny, be resolved by controlling a hitherto unknown experimental factor. With that factor controlled, the material could be subsumed under the law.

Of course, the procedure of using tables to support general claims, and to compare measured and calculated (or deduced) values, was not new. It had been used in astronomy for centuries and in physical sciences since at least the seventeenth century, with Boyle arguing for the inverse relation of volume and air pressure, for example (the relation later called "Boyle's law": Boyle 1662, 59sqq.). Brewster, however, pursued the strategy with particular intensity, always basing it on comprehensive experimentation.

Given my focus on the inductive process, it is significant that Brewster went a step further. To underscore the law's reliability and precision, he undertook to evaluate the discrepancies between calculated and measured values. While there existed no established procedure at the time to quantify the agreement or disagreement of those values—mathematical error analysis came only later—Brewster still wished to understand them in more detail. He accordingly discussed them in various ways. First, he took a quantitative approach, for one: he added the absolute values of the discrepancies in his measurements and calculated a mean discrepancy of 15′ of an arc. Moreover, he found an asymmetry: the total amount of negative discrepancies was roughly twice that of positive ones. This evaluation of error may seem quite crude to us, but we must remember that he performed it at a time when error analysis in physical measurements had not (yet) been refined. This was true both in Britain and in Paris, where the program of precision measurement had its stronghold. The method of least squares, presented by Gauss in 1809, had been developed and used only in astronomy.

Second, and with greater intensity, Brewster focused on the discrepancies' possible sources. To explain them generally he pointed to the difficulties of measuring both the index of refraction, which constituted an important numerical factor in the law, and the angle of minimum intensity of the reflected beam, or the polarizing angle. We might surmise that he attributed the mean discrepancy or error of 15′ to these two difficulties, and to the ensuing uncertainties. But this would not have explained the asymmetry between positive and negative discrepancies. For this reason he drove his analysis further and identified two specific sources of uncertainty in measuring the polarizing angle: the practical conditions of observability, and the variations of the angle with color combined with the varying intensities of different colors. Both factors, he concluded, favored the observed tendency to negative discrepancies. With these considerations he could at least qualitatively account for the asymmetry between positive and negative discrepancies.

(As a side note, we might see here a first intimation of what was later called the difference between statistical and systematic errors. Brewster gave only a general explanation for the occurrence of the mean error, but a much more specific one for the asymmetry between positive and negative discrepancies.)

Based on these results, Brewster understood more deeply why and how errors occurred in his measurements. He therefore trusted the empirical law even more now. He emphasized that "the law of the polarisation of light by reflexion [had been] thus experimentally established" (130). In modern terms, this is a significant case of inductive generalization: the researcher knew about the boldness of the inductive step and did everything he could to justify it as much as possible. Key points here were fully grasping *all* relevant experimental conditions and precisely controlling them. In addition, there came at least a qualitative understanding of the remaining "errors," or the deviations between measurement and the law's predictions. Step by step he had succeeded in overcoming those challenges, and hence was able to include even those cases that had not initially fit the law.

At that point Brewster was so confident of the law's validity and generality that he made a most significant epistemic switch: he changed the status of the law in the text from an empirical rule to an unquestioned scientific principle. "It will thus be seen," he wrote, "that the subject assumes a scientific form, and that we can calculate *a priori,* the result of every experiment" (130). While he did not explicate the phrase "scientific form," the subsequent text makes his meaning clear: he no longer regarded the law as a matter of empirical doubt but instead ascribed to it a fundamental degree of certainty. It was certainty so great that the law could, from now on, be an unquestioned starting point for all further investigations. Brewster's change was also manifest in the text's structure: from that point onwards he arranged it in a Euclidian manner, with numbered propositions followed by a sort of proof. The proofs were no longer experimental, but rather just gave the "geometrical consequences" (130) of the law he was now using as a principle.

In sum, we see an impressive pathway. It begins with carefully conducted individual experiments and brings them together in a series, and then rigorously analyzes the relevant experimental conditions. It also offers at least a qualitative understanding of the remaining "errors," or deviations between experiment and expectation. All this leads to a general empirical law. Most strikingly, the end involves an epistemic step whose boldness cannot be overstated—Brewster was so confident in the validity of the empirical law that he raised its status to that of a principle. Thereafter he treated it like a geometrical axiom, and used it as a physical principle for all sorts of geometric deductions. As such, in his mind at least, the principle was no longer subject to empirical test; it was to be taken as absolute, as an axiom. We see the pathway from provisional law hypothesis to full and absolute certainty. I know only few instances in the history of empirical science where a researcher consciously went as far as this last step. Kepler, with whom Whewell compared Brewster, provides a case from astronomy. Crucial elements of the pathway include the procedures of broad experimentation and leaving nothing out: in the included factors, in the breadth of experimental materials, and in analysis of the remaining discrepancies or errors between expectation and experiment. Every step was based on careful experimental scrutiny—highly controlled, and rigorously carried out.

4.5 Brewster's Reflections on How to Support Induction

Before discussing this procedure in a wider context, I shall examine Brewster's own methodological reflections. He was not an epistemologist and did not give methodological rules, but we can still discern his approach. We see it in his practice, in scattered side remarks, and (indirectly) in his criticism of others' procedures.

One striking example is his analysis of previous researchers' failure. After reporting Malus' claim of the non-relation of polarizing angle and optical properties of the materials, he analyzed the background of that failure:

> This premature generalisation of a few imperfectly ascertained facts, is perhaps equalled only by the mistake of Sir Isaac Newton, who pronounced the construction of an achromatic telescope to be incompatible with the known principles of optics. Like Newton, too, Malus himself abandoned the enquiry; and even his learned associates in the Institute, to whom he bequeathed the prosecution of his views, have sought for fame in the investigation of other properties of polarised light. (126).

The critique occurs on many levels. The facts had been too few and they had not been well ascertained; as a result, the generalization was premature. From these points we can see what he thought of as a good, or mature, generalization. The criteria would be:

1. *Many* facts or experiments are needed.
2. Each fact must be *well ascertained*.

The example I have discussed illustrates these points, and how he used the requisite facts to generalize. I shall return to this in a moment.

It is also interesting to note that he included a social aspect: he criticized Malus' generalization as premature, but also criticized others for accepting it too easily. They did not care, he thought, since more "fame" could be gained elsewhere. One could assume that, in Brewster's view, there was not much fame to be won in Paris by the meticulous work it would have taken to improve the earlier failures. To give it yet another twist: looking for fame, perhaps particularly in Paris, might sometimes work against the quality of experimental work and the control of generalizations associated with it. This could be true both for the researcher himself and for his academic fellows, at least if the local academic culture was strongly shaped by specific ideals (such as mathematization) at the expense of others (such as experimental broadness). This remark, concerning the impact that local academic culture (as we might call it) and competition for fame had on the research process, has become a pressing topic in our times. It strikes me as a remarkable observation and critique in Brewster's period. That Brewster made the remark with Paris in mind had probably to do with the historical situation: academic physics in Paris, much more so than elsewhere in Europe, had a dominant epistemic ideal. Even those who no longer followed a strict Laplacian program shared the ideal of mathematization, often at the cost of broad experimentation. There was little chance for visibility in Paris physics without following that ideal. It should also be noted that Brewster, despite his critique of Malus' premature conclusion, expressed deep respect at the end of the paper for Malus as a productive researcher (159).

We must also address the role of theory in Brewster's experiments, as part of his epistemic approach. On the one hand, Brewster claimed to do his experiments independently of any "hypothetical assumption" (158), probably having in mind the debate between the wave and corpuscular theories of light. Indeed, his experimental reasoning did not discuss that question at all. He did not position his findings within that debate, nor do we see his experiments designed with the debate in mind. As Hacking and his co-author Everitt famously characterized it, Brewster just analyzed "how light behaved" (1983, 157). Even if such an expression may sound naïve—there is no 'innocent' analysis of how things behave—it highlights the absence of theory in guiding experiments. Brewster emphasized that optics could advance only "when discovery shall have accumulated a greater number of facts, and connected them together by general laws" (158). Only then could it invent "better names" (158), that is, a more fitting terminology, and "speculate respecting the cause of those wonderful phenomena" (158–159). He thus gave the epistemic process a clear sequence: first facts, then laws, and after that, theories about physical causes like waves and particles.

At the same time, however, as an admirer of Newton (he published a biography in 1831 and more papers on him in 1855), Brewster was convinced of the corpuscular theory. To put it more precisely, he was convinced of the "selectionist' approach to light that Newton had developed from the background of the corpuscular view.[5] At its core were several assumptions: that every beam (or "pencil") of light could be understood as a multitude of individual rays; that all its properties could be reduced (in the meaning sketched above) to the properties of those rays; and that the interaction between a beam and a surface could always be understood in terms of selecting certain rays from the multitude of the beam. As Buchwald has pointed out, Brewster's commitment to this framework did not affect his experimental design, but it manifested in the terminology he used to describe experiments and results. He often spoke of rays and used that framework without much discussing it. He obviously was not aware of all the philosophical baggage such an approach brought with it. Sometimes it was difficult to formulate his findings within that framework, and the result could be contorted expressions that Buchwald described as "hodgepodge" (Buchwald 1989, xix, also p. 259 or 449). However, Brewster did not question the framework.[6]

This is a case of an experimental approach not oriented toward theory or driven by theoretical goals, even while others around Brewster were obsessed by them. At the same time, it was not fully separate from Brewster's own theoretical preferences. Those preferences left linguistic traces in concepts and terminology, and Brewster did not choose or develop a more neutral language. While the above quotation shows Brewster sensing the need for a more appropriate language, he did not invent one. This inflexibility for basic concepts makes Brewster's case differ from others, and most significantly from Faraday's, in Brewster's own day. Faraday knew the importance and laden-ness of terminology, and kept it as flexible as possible (Ross 1961).

---

[5] I rely here on the excellent analysis given in Buchwald (1989).

[6] Buchwald (1992) gives a profound analysis.

4.6   Dimensions of Experimental Control

While Brewster himself scarcely spoke of rigor or control—almost no one in the physical sciences did at the time—we can use these analytic concepts to understand and contextualize the historical case. The case indicates that experimental control can be exerted in different dimensions and that it comes in degrees. With Brewster's own experimental practice and reflections thereon in mind, I identify four dimensions with which we can characterize his strict inductive procedure via control and rigor.

The first dimension concerns the reliability and precision, in the sense of precise and well-ascertained numerical outcomes, for *each and every individual experiment or measurement*. As Brewster's first book made clear, he regarded this criterion as the foundation for any reliable experiment in optics, and in this respect he criticized Malus for being too sloppy. Every optical experiment must be controlled carefully to allow for the utmost reliability and precision, both in arranging the apparatus and in conducting measurements. Interestingly, Brewster did not mention simple repetition of experiments and measurements; we do not know whether his measurement results, like those given in the table above, were the outcome of single measurements or multiple. In his time discussions about those issues were not happening, although in astronomy they were about to start. Given the difficulties Brewster describes for measuring the polarizing angle, however (129–130, see my discussion above), it is plausible to surmise that he did, at least sometimes, measure more than once. Whether he obtained differing results in those cases, and how he might have calculated the final value for the table, we do not know.

However, Brewster's silence on the issue of repeating measurements and observations indicates a more general point: in most of the physical sciences of his day, repeating experiments was not an important issue. The focus was on controlling relevant experimental conditions so carefully that the outcomes were well-determined and stable even when repeated. This resonates with what Caterina Schürch (this volume) reports as the methodological reflection of Albrecht Thaer in 1809. Thaer noted a difference between those sciences that could fully control their experimental subjects in the closed space of the laboratory—he was thinking of chemistry and perhaps also of physics—and those that could never achieve that control. The most obvious of the latter would be those involving living beings, like plants or animals. The corresponding experimental strategies were described differently: in the first group, "completely perfect and pure experiments" could be performed, probably without needing repetition or comparative experiments. But the second group needed that method. While in physics, with increasing importance of precision measurement, such a view on experimentation would change in the decades to come, it might still have been possible around 1810.

The second dimension of control in Brewster concerns the goal of knowing about *all relevant* experimental factors, i.e. all those that affect the result, and to be able to control them, i.e. is to keep them constant or to vary them at will. A puzzling moment in Brewster's experiments occurred when he realized that he obtained different outcomes for the polarizing angle, even when he used the same specimen of glass but observed different surfaces of it. This result made him aware that the outcome was determined not only by the type of glass and its known optical properties. There had to be another, unknown factor, belonging to the different surfaces of the glass, even if all shared the same refractive index. He identified that factor successfully and thus regained full control of the experimental situation—that is, he knew all the experimental parameters required to determine the polarizing angle. As for terminology, Brewster did not speak of an "error" at all in this situation; rather, he spoke of "anomalies." Such anomalous results should not occur once the experimenter knew all the relevant experimental parameters.

Of course, this dimension is in one respect very common to experimental work. The strategy of varying parameters systematically, so important in experimenting, intends to discover which experimental parameters are relevant for the effect in question, and which are not. Coko (this volume) provides another striking example and explicates the strategy. What has seldom been studied, by contrast, is another way researchers might become aware of the problem: by obtaining results that are "anomalous" in Brewster's sense, or results that should not occur if the experimenter already knew all relevant parameters. When confronted with such results, experimenters might wonder about and initiate the search for unrecognized but relevant experimental factors.

A third dimension of control concerns error analysis. Brewster spoke of "errors" with a specific meaning: they were discrepancies between the results expected from the (perhaps still hypothetical) law, and those obtained from actual measurement. As I described above, he discussed possible sources of those discrepancies and arrived at least at a qualitative understanding of their occurrence and distribution. It is important to note that the factors he identified were all based on actual procedures and the conditions of observation and measurement. As I have suggested, he needed this understanding to take the bold inductive step after that discussion: to promote the law from an empirical statement to a principle that would no longer be subject to empirical uncertainty. Understanding error sources enhanced his control over the experiment and so was essential for induction.

We ought to remember that these three dimensions of experimental control were not unfamiliar at the time. Both the ideal of precision measurement and of analysis on measurement error had originated centuries earlier in astronomy, where Tycho Brahe is a striking example. Both started to be introduced into the experimental sciences during the final decades of the eighteenth century. Nevertheless, they were still not common in Brewster's day. Even in Paris, where a mathematical approach to physics had been thriving[7] and hence the issue was most pressing, there were no

---

[7] For a "locus classicus" see Robert Fox's characterization of Laplacian physics (Fox 1974), and Norton Wise's collection on the "Values of precision" (Wise 1995).

common procedures for analyzing possible measurement errors.[8] In his optical experiments, Brewster knew what could be achieved with reliable precision and error analysis, and he was among the first to practice them in Britain. What he criticized in his Paris colleagues was not the lack of control, but the degree, insufficient in his eyes, to which researchers had implemented the three mentioned dimensions in optical research on polarization. To use our terms here, his criticism concerned insufficient rigor in implementing control procedures. That insufficiency itself was, as he suggested indirectly, probably attributable to the heated atmosphere in Paris academia, which did not reward such rigor. His own optical research, by contrast, provides a striking case of the success of those ideals. It appears not only in the polarizing angle but also in other achievements as well, including what came to be called "Fresnel's formulas." On the empirical side, those were the outcomes of Brewster's meticulous measurements.

However, Brewster's case also points to a fourth dimension by which experimental research can be well-controlled and rigorous. It deals less with individual experiments and more with how to arrange them as a group. What Brewster did was use the same experimental procedure—measuring the angle of polarization—and apply it to as many materials as he could (as long as they were appropriate; they needed reflecting surfaces, for example). Determining the polarizing angle was in itself a procedure far from trivial, and it required strict control in the three dimensions listed above. But what he did (and required as part of his practice) was to use that procedure on a broad range of materials while leaving other parameters unchanged. Only with such variation, or so he claimed, could one build the inductive argument needed to formulate a law. In other words, it would be impossible to base the law on a small sample or individual collection of experiments. It could come only from a group of experiments that was well-structured, coherent, and as large as possible. Within a group like that, everything remained the same except one parameter that researchers systematically varied—in this case, the material to be analyzed. That variation made the group coherent and gave the central epistemic argument for the induction process. Not the individual experiment, but only the whole group, designed to be internally coherent, could serve as a basis for the inductive step.

This dimension of experimental control is hugely important, and I shall add some observations. First, presenting those experimental results in a table aligns with that dimension and its procedure: for each line in the table, the basic situation is the same. Only one parameter—the material—is varied, and for each variation there is a new line, with the parameter in the first column and the results in the others. To be sure, not all experimental groups in Brewster's research were presented in such

---

[8]To my knowledge, we still do not have a comprehensive picture of how those procedures made their way into physical and chemical experiment. Some of the articles in Wise (1995) touch on the topic; see also Hoffmann (2006). Astronomy is better studied here; for examples, see Schaffer (1988) and Hoffmann (2007). A workshop in Dresden (September 2021) on "Promises of Precision—Questioning 'Precision' in Precision Instruments," organized by Sibylle Gluch, made another attempt at the project, but no publications have yet resulted.

tables, as when there were no measurements or numbers involved. But the appearance of the tables suggests that a group of that kind had been created.

Second, and with respect to the history, it is obvious that this dimension of control was not present in Paris in his time. Indeed, the lack of such experimentation was one of the central critiques Brewster posed to his Paris contemporaries. And his intriguing remark about the local practices probably hit a crucial point: in Paris, with its intense atmosphere of mathematizing ever-new domains, such meticulous work was less honored than was finding new effects and mathematizing them.

Third, this dimension of experimental control is intimately connected to the epistemic goal of establishing regularities and laws from empirical (usually experimental) research. This happens in cases in which explanatory theory is either not available or deliberately kept excluded (e.g., because it is thought premature). While such a goal might resemble the general empiricist ideal, formulated repeatedly since Bacon, of basing scientific insight on broad empirical input, this one is much sharper: the type of scientific insight is clearly defined as laws, in contrast to explanatory theories. The procedure is also clearly spelled out. The empirical foundation is not just a collection of experiments, but a highly structured and well-ordered one, often in the form of an experimental group, as described above.

In the history of the physical sciences, we find many cases of just such a connection between an epistemic goal and this dimension of experimental control. I shall note further cases below. When and where exactly that connection had its first historical appearance is difficult to say, but it may already have existed in the seventeenth century in Mariotte (see Steinle forthcoming). Here we find a tradition more bound to specific epistemic goals than to local cultures. At the same time, the claim of basing laws on empirical findings has not always involved that specific type of control. There are many cases in which the argument for a law's empirical validity had a different structure, and the tradition into which Brewster's approach fits is not identical with the more general tradition of looking for empirical laws. Hence, In my final section I shall discuss these considerations in more detail.

4.7 Experimental Control and Empirical Laws

The epistemic goal of establishing empirical laws, in contrast to the search for causes or explanatory theories, has been formulated and practiced at least since the early modern period.[9] Those laws have taken different forms, including mathematical proportions or formulas like the sine law of optical refraction or Hooke's law of force of the elastic spring; they have also appeared in non-mathematical if-then statements, such as Dufay's law of electric attraction and repulsion. The process leading to those laws has often been connected to the idea of induction, and in many

[9] The concept and terminology of laws of nature has itself a complex history, but came into common acquaintance in the seventeenth century; see the contributions in Daston and Stolleis (2008).

cases experiment was the central means of research. However, views on how exactly to understand that induction process, and how to conduct and control experiment, differed widely. I shall sketch just two specific examples, which probably lie at the two ends of a spectrum.

One can be seen in Brewster's research. It has the general plan of creating a series (or several series) of closely connected experiments, covering as much empirical ground as possible. It also involves systematically varying parameters as the core of the experimental procedure. Only sufficiently broad arrangements were regarded as solid bases for both the inductive step and for the law. In these cases, theoretical explanations—or even just strong conceptual frameworks—were typically not available, or were deliberately excluded from the process (as happened with Brewster). The experimental approach is inherent in what I have elsewhere described as exploratory experimentation (Steinle 2016, ch.7, among others). As prominent historical cases, one could include here, among others, Hooke's law of 1678 (although we have little documentation for how he arrived at it), or Dufay's research on electricity. Dufay had explicitly postponed all questions of theoretical explanation and focused solely on laws, whereby he had formulated the law of electric attraction and repulsion, among others (1742; see Steinle 2006). One might also include the law of definite proportions in chemistry, formulated by Proust in 1797, or Faraday's research on electromagnetic induction with the resulting law in 1832 (see Steinle 1996). This approach was explicitly addressed by different authors; d'Alembert spoke, perhaps with someone like Dufay in mind, of the need to "multiply" (vary) phenomena in experimental physics, and to make them "a chain with as few missing links as possible" (d'Alembert 1756, 301). Faraday described the core of his experimental procedure in a letter to Ampère as "facts closely placed together" (James 1991, letter 179). Brewster might have been the first to follow such a procedure in a domain based on precision measurement.

On the other hand, we find very different constellations, viz. those in which the law was strongly suggested by more general considerations. It was often framed by an overall theory and then "confirmed" by a small number of selected experiments. One prominent case might be Coulomb's force-law of electric repulsion (1785; see Heering 1994): in support of this law he published exactly three experimental data from one measurement with his torsion balance. Another case might be the law of electromagnetic action, presented by Biot and Savart in 1820. It was the result of few but highly delicate experimental measurements (see Steinle 2016, ch.3). We could also add Malus' research: while he shared Brewster's goal of establishing a law for the polarizing angle, he tested only two materials and gave up when one did not give the expected result. The idea of widening the scope and including more materials had obviously not been part of his approach for an empirical law.

In all of these cases, experimental control was very different from control in the first group. While these strongly emphasized the precision and reliability of one or few experiments, there was no intent to embed them in a broader field of connected experiments. The very idea of using a single experiment or an otherwise small sample as "proof" of a law or a general statement had been most prominently presented by Newton, in the first book of his 1704 *Opticks*. We see the same even earlier in discussions about his 1672 "new theory" of light and colors, where he insisted on setting aside further experiments and put all the weight on the *experimentum crucis* (letter to Oldenburg of 16 May 1676, in Turnbull 1960, 79). Induction was understood here in a very different way, one much less exploratory and systematic. The focus of experimental control was likewise substantially different.

To illustrate these differences I shall use a case in which we see the inductive approach shift in a single researcher within a short period. The type of experimental control also shifted. The case is instructive because we also see how these two approaches, and the shift between them, connect to specific epistemic goals (which can also switch). The episode concerns A. M. Ampère's reaction to the surprising discovery of electricity's action on magnetism, communicated by Ørsted in July 1820.

I have elsewhere elaborated the case in more detail (Steinle 2016, chs. 3 and 4), and shall here focus on just one aspect. The new effect challenged established thinking because it involved complex spatial issues; at first it was impossible to grasp it in traditional terms of attraction and repulsion, i.e. with the concept of central forces as it had been so successfully mathematized in Paris. In this situation of deep conceptual uncertainty, Ampère started out looking for laws—he also spoke of "general facts" ("faits généraux"), to which all the other phenomena should be "reduced." To find them he performed broad experiments with relevant instruments, and the centerpiece of experimental control was a systematic and broad variation of experimental parameters. With his core result he could formulate two "general facts," which gave the necessary and sufficient conditions for electromagnetic action to occur. He also detailed the direction of that action, captured in what was later called "Ampère's swimmer rule."

Before finishing, however, he abruptly changed his research agenda. Not only had he discovered a totally new effect—the interaction of currents without magnetism—but he became also quickly convinced of two things. First, this new effect could explain also electromagnetic effects and, second, its ultimate cause was the interaction of infinitesimal current elements, much in the mode of central forces. All his effort then turned toward demonstrating the first point and, for the second point, toward finding a mathematical law for that force. His experimental approach therefore changed completely: he designed few and very specific experiments for just these two purposes. When after great pains he succeeded, those few experiments corroborated not only the general thesis but also the specific mathematical law that he had framed from various non-experimental considerations. When he presented the law to the Paris academy in December 1820, its empirical supported consisted of a few different and well-selected experiments. Of his former approach, which covered a broad range of experiments and established "general facts," nothing was left.

The episode strikingly shows the connection between experimental procedure and epistemic goals. It also nicely illustrates the general preferences among Paris scientists of the period, which Brewster had so sharply criticized: Ampère had left his broad, exploratory work at a point where he knew it was not finished and had many questions outstanding. However, the approach of formulating a mathematical law for the new electrodynamic action was much more promising in the heated Paris environment than an approach of further solidifying the laws (or "general facts") of electromagnetism, or of broadening their empirical basis.

4.8   Epilogue

Returning to the introduction, I hope to have illustrated and substantiated three claims. We see experimental research in Brewster's case that is highly controlled and rigorous throughout, but we can also differentiate between at least four dimensions of experimental control. These are (1) securing every individual experimental outcome, (2) embracing all relevant experimental factors and leaving none out, (3) rigorously analyzing the sources of observation error and measurement error, and (4) creating a whole field of closely connected experiments to provide the central means for supporting a law. Brewster's story also makes clear how the specific invocation of experimental control relates to the epistemic goals in question; I have mentioned other historical cases to develop that point further. Finally, with a focus on a specific historical context, we see how, in nineteenth century experimental optics, two different traditions of experimental control and rigor united, resulting in remarkable optical achievements.

Acknowledgments Many thanks go to all contributors to this volume, from whom I learned much in our discussions. I am deeply indebted in particular to Theo Arabatzis, Tawrin Baker, and Vasiliki Christopoulou for their careful reading of a former version of this paper and their intriguing and most helpful comments (not all of which I could live up to, I fear). Jutta Schickore raised intriguing questions in her reading of the prefinal draft and I am deeply grateful for her sharp analysis, which made me rethink and hopefully clarify important points. Finally, many thanks to Jutta Schickore and Bill Newman for organizing and bringing together such a constructive and stimulating discussion environment.

References

Boyle, Robert. 1662. *A Defence of the Doctrine Touching the Spring and Weight of the Air*. London.
Brewster, David. 1813a. *Treatise on New Philosophical Instruments, for Various Purposes in the Arts and Sciences, with Experiments on Light and Colours*. Edinburgh: Murray.
———. 1813b. On Some Properties of Light. *Philosophical Transactions of the Royal Society*: 101–109. https://doi.org/10.1098/rstl.1813.0016.
———. 1814a. On the Polarisation of Light by Oblique Transmission Through All Bodies, Whether Crystallized or Uncrystallized. *Philosophical Transactions of the Royal Society* 104: 219–230. https://doi.org/10.1098/rstl.1814.0013.
———. 1814b. On new Properties of light exhibited in the optical Phenomena of Mother of Pearl, and Other Bodies to Which the Superficial Structure of that Substance can be Communicated. *Philosophical Transactions of the Royal Society* 104: 397–418. https://doi.org/10.1098/rstl.1814.0020.

————. 1815. On the Laws Which Regulate the Polarisation of Light by Reflexion from Transparent Bodies. *Philosophical Transactions of the Royal Society*: 125–159. https://doi.org/10.1098/rstl.1815.0010.

————. 1841. On the Phenomena of Thin Plates of Solid and Fluid Substances Exposed to Polarized Light. *Philosophical Transactions of the Royal Society* 131: 43–58. https://doi.org/10.1098/rstl.1841.0007.

Buchwald, Jed Z. 1989. *The Rise of the Wave Theory of Light: Optical Theory and Experiment in the Early Nineteenth Century*. Chicago, IL: University of Chicago Press.

————. 1992. Kinds and the Wave Theory of Light. *SHPS* 23: 39–74.

Cantor, Geoffrey N. 1975. The Reception of the Wave Theory of Light in Britain: A Case Study Illustrating the Role of Methodology in Scientific Debate. *Historical Studies in the Physical Sciences* 6: 109–132.

————. 1983. *Optics after Newton: Theories of Light in Britain and Ireland, 1704-1840*. Manchester: Manchester University Press.

d'Alembert, Jean-Baptiste le Rond. 1756. Experimental. In *Encyclopédie, ou Dictionnaire raisonné des Sciences, des Arts et des Métiers*, ed. Denis Diderot and Jean L.R. d'Alembert, 298–301. t. 6, Paris, 1756.

Darrigol, Olivier. 2012. *A History of Optics from Greek Antiquity to the Nineteenth Century*. Oxford: Oxford University Press.

Daston, Lorraine, and Michael Stolleis, eds. 2008. *Natural Law and Laws of Nature in Early Modern Europe. Jurisprudence, Theology, Moral and Natural Philosophy*. Aldershot: Ashgate.

Fox, Robert. 1974. The Rise and Fall of Laplacian Physics. *Historical Studies in the Physical Sciences* 4: 89–136.

Hacking, Ian. 1983. *Representing and Intervening: Introductory Topics in the Philosophy of Natural Science*. Cambridge: Cambridge University Press.

Heering, Peter. 1994. The Replication of the Torsion Balance Experiment. The Inverse Square Law and its Refutation by Early 19th-century German Physicists. In *Restaging Coulomb: Usages, Controverses et Réplications Autour de la Balance de Torsion*, ed. Christine Blondel and Matthias Dörries, 47–66. Firenze: Olschki.

Hoffmann, Christoph. 2006. *Unter Beobachtung: Naturforschung in der Zeit der Sinnesapparate*. Göttingen: Wallstein.

————. 2007. Constant Differences: Friedrich Wilhelm Bessel, the Concept of the Observer in Early Nineteenth-Century Practical Astronomy, and the History of the Personal Equation. *British Journal for the History of Science* 40 (146): 333–365.

James, Frank A.J.L. 1984. The Physical Interpretation of the Wave Theory of Light. *The British Journal for the History of Science* 17 (1): 47–60.

————., ed. 1991. *The Correspondence of Michael Faraday, Volume 1, 1811 – December 1831, Letters 1 – 524*. London: Institution of Electrical Engineers.

Morse, Edgar W. 1973. Brewster, David. In *Dictionary of Scientific Biography*, ed. Charles Gillispie, vol. 2, 451–454. New York: Charles Scribner's Sons.

Ross, Sydney. 1961. Faraday Consults the Scholars: The Origins of the Terms of Electrochemistry. *Notes and Records of the Royal Society of London* 16: 187–220.

Schaffer, Simon. 1988. Astronomers Mark Time: Discipline and the Personal Equation. *Science in Context* 2 (1): 115–146.

Steinle, Friedrich. 1996. Work, Finish, Publish? The Formation of the Second Series of Faraday's 'Experimental Researches in Electricity'. *Physis* 33: 141–220.

————. 2006. Concept Formation and the Limits of Justification. 'Discovering' the Two Electricities. In *Revisiting Discovery and Justification. Historical and Philosophical Perspectives on the Context Distinction*, ed. Jutta Schickore and Friedrich Steinle, 183–195. Dordrecht: Springer.

————. 2016. *Exploratory Experiments: Ampère, Faraday, and the Origins of Electrodynamics*. Pittsburgh, PA: Pittsburgh University Press.

———. forthcoming. Phenomena and Principles: Analysis-synthesis and Reduction-deduction in 18th-Century Experimental Physics. In *Analysis and Synthesis*, edited by William Newman and Jutta Schickore (working title).

Turnbull, H.W., ed. 1960. *The Correspondence of Isaac Newton, Volume II: 1676–1687.* Cambridge: Cambridge University Press.

Whewell, William. 1837. *History of the Inductive Sciences: From the Earliest to the Present Time.* Vol. 2. London: J.W. Parker.

———. 1859. *History of the Inductive Sciences: From the Earliest to the Present Time.* Vol. 2. 3rd edition with additions. New York: Appleton.

Wise, Norton M., ed. 1995. *The Values of Precision.* Princeton, NJ: Princeton University Press.

Friedrich Steinle is Professor of History of Science at the Institute of History and Philosophy of Science, Technology, and Literature at the Technical University Berlin. His research interests include history and philosophy of experimentation, concept generation, and the history of the study of electricity and colors.

Chapter 5
Carl Stumpf and Control Groups

Julia Kursell

In the fall of 1917, a group of students visited the Institute of Psychology at the University of Berlin. During their lectures in psychology, they had been invited to participate in an experiment. Explanations were provided on the spot. The students had to enter a booth mounted in one of the Institute's rooms, where they found the opening of a tube connecting the booth to the adjacent room. Through that tube they would hear sounds, which they were supposed to judge. Did they recognize any vowel? And what did they think of the vowel's quality? Such were the questions they received in advance.

In a paper read to the Prussian Academy of Sciences in 1918, the head of the Institute, philosopher Carl Stumpf, commented on the prior knowledge—or rather ignorance—he sought in these experimental subjects. He also pointed to the difficulty of judging sound without previous information:

> They had no idea about the entire setup and its purpose. They were only told that they would hear vowels. Because a sound so short and without the characteristic beginning [that was cut off from the transmission] is so ambiguous, such previous information is necessary to make any interpretation possible. (1918, 353)[1]

The purpose of the experiment was to test the sound quality of synthetically produced vowel sounds. With a gigantic structure—the so-called interference device (*Interferenzeinrichtung*) for sound analysis and synthesis—occupying almost all of the rooms of the Institute, it had become possible to emulate the sound spectra of

The author wishes to thank the editors of this volume, all participants of the workshop and especially the two readers Christoph Hoffmann and Klodian Coko for their support and insightful commentaries.

[1] Translations are mine unless otherwise stated.

J. Kursell (✉)
Faculty of Humanities, University of Amsterdam, Amsterdam, Netherlands
e-mail: J.J.E.Kursell@uva.nl

© The Author(s) 2024
J. Schickore, W. R. Newman (eds.), *Elusive Phenomena, Unwieldy Things*,
Archimedes 71. https://doi.org/10.1007/978-3-031-52954-2_5

vowels so convincingly that uninformed subjects were likely to recognize them. This was exactly the role of the visitors: they were test subjects in what Stumpf called "uninformed experiments" (*unwissentliche Versuche*). An Institute staff member prepared the synthetic production of the vowel under scrutiny in accordance with previously determined data, and then sent it to the booth in random alternation with the sound of a singer in another room. The students' answers helped the researchers determine whether the synthetic vowels withstood a comparison. Stumpf explained the rationale of this quality check:

> Due to the tendency to think of the synthetic vowels as reaching truth to nature as soon as a slight resemblance has been achieved, I worked not only with alternating observers, but I also systematically carried out uninformed experiments the statistics of which I compiled. (Stumpf 1918, 353)

As it turned out, the uninformed subjects did not reject the quality of the artificial vowels as deficient. On the contrary, they often found them more convincing than the ones produced naturally. More interesting for the present chapter, though, is the role of the uninformed experiment in Stumpf's methodology. As the quotation above reveals, Stumpf used a comparison between the presence and absence of a concrete condition of judgment, which allowed him to control the observers' bias in new ways. Whether or not the goal of synthesizing vowel sounds had been reached could be determined only through subjects who were ignorant of that goal.

The experiment with uninformed subjects was part of a setup that involved control on several levels. First, the available data about frequency components in vowels were compared with data extracted with a new device for sound analysis. These data were then recreated synthetically and compared with the original vowels by trained observers. Finally, the uninformed subjects were exposed to the comparison between synthetically and naturally produced vowels. In all steps, the comparison between independently produced sets of data was central. This is in line with the etymology of "control" tracing to the French "contre rôle" or counter roll: a second, independent list to be compared with a first. The term emerged in processes of administration and was soon used in the context of scientific experimentation. Similar to what historian of psychology Edwin G. Boring (1954) states for the English word "control," its German equivalent gained currency in the first half of the nineteenth century.[2]

German writer Johann Wolfgang Goethe, for instance, used the word for administrative matters and made the character Odoard in his novel *Wilhelm Meisters Wanderjahre* (1829) do so too when explaining the supervision needed to instigate an agrarian reform.[3] The word is not mentioned in the dictionary of the German

---

[2] Boring also mentions the use in German psychophysics. For a close reading of this paper see Schickore (2019).

[3] This is in line with the emphasis on administration in the entry "Kontrolle" of the term's history given in the German dictionary *Meyers Großes Konservationslexikon* from 1905 to 1909. This entry draws back to French uses of the seventeenth century and mentions that the German equivalents "Gegenschreiber" (counter-writer) and "Gegenbuch" (counterbook), as well as the direct borrowing "Kontrolleur," appeared in mining contexts "a long time ago." See *Meyers Großes*

language initiated in 1838 by the two brothers Johann Jacob and Wilhelm Grimm (Deutsches Wörterbuch 2021), even though that project grew to comprise sixteen volumes during 123 years of collecting and editing. However, by the time Stumpf published his first book on the psychological origin of spatial representation (Stumpf 1873), both the verb *controliren* (to control) and the noun *Kontrolle* were established. In later publications, Stumpf used the terms more and more frequently, consistently referring to instances of comparison by which the calibration of experimental setups could be checked and the validity of findings confirmed.

In Stumpf's vowel experiments, several functions of control, as well as several strands of its history, intersect. The present chapter will discuss them under the umbrella of the term "control group." Control groups delineate the process of experimentation in experiments with human subjects. They constitute what the "other things" are, the *ceteris paribus* that are supposed to remain unchanged when the main group in the experiment undergoes a certain intervention. Control groups or "unbiased comparison groups" (Chalmers 2001) can consist of subjects who do not know about the aims of the experiment; they help to conceal from those performing the intervention on whom they perform it, thereby counteracting bias; and they often consist of randomly chosen subjects, thereby counteracting bias in the researchers who might otherwise privilege a certain group without noticing. All of these functions help to create a group of subjects that remains "unchanged" in comparison with the experimental group through measures of blinding and randomizing.

In the notion of the control group as instantiating a gauging standard opposed to the experimental group, historian Trudy Dehue (2005) has identified two important assumptions. These assumptions have been taken for granted in the contemporary notion of the control group, but came about only gradually during the nineteenth century. First, the groups had to be understood not as consisting of individuals but as representing "populations," so as to—the second assumption—make them susceptible to statistical treatment. In addition, this chapter points to yet another genealogy of control groups, namely in an experimental logic pairing two states, the positive and negative, of the same condition. To reduce variation this way Stumpf needed neither a notion of population nor the law of large numbers. This chapter will discuss how he devised his method, conceiving of himself as a philosopher who understood the psychology he contributed to as a subdiscipline of philosophy.[4]

Just as much as the vowel study required control, the comparison of the subjects' judgments produced further insight into Stumpf's other, perhaps main, subject: the study of judgment itself. Starting out as a philosopher who also integrated collections of individual judgments, his method became experimental and eventually led him to seek judgments more and more systematically. During this search he refined his theory of judgment and explicated it more fully.

Konversationslexikon (6th edition of 1905–1909), digitalized in "Wörterbuchnetz des Trier Center for Digital Humanities," version 01/23, <https://www.woerterbuchnetz.de/Meyers>, accessed 03 Sept. 2023.

[4] On this, see, for instance, Kaiser-el-Safti (2011) and Martinelli (2015).

The work on phonetics can be seen as a culmination of this development. Stumpf's ingenuity in devising measures of control (*Kontrollmaßnahmen*) involved practical matters in unique ways, allowing me to ask how the fleeting nature of sound prompted functions of control within his research on aural judgment. To unfold the peculiar way in which Stumpf's experimental practice combines control and judgment with his own understanding of what logic should be, this chapter proceeds in three steps, corresponding to three sections of this paper. The first section introduces the experimental setup of the vowel study. It has two parts: the first introduces the workings of the setup, the second focuses on the experiment with uninformed subjects. The second section reviews how Stumpf's method of comparing judgments evolved from his first experiments on auditory judgment after 1873. This section partly confirms the findings of Dehue, while also showing the way toward his theory of judgment. The third section discusses that theory. These three steps will help me to discuss the notion of a control group as an operation rather than as a term. By this means I hope to contribute to the project of researching the history of what becomes a term of art at a given moment in time.

5.1 The Interference Device

5.1.1 Measures of Control for Acoustic Experimentation

In 1926 Stumpf published a book titled *Speech Sounds: Experimental-Phonetic Studies with an Appendix on Instrumental Sounds* (*Die Sprachlaute: experimentell-phonetische Untersuchungen nebst einem Anhang zu Instrumentalklängen*). The book's main part detailed the work with the interference device. When its construction began in 1913, the device's scale was unprecedented in acoustic research. It comprised two independent systems of tubes, one serving to analyze and the other to synthesize sound. The operative principle was interference: from actual sound waves propagating through the tubes, single frequencies were subtracted by adding vertical spikes to the main tube. The length of the spikes was calculated so as to project the reverse pattern of rarefaction and compression onto the partial wave in question, thus canceling out that frequency component in the overall sound. Potentially, all frequency components could be canceled from the incoming sound with all spikes added to the main tube. The spikes could also be inserted separately, enabling the researchers to generate various configurations to test.

The other part of the structure, the synthesis system, also used interference. There, periodic sound was purified by interference as described above, so as to obtain simple tones consisting of a single frequency. From these simple tones the "synthetic" sound was composed. For instance, the pattern of frequencies resulting from the analysis could be recreated. For these, the simple tones resulting from the purification were joined into a single tube at a place best understood as the device's control room. Both parts of the device provided a spot for an observer in this room. The synthesis structure also allowed some limited manipulation of the incoming simple tones; tones could be selected and their intensity changed by means of mechanical devices. The tones could be dampened or fully eclipsed with the help of clamps around rubber fittings that were fixed to the ends of each tube. The rubber fittings eventually merged into a single tube, so as to propagate the recreated component pattern to a spot of observation or the booth where the uninformed experimental subjects were located.

"The more finely a method of investigation operates, the more complicated the devices used must be," Stumpf wrote, when explaining the needs of acoustic experimentation in the introduction to his book (1926, 8). His own technical setup certainly met this criterion. Although the interference device did not yet involve electrical transduction of the sounds, operating instead on acoustic sound propagating through the system, it opened up many new procedures that would become standard features when psychoacoustics labs of the interwar period began using electronic technology on a large scale. One feature of particular interest here is connected to the dimensions of the Berlin device. The tubes propagated actual sound waves, the size of which, for human hearing, ranges between two meters and some millimeters. A plotted floor plan Stumpf added to his publications shows the setup (Fig. 5.1). The entry of natural sound into the system happened at a distance, in different rooms. No sound source was located in the control room, and there the observer could only listen to the tubes' outputs. For the first time, seeing and hearing were systematically disconnected.

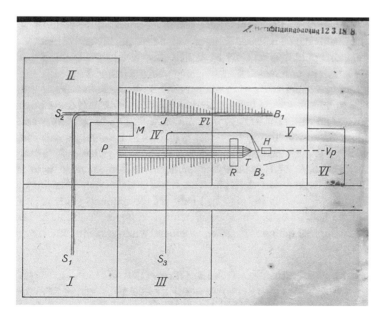

Fig. 5.1 Plotted schema of the interference apparatus of the Institute of Psychology, here taken from a blueprint for Stumpf (1918). The schema was used again in Stumpf (1926) on p. 44. Stumpf Papers, with the kind permission of the Ethnologisches Museum, Staatliche Museen zu Berlin

While earlier experiments attempting to do the same, such as Stumpf's own experiments from 1910 on recognizing the sound of musical instruments from an adjacent room (Kursell 2013), remained incidental, that disconnection became a basic feature in the new device. This feature included the booth for placing uninformed subjects. The method to investigate sound thereby became not only more fine-grained and more complicated, but also more stable, while leaving sound to auditory observation alone. As we shall see, Stumpf took this disconnection to enable also the disconnection of previous knowledge and observation.

The method of interference itself raised the quality of the data on vowel sounds to such a level that the sounds became an object of interest in their own right. In this regard, the work built on acoustic experimentation as it had been introduced into the field by Hermann von Helmholtz. Helmholtz was the first to assume that the ear analyzes sound by reacting to the frequency components selectively. Although he could not determine which mechanism exactly was responsible for reacting only to the frequencies present in a sound—his own "resonance theory" was proven wrong by György Békésy 50 years later—he did everything he could to test the usefulness of this hypothesis.[5] He was also the first to build an apparatus "for the artificial construction of vowels," often referred to as the first synthesizer. This instrument allowed him to re-instantiate frequency patterns he had determined before, using sets of resonators—hollow spheres with two openings that would react to a single frequency and were held to the ear, while listening to, e.g., a sung vowel. His synthesizer provided only eight and later twelve frequencies to choose from, and the resemblance to actual vowels was weak.[6] Yet, with the help of a keyboard allowing him to manipulate the strength of each frequency separately and in quick succession, he could enhance the slight differences in the sound of the patterns. A trained pianist, he could change between frequency patterns quickly and distinctly. A minimal distinction could thus be claimed, which was sufficient to confirm that the ear somehow in fact discriminated the frequency patterns in question.

Although a resemblance to actual vowels was not strictly necessary for Helmholtz's claim, it greatly helped the rhetoric: no other sound could be described in written text so easily for so large a community. Stumpf himself fully assented to it. He summarized the history of vowel synthesis in his book, mentioning many testimonies of experimenters who did not manage to obtain convincing vowel sounds with replicas of Helmholtz's apparatus. He corresponded with physicist Felix Auerbach, who reported that he recognized a vowel only occasionally when configuring the apparatus in order to finely set the required values (1926, 167). While Stumpf held a position in Munich before coming to Berlin, he had access to Helmholtz's original tuning forks, then stored at Deutsches Museum. He could not

---

[5]After scarce attention since the call in Hiebert and Hiebert (1994) to address music in Helmholtz's work, the literature on Helmholtz's acoustic research has been vast during the past 20 years. For general accounts, see Steege (2012) and Kursell (2018) and for more specific questions, see Jackson (2006), Pantalony (2009), Hui (2013), and Hiebert (2014). Among earlier studies, Scherer (1989) is notable.

[6]On this, see Pantalony (2005, 2009).

use them, however, as one central element, the interrupter-fork, was missing. This did not shake his assumption that Helmholtz did hear vowels: "After all: a Helmholtz cannot be lured with fancies" (168).

Stumpf's own vowel studies no longer had the function of the *experimentum crucis* that was to decide whether the ear can be said to "analyze" sound. If, for the testing of a hypothesis, it was sufficient that the sound leaned toward a noticeable distinction, Stumpf read the description to the letter and embedded its two components—analysis and synthesis—into a rigid experimental architecture with new points of fixity and openness. Analysis was delegated to the interference apparatus, which provided data about vowels. The synthesis, as in Helmholtz, was supposed to re-instantiate them, but in this new method the re-instantiation would serve to control the data quality and not to test the validity of the connection of analysis and synthesis as such. The analyzing ear was taken for granted in Stumpf's setup. The focus then moved to the mind.

Indeed, the mind was at stake in Stumpf's systematic manipulation of prior knowledge. The interference device offered multiple possibilities to situate human observers, but these observers also played a crucial role in making the device function. They monitored change in the analysis structure and determined the strength of the purified tones to be combined in the synthesis structure. They also verified quality in the uninformed experiments. The obvious reason for this human factor in controlling experimentation was that Stumpf could not measure intensity. Only with a concept of sound as energy could amplitude be measured, but such a concept came about only with electroacoustics (Wittje 2016). In Stumpf's apparatus, all the researchers could do was estimate the strength of a component. The ear remained the judge in matters of acoustics, as Stumpf never stopped insisting.

The two spots for observers in the control room shown on the floor plan (Fig. 5.1, room V) indicate two modes of judging sound: observation of a process, and comparison among results. In the analysis structure, sound could enter at three points: room I or II, with S indicating the position of a singer, and at a third point for whispered vowels (*Flüstervokale*), indicated in room IV with the letters Fl. As the use of interference depended on sound that could be kept constant for a certain amount of time, sung vowels worked the best for this setup. In rooms IV and V, then, the actual process of canceling out frequency components took place. For this procedure the spikes were opened one by one. The observer, located at point B1, monitored the change in sound and its overall intensity. After the sound had disappeared, the procedure was reversed and the sound rebuilt, now closing the spikes one by one, until the sound was transmitted unchanged through the tube again. Once again, the observer's task was to monitor the change.

On the other tube ending in the control room, marked B2, the observer had to take another action. This spot was connected to the synthesis structure initiating at a soundproof box (P) in which organ pipes were mounted. The pipes were driven by a motor in yet another soundproof box (M). Along the way though room IV, the pipes' sounds were the purified from overtones (all frequency components except the lowest or fundamental frequency). They then entered room V as simple tones. Here the sound intensity was regulated (R). The observer handled the clamps around the tubes' rubber fittings while monitoring the resulting sound changes. All components merged at T. The observer then had several options to induce comparison by choosing to listen to several tube endings. One ending transmitted the sound from room III, where a singer could be signaled to start producing the sound in question with an electric bell. Both the synthetic sounds and the singer's sound were brought to a switch that enabled the observer to choose between them to either listen to them him- or herself or to send them along to the booth.[7] There, the uninformed subject could hear them without suspecting their twofold origin.

5.1.2 Functions of Control in the Interference Experiments

From a table-top experiment with Helmholtz, the comparison between analysis and synthesis turned into the content of an entire Institute with Stumpf. Vowels were no longer supporting the rhetoric, becoming instead the object of analysis. Analysis and synthesis, in turn, could be carried out and observed in much greater detail, using the new procedure of the step-by-step canceling or adding of single frequency components with the help of the interference structure. Control was at the core of the structure's division into the two independent systems of tubes for sound analysis and synthesis. The division provided the researchers with corresponding sets of data for comparison. That comparison, however, could not skip the human ear, the final judge for whether a sound could be considered a vowel or not. Control was thus not encapsulated in an exchange between B1 and B2, but spilled over into other points in the setup as well.

Two basic categories of control stand out: the technology and the human observer. They prompted different regimes of control. On the one hand, the fine-grained analysis that was so important to Stumpf required a constant monitoring of the setup's functioning. Thus, *Die Sprachlaute* discusses technical problems at great length. The tubes distorted the sound, to begin with; this could be partly remedied using funnels for the singer, but the funnels had their own impact on the sound. The sound itself could not be controlled with the ear alone. Additional tools were needed, because even when below the threshold of hearing, a sound might nevertheless distort the devices' functioning. This was true, for instance, for the presence of unwanted components in the allegedly pure sounds used in the synthesis. Tuning forks with frequencies that deviated slightly from those of unwanted components were held in front of the openings so as to make them audible as beats in the forks' audible sound. Finally, the basic principle of the structure, sound canceling by interference, was difficult to handle. It could have side effects, such as the canceling of higher frequencies that fit into the same wave pattern or slightly deviating frequencies being reduced below the threshold of hearing within a certain range.

---

[7] Stumpf had female assistants and doctoral students at that time, such as Katharina von Maltzew, who later went to Soviet Russia.

On the other hand, the observer could not be trusted. Stumpf reports, again in great detail, about their failures, including his own. He writes about surprising observers by interpolating sounds that had not been agreed on beforehand into a series of tested items, such as a consonant in a series of vowels. He constantly compares humans and devices using the same term "Einstellung" for both. One observer, for instance, stubbornly "recognized A" when a "whispered Ö" was spoken into the entry point for whispered vowels, although Stumpf assures the reader that the "Ö" by then was fully recreated in a process of building the components with the analysis device. He also expresses his amazement about "untrained" observers, such as a group of students and staff from the university department of modern languages, to whom he introduced the workings of the setup:

> One day, I demonstrated the change of vowel sounds with interference tubes to a group of members from a seminar for modern languages, among them a lecturer. The vowel Ö was being deconstructed, and long after it had transformed into a pure and even dark-shaded O, the first observer [from this group], a lady, insisted in still hearing Ö. This assessment was taken up by all ensuing observers, who had heard her assessment. I almost began to doubt my own ears, until a reliable staff member, Dr. Wertheimer, was called and immediately and without previous information recognized O. (Stumpf 1926, 51)

The cameo of Gestalt psychologist-to-be Max Wertheimer is an aside in this anecdote about the distinction of trained versus untrained observers. At the time, Wertheimer was working with Stumpf, and Stumpf praised his fine ear. However, even the best observers kept failing in a specific task:

> There is one point at which even the most trained observer is exposed to a constant psychological influence: the results of decomposition and re-composition consistently deviate, as the stages of transformation are situated at a lower point [of the acoustic spectrum] during re-composition than during decomposition. (Stumpf 1926, 51)

Starting from the fully present transmission, the observers were ready to note any small change, whereas the opposite direction—the re-composing of the vowel from its lowest partials—prompted them to recognize a reappearance of the vowel at the earliest moment. As a result, recognition was lost and gained at different points in the two directions of the process. The reaction of human hearing to language was not like a measuring device, but rather was sensitive to immediate context.

Stumpf's comment on this deviation demonstrates that his psychological interests were not absent while generating data for phonetics. "That difference can only have psychological causes," he noted, drawing an analogy to the difference in the threshold of audibility when a sound source was moved toward or away from the ear. He explained the deviating points of loss and recovery of the vowel's "specific character" in the same way. "To this diverging behavior," he concluded, "one is submitted even with a high degree of training and even when de- and reconstruction succeed each other immediately" (52). As these observations demonstrate, the measures of control generated their own surfeit of research findings.

The experiment with the uninformed subjects can be seen as a counterpart to the observer comparing natural and synthetic sounds. When recounting it in the book, Stumpf added a detailed description of the procedure, beginning from what has remained the practice in psychological experimentation ever since: "To an invitation during my psychology lecture to participate as observer in my vowel studies, 30 students, both female and male, reacted" (182). He first tested these students with regard to their general ability to recognize any sound in the transmission, using only vowels produced by a singer. Of these, eighteen succeeded and were invited to the actual tests. Each vowel was tested with five series of ten pairs of vowels, one natural and one synthetic, in predetermined but randomly chosen sequences. The experimental subjects were instructed as follows:

> You will hear vowels of very short duration. Ask yourself when you hear the first of them, which vowel it is and whether its transmission is good or deficient, and if the latter, in which regard, e.g., E too much towards Ä. When you hear the second, ask yourself whether it is the same and if so, whether it sounds better or less good than the one before and why. Then you will always hear pairs that you should compare. Anything remarkable should be noted. (Stumpf 1926, 183)

Although the subjects most often did not follow the instruction to compare pairs, Stumpf found the results to be sufficient for his purpose. Figure 5.2 shows the notes by one of the subjects from October 22, 1917. Reacting to samples of the vowel "Ö," "Fräulein Cassirer" wrote down on the left side which vowel she thought she had heard. The experimenter added on the right with the letters "k" (*künstlich*, "artificial")

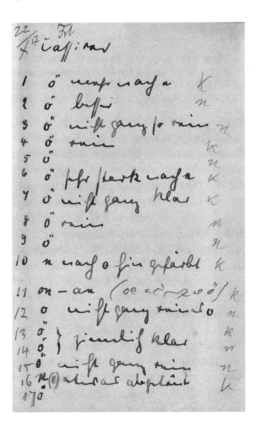

Fig. 5.2 Notes by a test subject (ink) in an "uninformed experiments" on the vowel "Ö", comments in pencil by Carl Stumpf. Stumpf Papers, with the kind permission of the Ethnologisches Museum, Staatliche Museen zu Berlin

and "n" (*natürlich*, "natural") whether the sample was naturally or artificially produced, also providing other necessary information for laboratory purposes, such as the
date, numbers where she forgot them, or which vowel she meant when her handwrit-
ing was bad. In between, one reads Miss Cassirer's comments, such as "more towards
e," "better," "not fully clear," and "pure," all conveying her estimate about the quality
and distortions in what she discerned. These remarks were fully in line with what
Stumpf had asked for: "anything remarkable should be noted."

Stumpf was pleased with the outcomes of the experiment with the eighteen students. In his usual way, he commented with a subjective tint:

> Often the exp. subjects stated somewhat depressed at the end of a series of samples that they
> had not found any significant differences, that they had always heard the same vowel, which
> I took note of not without some hidden pleasure. From all their comments it was clear that
> also during the experiment the subjects had no clue that natural and artificial vowels were
> presented alternately. (Stumpf 1926, 183)

The experiment confirmed the expectation that the synthesis plausibly reproduced
vowel sounds, based on the data generated in the analysis. It also introduced new
methodological components, such as blinded testing, random samples, and a statistically relevant number of answers to avoid individual bias:

> Individual propensities were showing up here as well. For one subject no A, whether natural
> or synthetic, was bright enough. Especially regarding A, the expectations indeed differ con-
> siderably among individuals. Another subject always heard the natural E to be closer to Ä,
> which might in fact be objectively not without a reason. It is exactly because of such small
> individual differences that a larger number of subjects had been involved. (Stumpf
> 1926, 183)

To sum up the explanation of how control guided phonetic experimentation here, we
can say that the ephemerality of sound, the inability to measure amplitude, and the
subjectivity of auditory observation were tackled with a triangle of control instances:
first, by monitoring the manipulations in the analysis device in deconstruction and
reconstruction; second, by comparing the resulting data with their re-instantiations
with the synthesis device; and third, by presenting the synthetically recreated sounds
to the uninformed subjects.

The interference device embodied strategies of granting independence for the
gathering and comparing of data in its architecture. It provides rich insights for
disentangling functions of control, as they were discussed by Jutta Schickore for the
life sciences around 1800 (2021a, b). While such a close analysis could instantiate
what is subsumed under notions such as Hans-Jörg Rheinberger's "technical object"
(1997, 2023), it is important to note that audition, to some extent, placed the emphasis not on the counterpart notion of the epistemic thing, which describes the moment
when the research object, from hindsight, can be understood to have guided the
process of experimentation. Instead, Stumpf's experiments on audition dealt with
the defaults that subjects fell back on when placed in a situation of ignorance. For
disentangling the distributed action in terms of functions of control, it is therefore
important to expand the analysis to the ways in which this research took such
defaults into consideration. This takes the chapter back to Stumpf's earlier work on
the psychology of auditory perception and cognition.

5.2   Comparing Judgments

In the triangle of control instances, the experiment with uninformed subjects presents a region of overlap between two competing interests. The subjects acted, as Stumpf writes, like rabbits and frogs, whom the experimenter does not query how they experience what is done to them. However, the opposite occurred in this uninformed experiment: the subjects were asked to take down their experience. The commentaries are literally inserted between the columns of data, pointing to another area of Stumpf's interest: the study of judgment.

He had pursued this study since his first appointment as professor of philosophy at Würzburg University, where he succeeded his former teacher Franz Brentano in 1873. At Würzburg he began researching the psychology of auditory perception and cognition or, in his own terms, tone psychology. He would later publish two volumes with the title *Tonpsychologie*. The first volume, appearing in 1883, dealt with sensory judgment more generally and the judgment of single and successively heard tones more specifically. The second, from 1890, discussed the judgments of two tones given simultaneously and the theory of fusion that Stumpf would remain known for. The two volumes together grew Stumpf's reputation as an experimental psychologist. His renown procured him further positions as professor of philosophy, first in Prague, then at Halle and Munich Universities, and eventually in Berlin, where he took up the position of chair of philosophy in 1894.

Back in the mid-1870s, Stumpf's interest in the general reliability of tone judgments prompted him to invite people to his home who said they had no talent for music. He described this endeavor in the first volume on tone psychology as follows:

> At first only with the intention of getting a more definite idea of the degrees of unreliability that occur in judgements about tones, years ago when I was in Würzburg, […], I asked several people – otherwise well-educated and normal in hearing, but very unmusical – about their judgement as to which of two tones is higher. These people were: Miss C., completely unmusical according to her own statement and those around her; Dr. K., who assured me that he has no clue about music; W., man of private means, who is not disposed to music and ignorant about it; S., man of private means who is, according to his own statement, able to retain easy melodies, but hostile to the violin, and almost never engaged in music in his youth; finally, the students Be. and Bo. I preferred the question "which tone is higher?" to that of "equal or different?", for I believed this would give me insight into the general conditions of the qualitative judgement. (Stumpf 2020 [1883], 201, translation slightly modified after Stumpf (1883)).

The reliability—or, as phrased here, the unreliability—of judgments was the object of this experiment, and the experiment later developed into a full-fledged method. What Stumpf calls "conditions of judgment" (*Urteilsbedingungen*) could be manipulated by contrasting two complementary conditions: the judgments of those who do, as opposed to those who do not, have a specific and well-defined predisposition, precondition, or, as he would say in *Die Sprachlaute*, "setting" (*Einstellung*) for making a judgment.[8] The question which of two notes was higher, as opposed to that about just noticeable differences, targeted the subjects' ability to find their way in the system of Western tonal music. To grasp pitch in that type of music means to subsume the spectrum of a periodic sound under one value and to understand this value as a tone or note that can be situated within a scale. For this one must be able to align the values on the basis of the parameter of "height," which the English language subsumes under the concept of pitch, but which in German appears in the compound *Tonhöhe*. A listener who could not grasp pitch in this specific sense, even if sensitive to sounds or tones being different among themselves, would not understand the rules of tonal harmony and counterpoint. Later musicality tests would continue to use this question.[9]

It is important to note that the judgments of the unmusical were interesting exactly because they were true without being correct. These subjects opened the possibility of working with false judgment in controlled ways, namely as inside or outside a conventional symbolic system. They allowed Stumpf to draw the distinction not in the physicality of the subjects' hearing, but in their access to a particular and very specific set of rules whose application also relies on a subject's exposure and training. While the answers of the unmusical subjects were perhaps not random, they did not match with a particular system in which they, for whatever reason, did not participate. But that system was also not the experiment's main interest, because the intent was to discover the extent to which subjects participate in and have access to any such system. In fact, Stumpf later changed the object of investigation, but he always searched for what he called "psychic functions" at play in accessing these systems, and confronted subjects with tasks that presupposed access. The first task Stumpf explored systematically was the judging of simultaneous sounds. Later he attempted to find out whether the confrontation with musical systems other than Western tonal music could be tackled in a similar way, but by using phonographic recording. Finally, he turned to vowel sounds and would oppose observation with uninformed experiment.

In all these experimentations, false judgment is a recurrent feature. Stumpf explained in the preface to the first volume of *Tonpsychologie*:

> The physicist seeks the motives of false judgements only in order to eliminate them. The physiologist as such is perhaps concerned with them for his speculations concerning unknown processes in the brain. To the psychologist, they are essential in that they help him elucidate the coming-about and conditions of judgements as such. In unpractised observers, whom the physiologist rejects from the outset, he studies the influence of practice; and in unmusical people, he studies the conditions of musical feelings. (Stumpf 2020 [1883], lxii, translation slightly modified).

---

[8] The German word "Einstellung" means both attitude or mindset, and calibration. The metaphorical transfer between humans and machines does not replicate easily in English. Psychologists and phenomenologists chose "attitude" to translate the word, but their choice does not match technical setups. On attitude and music, see Steege (2021); on "Einstellung" in Stumpf, see Kursell (2021).

[9] See, for instance, Honing (2018).

It was "precisely the curious differences between musical and unmusical natures" (Stumpf 2020, lxi), he added, that supported his research. In other words, false judgments and unmusical subjects provided a key to reducing the complexity of judgment in the realm of music. Experiments with unmusical subjects became a cornerstone in Stumpf's experimental method. Next to observing how the unmusical judged two tones presented in sequence, he also exposed them to simultaneous tones. This became the main topic in the second volume of *Tonpsychologie* and a key to his influential concept of "fusion" (*Verschmelzung*, see Stumpf (1890)). The method for working on fusion proceeded in two steps. First, subjects were tested about their access to the concept of pitch as explained above. A person who could not answer the question which of two notes was higher would normally be disqualified to judge anything musical. In Stumpf's setup, however, they qualified for further experimenting, as he needed them for working with two groups: one of them "musical," the other "unmusical." Then he exposed both the unmusical and musical subjects to two simultaneously played musical notes. He chose intervals that differed with respect to their consonance and dissonance, speaking in terms of Western tonal harmony.

The question Stumpf asked the subjects in the experiments on tone fusion was not whether they heard a consonance or dissonance, but whether they heard one or several tones. Again, that question is remarkable in how it reduced the complexity of the potential musical background in the subjects' answers. The answers of "one" or "many" situated the question below a level that already assumed Western tonal music for its framing. Much of the charm of Western music depends on the melting or diverging in simultaneously produced voices—from the choir singing in unison to personalities on stage, like a Marquis de Posa and title hero Don Carlos in Verdi's 1884 opera singing in parallel thirds and sixths. Those intervals fuse just enough to show two distinct individuals joining in one movement. Music theory was lacking the vocabulary for such features, instead taking the notes on the page as a point of departure: they unambiguously showed whether one or two distinct voices or pitches had to be involved. The category of consonance and dissonance, then, addressed a classification of intervals, rather than their effects in context.

Stumpf's questions about tones did not depart from music teaching, or, as its elementary level was called in German, *Musiklehre*. Instead, he built on research by Hermann von Helmholtz also when it came to music theory. Helmholtz's book *On the Sensations of Tone as a Physiological Basis for the Theory of Music* (1863; first English translation 1875) was notorious for providing an explanation of the complementary notions of consonance and dissonance. Music theorists held Helmholtz's theory of beats that caused roughness in the frequency compounds of simultaneously given tones to favor dissonance, while not explaining the effect of consonance itself. Helmholtz replied in the preface to the third edition that he had never aimed at providing a natural foundation for Western music.[10]

---

[10] On Helmholtz's notion of consonance and dissonance in comparison to Stumpf's, see Kursell (2008).

Stumpf took another observation in Helmholtz's treatise as his point of departure, namely the amazement regarding the fact that any sound sources can be distinguished at all. Helmholtz used the metaphor of waves on the surface of water to describe the problem. Looking at a water surface in motion, the eye can discriminate directions in the motion and sometimes even discern how many waves intersect in a spot and whence they come. The ear, in contrast, distinguishes only a small spot of such a surface and instead calculates the presence of waves like a mathematician (1954 [1877],[11] 36f.).

Interested in the mental operations involved in recognizing tones, Stumpf devised a question posing a simple alternative: do you hear one sound or many? This did not just shift a basic operation of psychophysics to a genuinely psychological task. By asking this question to his two groups of subjects, he also avoided having the subjects—those with and those without a musical background—depend on the vocabulary and knowledge of music for their answer.[12] He then varied the stimuli, always choosing two musical tones but changing their distance or, in other words, the musical interval the tones constituted. The musical intervals, while providing the choice of stimuli, were thus emptied of their musical meaning. Within the system of tonal music (i.e., the music of roughly the year 1600 until Stumpf's own time), the correct answer would always be two tones. All subjects, however, occasionally did not recognize an interval as a manifold. The answers of those subjects for whom musical theory was inaccessible, the unmusical (*Unmusikalische*) in Stumpf's words, in particular shed new light on the reactions from the other group. They tended to hear one sound as the degree of consonance, in musical terms, became higher.

The musically able subjects were, in turn, unable to distinguish the application of their musical knowledge. As a consequence, they could not separate their ability to identify tones as musical notes in certain defined relationships from an immediate sensation. They would, for instance, react to the distinction between consonance and dissonance as it is made in music theory, identifying the two tones accordingly as two consonant or dissonant notes. But they frequently did not identify two notes in the interval of an octave as "many" tones and thought instead they heard just one. Each group could thus be found lacking. The unmusical did not further analyze a multiplicity of tones, they only "sensed" the sound; the musical took the analysis to provide an answer to the question of "one or many," without realizing that they also depended on discerning the multiplicity in a hypothetically prior stage.

From these findings, Stumpf inferred that all subjects sensed two simultaneously given notes as one sound to begin with. The unmusical would remain in the state of that sensation. The musical, in contrast, would analyze the sound in accordance with the rules they had acquired. The immediacy with which the musical subjects reacted to the two notes being consonant or dissonant, in fact, operated as an obstacle for detecting the state of sensation. The analysis happened so fast that

---

[11] The reprint of 1954 is based on the 1885 translation of the German fourth edition of 1877.

[12] On how Ernst Wilhelm Weber construes an alternative question to investigate skin sensation as a means to understand the physiological anatomy of the nerves the skin conceals, and on how this modifies the standard account of psychophysics' history, see Hoffmann (2001, 2005).

they would not notice how they sensed the sound, except when the degree of fusion was exceptionally high. From this, Stumpf construed his notion of fusion, which was later developed into further-reaching phenomenological and Gestalt-theoretical assumptions in his own work, and in that of some of his disciples and colleagues.[13]

As the research on fusion shows, the experiment construed a judgment on sound in terms of simple alternatives operating on two levels. Both the question to be decided on—one or many sounds—and the conditions of judgment—with or without musical ability—were conceived this way. Another aspect is that the subjects could not see the sound sources. They reacted to the sound exclusively, although little was done to shield them from knowledge of a local distribution of the sound, for instance, as would be implied in the sound of specific instruments such as the organ or piano. The logical operation of combining two simple alternatives was at the core of the experiment, and it is this combination that marks a major step in Stumpf's formalization for his method of inquiry. If Stumpf had before collected statements in more informal ways, for instance, through writing letters to friends and colleagues or excerpting literature, he now began to work more systematically with experimental subjects.[14]

In 1885 another shift in his work occurred. Attending a performance by non-European musicians in Halle, Stumpf realized that he himself was now in the position of the unknowledgeable listener. The performance appeared to him like some howling and rattling, although he was convinced that this judgment was unjust. He seized the occasion to work with two of the musicians, the singer Nuskilusta and another whose name is not known. These two Nuxalk First Nation singers from British Columbia patiently auditioned with Stumpf in individual sessions. Stumpf did his best to make notes, as he wrote in a paper on this encounter in *Vierteljahrsschrift für Musikwissenschaft* (Stumpf 1886). However, he realized that neither his note paper nor his mind were up to properly marking distinctions relevant to the two singers. A second performance, then, already made a different impression on Stumpf: he meant to hear some singers deviating from what Nuskilusta had taught him. As he remarked tongue in cheek to his readers, unmusical individuals were not a privilege of Western music (Stumpf 1886, 421).

Between the encounter at Halle and the beginning of his research on speech sounds, Stumpf's work on auditory cognition explored the question whether music gave further insights into the mind making sense of it. He eventually founded the Berliner Phonogramm-Archiv, which was to become the largest collection of phonographic wax cylinder recordings worldwide (Ziegler 2006). However, enticing as the prospect might have been to have a multitude of musics to experiment with, all of which followed different implicit rules, recorded sound did not allow

---

[13] On Stumpf and fusion in the context of phenomenology, see, for instance, Rollinger (1999); on Gestalt psychology in the early years of the Berlin Institute of Psychology, see Ash (1995) and Klotz (1998); on Stumpf's perspective on the interpretations of his notion of fusion, see, e.g., Kaiser-el-Safti (2011).

[14] On the research that Stumpf based on collecting statements from acquaintances and colleagues, see Kursell (2019).

experimenting on judgment by comparing groups of initiated listeners with groups who were not. The turn to phonetics eventually brought experimental subjects back into the Institute. As I shall argue below, Stumpf's interest in judgment now also migrated into the material structure he devised for his experimentation. Between his work on music and the study of language sounds, he published several philosophical papers dealing with, among other things, judgment as an epistemological and cognitive problem. With the experimental setup for his phonetics research, then, he practiced a rigorous and controlled way of judging that offered new perspectives on how to provoke judgment for the purpose of empirical scientific investigation.

5.3   A Two-Level Practice of Judging Judgment

The first experiment with unmusical subjects marks the instantiation of what could be called a "practical epistemology," in Stumpf's own terms. He coined the term in a lecture on logic held at Halle and preserved among the papers of Edmund Husserl. Husserl took notes in 1887 and received a printed version, a so-called "Diktat" (i.e. a text to be dictated), in the following year, 1888 (Fisette 2015a, b; Rollinger 1999, 2015; Schuhmann 1996). While the lecture is considered to lean heavily on those of Stumpf's own teacher Brentano, the term "practical epistemology" is considered to be his own (Rollinger 2015, 77; Schuhmann 1996 on Stumpf's dependence on Brentano more generally). It expresses his opposition to a merely formal approach to logic and asks about uses of logic. Logic is defined in the beginning of the lecture as *Kunstlehre*—to be translated, following Rollinger (2015), as the "instruction to practice an art," namely the art of correct judgment. The lecture on psychology of the winter semester 1886–87, equally preserved in Husserl's notes, comes back to this understanding of logic:

> Logic must go back to the essence of judging, to the different classes of judgments, the expression of them in language, which is indeed also a psychological function. It must also sort out different motives of judging, attend to motives of feeling, habits, exhibit the origin of prejudices, etc. A logic that would abstain from this, a purely formal logic, would otherwise be useless from the outset. (Rollinger 2015, 83 trans. slightly modified)[15]

This reads like an outline to Stumpf's work on auditory cognition all the way through, from the first experiments about the reliability of judgment after 1873 up to 1926, when he published his book on speech sounds. Logic, for Stumpf, was not an aim in

---

[15] The German is also given in Rollinger (2015, 83, n. 19): "Die Logik muss zurückgreifen auf das Wesen des Urteilens, auf die verschiedenen Klassen der Urteile, den Ausdruck derselben in der Sprache, die ja auch eine psychologische Funktion ist. Sie muss auch verschiedene Motive des Urteilens auseinanderhalten, die Gefühlsmotive, Gewohnheiten beachten, die Entstehung der Vorurteile aufzeigen etc. Sonst würde eine Logik, die davon absähe, eine rein formale Logik, von vornherein nutzlos sein." Rollinger emphasizes that Husserl borrowed the notion of "antipsychologism" from Stumpf and not only exempted his mentor from it, but also "allows for logic as Kunstlehre to be dependent on psychology and only argues that a pure logic, strictly as a theoretical discipline, must be seen as free of all psychology." See also Textor (2020).

itself, but as a *Kunstlehre* it had a purpose. The lecture discussed and dismissed other purposes, such as defining logic as either concerned with *thinking*, which Stumpf declares instead to be a matter of psychology, or with *concluding*, which for Stumpf would make it a task of assessing knowledge by means of proof. As the art of practicing correct judgment, logic shared the concern about judgment with psychology. Whether "useful" in the sense of "practicing the art" or useless, Stumpf was critical about the idea that psychology would explain logic or embed it into its own study of the mind. Instead, his interest in judgment overarched two parallel activities: the study of judgment, and the elaboration of methods for doing so. Pushing this further, one could say that Stumpf's practice of logic included experiment.

In the light of these deliberations, the first volume of *Tonpsychologie* presents a parting of ways. Discussing sensory judgment more generally, Stumpf mused on what psychophysics added to grasping the reliability of judgment:

> It would, incidentally, be *a priori* conceivable that yet another constant would have to be added to the specified conditions of the subjective reliability for each individual. […] If we assume that all previously specified conditions are maximally favourable for a judgement about the equality of two impressions, the question would be whether we would in this case notice every difference, be it ever so small. If not, there would be a threshold that the difference in sensations would have to cross over in order to be discerned as such. This threshold would not have to be dependent on the aforementioned and empirically familiar changeable conditions, but should rather be noted as a peculiarity of the mental (central) organism, as a constant coefficient of discrimination (more generally: of judgement), perhaps variable between individuals. The question, however, can hardly be decided experimentally, for there is, strictly speaking, simply no maximally favourable state for those empirical conditions. They can rather by their nature operate more favourably into infinity. (2020, 21, trans. slightly modified)

Psychophysics was thus caught up in not having and never reaching ideal conditions for experimentation.[16] The quantitative premise that could be tied to what Stumpf identified as its main type of question—same or different—would never be accessible to the ideal conditions it presupposed. More importantly, Stumpf needed a threshold of a different nature. He could not accept pitch to be a homogenous parameter. In *Tonpsychologie* he argued that, at least for those trained in Western tonal music, pitch implied values separated by thresholds beyond which recognition tilted toward one or the other of two neighboring values; it did not imply a fine-grained but compact line between any two values. What is more, the highly developed ability to distinguish pitch in musically trained individuals did not concern the mere question of same or different, but what in music teaching was called "intonation," that is, the possibility of indicating a value's closeness to an intended "correct" value. Experimentation that disregarded these features in the musically trained mind was flawed from the outset.[17]

Stumpf's own practice instead proposed to ask what he called "qualitative" questions. Recall the experiment with the unmusical, where he explained, as quoted above, that he "preferred the question 'which tone is higher?' to that of 'equal or

---

[16] The unattainable ideal of experimental conditions has been discussed with the example of Wundt's experiments on reaction time, in which the subject becomes the main obstacle for assessing the subject's reactions (Schmidgen 2014).

[17] On the debate more generally, see Hui (2013).

different?'," for he believed this would give him "insight into the general conditions of the qualitative judgement" (2020, 201). The qualitative method could not operate with the core of psychophysics, or a parametrization of sensation in correlation with measured stimuli. The field that interested Stumpf lacked such homogeneity. He described the way in which the realm of pitch was organized for those educated in Western tonal music like a land surveyor's perspective: nineteenth-century music required a standpoint and, seen from there, the recognition of signposts, rather than a parametrization of pitch.

Stumpf's own two-level comparison of judgments harmonized with the development of his logic from Brentano's. From the lectures Stumpf heard from Brentano between 1865 and 1868, he could take a definition of judgment stating that all judgments are "reducible to positive or negative existential judgments" (Schuhmann 1996, 111). Brentano furthermore used a distinction of matter and content, as Karl Schuhmann has explained in a paper on Stumpf as a disciple of Brentano:

> The whole complex of presentations underlying the judgment [Brentano] called the judgment's matter and the act of affirmation or negation he termed the judgment's form or quality. Further he posited the judgment's content which he defined as that which is accepted or rejected in the judgment (the immediate target of affirmation or negation, as it were). Such judgmental contents are linguistically expressible in infinitival clauses or in that-clauses. This notion of a content allowed Brentano also to explain so-called indirect judgments of the type 'it is possible, necessary, true, wrong that ---' by referring to their content. Thus the judgment 'it is possible that A exists' has as its presentational matter A and as its content the possibility of A's existence. (Schuhmann 1996, 111)

The distinction between preparing the way in which the matters to be judged were phrased, and the ensuing positive or negative judgment, obviously appealed to Stumpf, who was studying law when he heard Brentano lecture for the first time. He later formulated himself the two steps in judging, taking over notions from Bolzano and his other former mentor, Lotze. Stumpf proposed calling the content "Sachverhalt" (i.e., state of affairs). This notion stemmed from German legal practice, where it described the preparation of the file encompassing everything that the judge was entitled to take into consideration for the judgment: "what is not in the file, is not in the world." This practice implied the separation of two steps in judging. The matter to be judged was first prepared and documents gathered, so as to be presented as the "state of affairs" in the file. The final judgment, then, answered to the state of affairs, not to the matter beyond the confines of the court. In Stumpf's time, the written file had been replaced by hearing statements before court.[18]

This foundation for the distinction between matter and content or state of affairs implies that the question of whether or not a statement is true or false cannot reach any rationalization beyond the content of the judgment. Stumpf could find support for this stance in Brentano. "According to Brentano," Arkadiusz Chrudzimski writes,

---

[18] The Latin phrase is quoted in Vismann (2011). On the philosophical notion of Sachverhalt as embedded in German legal practice, see Smith (1978). On how Stumpf's notion of Sachverhalt evolves from gathering statements to experimentation and on how the development of an actual hearing in court may have impacted Stumpf's interest in acoustics, see Kursell (2022).

"a judgment is not true when it coincides with a part of reality, but when it could also be made by someone who judges based on evidence" (Chrudzimski 2015, 178). This "epistemic notion of truth," he continues, entailed that Brentano not only could dispose of propositional truth makers, but of any kind of truth makers.[19]

Stumpf's method of studying judgment privileged instances in which considering a judgment's truth as the basis for further elaboration is irrelevant.[20] He took the subjects to be judging as best they could, based on their individual epistemic situatedness or conditions of judgment (*Urteilsbedingungen*). He then compared the outcomes of those judgments by looking at more than one individual. He thereby formed what is in focus for this chapter: groups who share elements of that epistemic situatedness. Rather than defining those elements, however—and this is central to his method's foundation in Brentano's logic—the shared element was reduced to being on one side of a yes-or-no alternative.

The members of the group shared that they all did not have some feature defining a second group. That feature could be very simple, such as that group A can tell which of two tones is higher while group B cannot do so; group A does know what sounds are used in the experiment while group B does not; group A is familiar with such and such a regional musical practice while group B is not. As is apparent, statistical relevance was not a defining feature, nor was random choice of the individuals: Stumpf was the only individual in the group of those not familiar with the music of the Nuxalk. The one defining feature was even used for categorizing one group without the second being investigated in a paired setup: the uninformed individuals were not systematically compared to informed subjects. That is to say, the control group, in Stumpf's case, emerges directly from a logical operation.

The interference device with its distributed architecture gave this logical operation a new, material shape. The judgments Stumpf studied were no longer based on genuinely invalid premises, but rather on arbitrarily induced premises invalidating the judgment. If, for the informed observers, the acoustic topography of language sounds was what they should observe in detail, the uninformed were supposed to resort to everything they were left with in the state of an induced lack. Their notes from the listening task display this function. While the controlled ignorance cut the subjects off from correct judgments of the sounds' origins, this ignorance not only allowed them to judge the sound in an unbiased way, the comments they added on the notes also made them explicate on which other conditions they fell back for their judgments. Framed by the task handling their controlled ignorance, they added insights on their motives, habits, and prejudices. In short, Stumpf's method can be summarized as creating situations in which subjects who could or could not judge truthfully were confronted with objects that were prepared for response in a controlled way, rather than a truthful way. What began as his interest in "false"

---

[19] See also Rollinger (2015, 80), on truth in Stumpf's logic.

[20] For instance, in Stumpf (1906b, 50): "Die Eigenschaft, um derentwillen wir von notwendigen Urteilen im logischen Sinne des Wortes sprechen, ist nicht [die] psychologische, reale Notwendigkeit. Sie ist eine immanente Eigenschaft des Urteils in Hinsicht seines Inhaltes, als des Sachverhaltes. Diesem kommt sie zu, nicht dem Urteilsakt." See also Stumpf (1906a).

judgment developed into a method investigating judgment based on the distinction between two alternative and mutually exclusive conditions of judgment.

5.4 Conclusion

This chapter has discussed the roles that groups of subjects played in Carl Stumpf's experimental practice or—stretching the etymology somewhat—their roles and counter-roles. Stumpf began working with groups of subjects long before he researched vowels, in his research on judging musical tones. Although those groups do not match the requirements identified by Dehue and spelled out in research on the history of the control group more generally (Dehue 1997, 2001, 2005; Chalmers 2001), and although he did not use the term "control group" even in 1926, by the time the notion gained currency, his research practice has shed new light on it. More specifically, Stumpf transfers one basic feature of control experiments to his psychological investigation: he reduces the claim that "all other factors remain the same" to a simple alternative that he eventually could control arbitrarily. Rather than taking all sentiments and feelings of music-listening into account, he split all music listeners into two groups, according to the criterion of whether they could or could not distinguish the higher of two notes. In this case, his homogeneous Middle-European population easily granted that his subjects would share many features otherwise. The musical and unmusical subjects, for instance, were all eloquent, had access to erudition, were exposed in various ways to music, etc. The uninformed subjects in the vowel experiment were first tested with regard to their general ability to react to the apparatus. Individuals who would not have accepted the transmitted sounds to begin with were thus not admitted to the experiment. In other words, the functioning of the logical operation at the core of this method had to be carefully handled, even though its explicitness varies greatly.

The story this chapter has been telling about the history of experimental psychology diverges considerably from the standard narrative emphasizing psychophysics, and in particular from the telling of Edwin Boring. Stumpf's notion of psychology welcomed experimental methods while rejecting the exclusive methodological choice of correlating sensations to outer stimuli, which had made psychophysics the center of psychology's alleged auto-historiography as instantiated by Boring (e.g., Boring 1929, 1942).[21] Instead, Stumpf's psychology was based on an immanent approach and his main object of inquiry was judgment. For his practical epistemology, he worked with two parallel strands of developing psychological methodology. In his experimental work, he provoked judgments that he held against the conditions in which they were made. Step-by-step, this method took on a systematic character and culminated in the setup for the vowel experiment, which went so far as to induce

---

[21] That Edwin Boring started out from the 1930s electrified psychological laboratory, when he made researchers like Helmholtz and Stumpf predecessors of, e.g., S.S. Stevens, has not yet been considered sufficiently. See, e.g., Boring (1938), and cf. Kursell and Schäfer (2018).

ignorance. This method turned this intervention via induced ignorance into a methodological device that became customary to the notion of control group as it has been discussed here. In parallel, he developed his own notion of logic as a practice. This theoretical backing, despite not proceeding at the same pace, remained constantly connected to his practical work. He anticipated this making him drift away from logic proper, when he wrote in the preface to his book *Die Sprachlaute* (p. v.): "The philosopher who will pick up this book, will shake his head in incomprehension and lay it aside again quickly."

References

Ash, Mitchell. 1995. *Gestalt Psychology in German Culture 1890–1967: Holism and the Quest for Objectivity*. Cambridge/New York: Cambridge University Press.

Boring, Edwin G. 1929. *A History of Experimental Psychology*. New York: Appleton-Century.

———. 1938. Perspective. In *Hearing. Its Psychology and Physiology*, ed. Stanley Smith Stevens and Hallowell Davis, v–vii. New York/London: Wiley/Chapman & Hall.

———. 1942. *Sensation and Perception in the History of Experimental Psychology*. New York: Appleton-Century-Crofts.

———. 1954. The Nature and History of Experimental Control. *The American Journal of Psychology* 67 (4): 573–589.

Chalmers, Iain. 2001. Comparing Like with Like: Some Historical Milestones in the Evolution of Methods to Create Unbiased Comparison Groups in Therapeutic Experiments. *International Journal of Epidemiology* 30 (5): 1156–1164.

Chrudzimski, Arkadiusz. 2015. Carl Stumpf über Sachverhalte. In *Philosophy from an Empirical Standpoint: Essays on Carl Stumpf*, ed. Denis Fisette and Riccardo Martinelli, 173–202. Leiden/Boston: Brill Rodopi.

Dehue, Trudy. 1997. Deception, Efficiency, and Random Groups: Psychology and the Gradual Origination of the Random Group Design. *Isis* 88 (4): 653–673.

———. 2001. Establishing the Experimenting Society: The Historical Origin of Social Experimentation. *The American Journal of Psychology* 114 (2): 283–302.

———. 2005. History of the Control Group. In *Encyclopedia of Statistics in Behavioral Science*, ed. Brian S. Everitt and David C. Howell, 829–836. Chichester: Wiley.

Deutsches Wörterbuch von Jacob Grimm und Wilhelm Grimm, digital version at Wörterbuchnetz of Trier Center for Digital Humanities, Version 01/21, https://www.dwds.de/d/wb-1dwb. Accessed November 6, 2021.

Fisette, Denis. 2015a. The Reception and Actuality of Carl Stumpf: An Introduction. In *Philosophy from an Empirical Standpoint: Essays on Carl Stumpf (Studien zur Österreichischen Philosophie, ed. Marco Antonelli, vol. 46)*, ed. Denis Fisette and Riccardo Martinelli, 11–53. Leiden/Boston: Brill and Rodopi.

———. 2015b. Archivalia: Introduction. In *Philosophy from an Empirical Standpoint: Essays on Carl Stumpf (Studien zur Österreichischen Philosophie, ed. Marco Antonelli, vol. 46)*, ed. Denis Fisette and Riccardo Martinelli, 423–432. Leiden and Boston: Brill and Rodopi.

Helmholtz, Hermann L.F. 1954. *On the Sensations of Tone as a Physiological Basis for the Theory of Music*. New York: Dover.

Hiebert, Erwin. 2014. *The Helmholtz Legacy in Physiological Acoustics*. Cham et al.: Springer.

Hiebert, Elfrieda, and Erwin Hiebert. 1994. Musical Thought and Practice: Links to Helmholtz's Tonempfindungen. In *Universalgenie Helmholtz. Rückblick nach 100 Jahren*, ed. Lorenz Krüger, 295–311. Berlin: Akademie Verlag.

Hoffmann, Christoph. 2001. Haut und Zirkel. Ein Entstehungsherd: Ernst Heinrich Webers Untersuchungen 'Ueber den Tastsinn'. In *Ansichten der Wissenschaftsgeschichte*, ed. Michael Hagner, 191–223. Frankfurt am Main: Fischer.

———. 2005. *Unter Beobachtung. Naturforschung in der Zeit der Sinnesapparate.* Göttingen: Wallstein.

Honing, Henkjan, ed. 2018. *The Origins of Musicality.* Cambridge, MA: MIT Press.

Hui, Alexandra. 2013. *The Psychophysical Ear: Musical Experiments, Experimental Sounds, 1840–1910.* Cambridge, MA: MIT Press.

Jackson, Myles W. 2006. *Harmonious Triads. Physicists, Musicians, and Instrument Makers in Nineteenth-Century Germany.* Cambridge, Mass., London: MIT Press (= Transformations: Studies in the History of Science and Technology).

Kaiser-el-Safti, Margret. 2011. Einleitung. In *Carl Stumpf: Erkenntnislehre,* ed. Carl Stumpf Gesellschaft, 5–45. Lengerich: Pabst Science Publisher.

Klotz, Sebastian, ed. 1998. *Vom tönenden Wirbel menschlichen Tuns. Erich M. von Hornbostel als Gestaltpsychologie, Archivar und Musikwissenschaftler.* Berlin: Milow: Schibri-Verlag.

Kursell, Julia. 2008. Hermann von Helmholtz und Carl Stumpf über Konsonanz und Dissonanz. *Berichte zur Wissenschaftsgeschichte* 31: 130–143.

———. 2013. Experiments on Tone Color in Music and Acoustics: Helmholtz, Schoenberg, and klangfarbenmelodie. In *Music, Sound, and the Laboratory From 1750 to 1980,* ed. Alexandra Hui, Julia Kursell, and Myles W. Jackson, vol. 28, Osiris, 191–211. Chicago: University of Chicago Press.

———. 2018. *Epistemologie des Hörens: Helmholtz' physiologische Grundlegung der Musiktheorie.* Paderborn: Fink.

———. 2019. From Tone to Tune: Carl Stumpf and the Violin. *19th Century Music* 43 (2): 121–139.

———. 2021. Coming to Terms with Sound: Carl Stumpf's Discourse on Hearing Music and Language. *History of Humanities* 6: 35–59.

———. 2022. Carl Stumpf and the Phenomenology of Musical Utterances. In *The Oxford Handbook of the Phenomenology of Music Cultures,* ed. Harris M. Berger, Friedlind Riedel, and David VanderHamm, C30.P1–C30.N24. Oxford University Press (online publication).

Kursell, Julia, and Armin Schäfer. 2018. Elektronische Musik für Radios von John Cage, Karlheinz Stockhausen und Michael Snow. In *Radiophonic Cultures,* ed. Ute Holl, vol. 1, 135–149. Heidelberg: Kehrer.

Martinelli, Riccardo. 2015. A Philosopher in the Lab. Carl Stumpf on Philosophy and Experimental Sciences. *Philosophia Scientiae* 19 (3): 23–43.

Pantalony, David. 2005. Rudolph Koenig's Workshop of Sound: Instruments, Theories, and the Debate over Combination Tones. *Annals of Science* 62: 57–82.

———. 2009. *Altered Sensations. Rudolph Koenig's Acoustical Workshop in Nineteenth-Century.* Paris, Dordrecht et al: Springer.

Rheinberger, Hans-Jörg. 1997. *Toward a History of Epistemic Things: Synthesizing Proteins in the Test Tube.* Stanford: Stanford University Press.

———. 2023. *Split and Splice: A Phenomenology of Experimentation.* Chicago/London: University of Chicago Press.

Rollinger, Robin D. 1999. *Husserl's Position in the School of Brentano.* Dordrecht/Boston/ London: Kluwer.

———. 2015. Practical Epistemology: Stumpf's Halle Logic (1887). In *Philosophy from an Empirical Standpoint: Essays on Carl Stumpf (Studien zur Österreichischen Philosophie, ed. Marco Antonelli, vol. 46),* ed. Denis Fisette and Riccardo Martinelli, 75–100. Leiden/Boston: Brill and Rodopi.

Scherer, Wolfgang. 1989. *Klavier-Spiele. Die Psychotechnik der Klaviere im 18. und 19. Jahrhundert.* München: Wilhelm Fink.

Schickore, Jutta. 2019. The Structure and Function of Control in the Life Sciences. *Philosophy of Science* 86 (2): 203–218.

———. 2021a. The Place and Significance of Comparative Trials in German Agricultural Writings Around 1800. *Annals of Science* 78 (4): 484–503.

———. 2021b. Methodological Ideas in Past Experimental Inquiry: Rigor Checks around 1800. *Intellectual History Review* 33: 267–286.

Schmidgen, Henning. 2014. *Hirn und Zeit: Die Geschichte eines Experiments.* Berlin: Matthes und Seitz.

Schuhmann, Karl. 1996. Carl Stumpf 1848–1936. In *The School of Franz Brentano*, ed. Liliana Albertazzi, Massimo Libardi, and Roberto Poli, 109–129. Dordrecht: Kluwer.

Smith, Barry. 1978. Law and Eschatology in Wittgenstein's Early Thought. *Inquiry* 21 (1–4): 425–441.

Steege, Benjamin. 2012. *Helmholtz and the Modern Listener*. Cambridge: Cambridge University Press.

———. 2021. *An Unnatural Attitude: Phenomenology in Weimar Musical Thought*. Chicago, London: University of Chicago Press.

Stumpf, Carl. 1873. *Über den psychologischen Ursprung der Raumvorstellung*. Leipzig: Hirzel.

———. 1883. *Tonpsychologie*. Vol. 1. Leipzig: Hirzel.

———. 1886. Lieder der Bellakula-Indianer. *Vierteljahrsschrift für Musikwissenschaft* 2: 405–426.

———. 1890. *Tonpsychologie*. Vol. 2. Leipzig: Hirzel.

———. 1906a. Erscheinungen und psychische Funktionen. In *Abhandlungen der Königlich Preußischen Akademie der Wissenschaften aus dem Jahre 1906*, 40 pp. Abh. IV.

———. 1906b. "Zur Einteilung der Wissenschaften." *Abhandlungen der Königlich Preußischen Akademie der Wissenschaften aus dem Jahre* 1906, Abh. V, 94 pp.

———. 1918. Die Struktur der Vokale. *Sitzungsberichte der Königlich Preußischen Akademie der Wissenschaften* 17: 333–358.

———. 1926. *Die Sprachlaute. Experimentell-phonetische Untersuchungen nebst einem Anhang über Instrumentalklänge*. Berlin: Springer.

———. 2020. *Tone Psychology, vol. 1: The Sensation of Successive Single Tones* (trans: Rollinger, Robin D.). London/New York: Routledge.

Textor, Mark. 2020. Stumpf between Criticism and Psychologism: Introducing 'Psychologie und Erkenntnistheorie'. *British Journal for the History of Philosophy* 28 (6): 1172–1180.

Vismann, Cornelia. 2011. Medien der Rechsprechung. In *and Markus Krajewski*, ed. Alexandra Kemmerer. Frankfurt am Main: S. Fischer.

Wittje, Roland. 2016. *The Age of Electroacoustics. Transforming Science and Sound 1863–1939*. Cambridge, MA: MIT Press.

Ziegler, Susanne, ed. 2006. *Die Wachszylinder des Berliner Phonogramm-Archivs*. Berlin: Ethnologisches Museum, Staatliche Museen zu Berlin – Preußischer Kulturbesitz.

Julia Kursell is Professor of Musicology at the University of Amsterdam. Her research interests include 20th and 21st-century composition, the history of musicology and the relation between music and science. She is co-director of the UvA's Vossius Center for the History of Humanities and Sciences (http://vossius.uva.nl).

Chapter 6
A "Careful Examination of All Kind of Phenomena": Methodology and Psychical Research at the End of the Nineteenth Century

Claudia Cristalli

6.1 Introduction

It was a warm night in Boston on June 26, 1857, and behind the shut windows and tightly drawn curtains of Apartment 12, on the third floor of the Albion, at the corner of Tremont and Beacon street, it was even warmer. Inside the apartment, in a room dimly lit by gas lamps, Professor Benjamin Peirce – the United States' foremost mathematician – was sweating profusely. He was crammed on a bench inside a wooden box, some musical instruments (two tambourines, a fiddle, a banjo, and a tin horn) gathered between his legs; two boys, tied with ropes, were sitting at his sides. The boys faced each other; the ropes fastened so their hands were behind their backs and their ankles together: they could not bend forward toward the Professor nor toward the musical instruments lying at his feet. The three were participating in a spiritual manifestation: the boys as mediums, and Professor Peirce as their control.

Bizarre as this setup might appear, the examination of psychical phenomena by a committee of experts was not an isolated case in the nineteenth century. The one just mentioned exhibited an incredible display of resources among both controllers and committed spiritualists. Besides Benjamin Peirce (1809–1880), the committee included the naturalist Louis Agassiz (1807–1873), the chemist Eben Norton Horsford (1818–1893), and the astronomer Benjamin Apthorp Gould (1824–1896). These last three had studied with some of the most eminent scientific men of Western Europe –Alexander von Humboldt, Justus von Liebig, and Carl Friedrich Gauss, respectively – before settling at Harvard and surroundings.[1] To ensure that such a

---

[1] Agassiz taught at the Lowell Institute, Horsford at the newly founded Lawrence scientific school, and Peirce and Gould at Harvard University proper.

C. Cristalli (✉)
Department of Philosophy, University of Tilburg, Tilburg, Netherlands

149
J. Schickore, W. R. Newman (eds.), *Elusive Phenomena, Unwieldy Things*,
Archimedes 71. https://doi.org/10.1007/978-3-031-52954-2_6

jury would have a chance of witnessing relevant manifestations, some of the most powerful mediums of the country had been invited, including the Fox sisters. The event was a response to a mediumship scandal that had happened earlier that year at Harvard University. The scandal involved a few faculty members and a Divinity student, the latter in the role of the fraudulent medium. To capitalize on the attention, the newspaper *Boston Courier* offered a $500 prize for successfully demonstrating spiritual manifestations. A certain Dr. H. F. Gardner took up the challenge (Moore 1972, 494–95).

The 3 days of experiments at the Albion were not the first attempts at investigating psychical phenomena in a controlled setting, nor were they to be the last ones. In the last decades, a growing body of literature has examined the connections of psychical research with ordinary science, highlighting the positivist and naturalist motivations of many researchers (Luckhurst 2002; Owen 2004; Noakes 2019) and their role within the process of institutionalization of psychology (Coon 1992; Plas 2012; Sommer 2012). Moreover, in a context in which physics and engineering were constantly expanding the boundaries of the possible, psychical phenomena often appeared just as plausible as the possibility of long-distance communication through the telegraph. Besides being suggestive, this analogy well represents the idea of a continuity between physical and psychical forces, to be investigated through scientific experimentation.

Other commentators highlighted how the rejection of psychical phenomena and of their scientific investigation was often motivated by religious sentiments, in so far as the belief in spirits communicating through table turnings and rappings was perceived as impious and/or materialistic (Moore 1972; Sommer 2018).[2] Even the journalist writing the *Boston Courier* report confessed his religious concerns when witnessing the Albion séances:

> …there are certain things of which I have no objection to say that I am afraid. I trust I am afraid to do anything deliberately, which would lessen my self-respect, – afraid of too intimate association with those, whom I do not conceive deserving of respect, – afraid of disobeying the instincts of my understanding, – afraid of violating the manifest law of God. I have never ceased to consider what was *professed* and *pretended* by the spiritualists as impious – what was *performed* by them as puerile. If the manifestations claimed by them, as within their power, actually corresponded with their pretensions, I should have no hesitation in ascribing them to diabolical agency… (Author Unknown 1859, 23; emphasis in original)

In spite of the rhetorical formulation of this nineteenth-century journalistic piece, it is clear that some elements of belief influenced the judgement of both parties involved. Belief indeed played a role in almost every case that I discuss in this chapter; however, it is interesting to note that belief was not, by itself, able to determine the type of investigation that the different actors chose to pursue, nor the way in

---

[2] The literature on these topics is now vast; it is also often country-specific, with, e.g., Bensaude-Vincent and Blondel (2002), Bower (2010), Plas (2012), and Hajek (2015) focusing on France; Luckhurst (2002) and Noakes (2019) on the United Kingdom and on the Empire; Sommer (2013) on Germany; and Moore (1972) and Sommer (2020) on the United States.

which they chose to control for the phenomena that they wanted to investigate. Gender and class also played an important role in structuring the setting of psychical experiments and in negotiating positions of authority (Taylor 1996; Owen 2004). Charles S. Peirce, son of Benjamin Peirce, when evaluating a collection of testimonies of telepathic phenomena, succinctly stated: "Women, children, sailors, and idiots are recognized by the law as classes peculiarly liable to imposition. If sailors' yarns are to be admitted, the reality of ghosts is put beyond doubt at once, and further discussion is superfluous" (Peirce 1887b, W6, 138).[3] All of these elements constitute the background of the present account, which focuses more narrowly on how different investigators came up with and defended different strategies of control. Such strategies were crucial to the interpretation of the phenomena under investigation, including to which branch of science (if any) they belonged.

"Psychical phenomena" is an umbrella term that referred generally to phenomena allegedly produced by spiritual forces or entities; in this chapter, we encounter four instances of these manifestations – table moving (Sect. 6.2), mediumistic communication (Sect. 6.3), "telepathic" communication (Sect. 6.4), and apparitions (Sect. 6.5). All of these instances offer the opportunity to capture different declinations of "control." First, "control" is here mostly about controlling the setting (i.e., the instrument(s), the people's perceptions, or the people's testimonies), rather than about controlling a result against a comparative trial (as it is most often the case for the chapters in this volume; see Schürch's contribution, for example). Second, in some cases "control" is an actors' term (see Sect. 6.4), a rare occurrence in this volume. Third, the actors presented in this chapter deploy two distinct but parallel (and ultimately intertwining) strategies of control, which are called – by the actors themselves – "experimental" and "historical."[4]

The experimental strategy is distinguished by its use of apparatuses; in this chapter, small tables are the simplest device for detecting and interpreting spirit messages. The historical strategy relies instead on collecting, evaluating, and publishing testimonies, and on attempting to statistically quantify their relevance. The experiments of Michael Faraday (Sect. 6.2) and Robert Hare (Sect. 6.3) illustrate control strategies in traditional experimental settings,[5] while Edmund Gurney's contributions to the Society for Psychical Research (SPR) show the complexity of the

---

[3] Henceforth, references to Peirce's works reprinted in the *Writings* will be as follows: W volume number, page number.

[4] While I focus here only on the anglophone context, the experimental/historical distinction was also present in France and Italy. Thus, in the 1890s Charles Richet and Camille Flammarion, convinced by the phenomena produced by Eusapia Palladino, would defend themselves from the allegation of "spirit communication" and claim their approach to be "essentiellement expérimentaliste" (Blondel 2002, 145). As research continued, though, they both came to appreciate that in this field, experiment is mixed with a type of investigation resembling that of the lawyer or of the historian (Blondel 2002, 151).

[5] Schickore (2019, 210) draws the distinctive meaning of "control checks" from the "controlling experiment" entry in Whitney's 1897 *Cyclopedia*. Interestingly, Whitney situates this definition in the context of chemistry ("in chem.," 1897, 1237), and Robert Hare, who adopted this notion of control in his experiments, was a chemist.

experimental-historical interaction from methodological and epistemological perspectives (Sect. 6.4). Finally, Gurney's controversy with Charles S. Peirce over
cases of spontaneous telepathic manifestations illustrates the challenge of controlling testimony within the historical method (Sect. 6.5).

6.2 Control as Education: Michael Faraday's Experiments on Table-Turning

In 1853, Michael Faraday (1791–1867), who at the time was already one of the
greatest scientific authorities in Europe, took the time and trouble to conceive
an experiment on the phenomenon of table moving. His research resulted in three
publications: a brief letter to the *Times* (appearing on June 28, 1853, Faraday 1853a);
a longer exposition for the *London Athenaeum* (July 2, 1853, Faraday 1853b); and
a description with figure for the *Illustrated London News* (July 16, 1853, Faraday
1853c; Fig. 6.1). While Faraday deplored in more than one instance the fact of having had to engage with such a topic, he was clearly aware of its importance both
from an epistemological and a moral perspective.

Faraday began by testing various materials in contact with the tables and the
participants' hands. He determined that interposing such materials between wood
and skin didn't interfere with the table's movements. He then conceived an experiment to ascertain the nature of the force operating on the table. This experiment was
fully within the methodological constraints of physical investigation, as he declared
in the opening paragraph of the *Athenaeum* letter:

Fig. 6.1 Illustration of
Faraday's table-turning
experiment. Source:
Illustrated London News,
July 16, 1853

the proof which I sought for, and the method followed in the inquiry, were precisely of the same nature as those which I should adopt in any other physical investigation. (Faraday 1853b, 801)

Participants in séances claimed (1) that the table would move *first*, and that the hands would follow; and (2) that their hands were laying still on the table's surface and/or pressing downwards only, while the table "turned" and hopped in any direction. According to Faraday, they did not give their testimony in bad faith:

The parties with whom I have worked were very honorable, very clear in their intentions, successful table movers, very desirous of succeeding in establishing the existence of a peculiar power, thoroughly candid, and very effectual. (Faraday 1853b, 801)

Their séances were "effectual," meaning that the participants were predictably successful in moving the table. However, Faraday could not detect a force that could justify the movement, whether electrical, magnetic, or otherwise. The only plausible option seemed to be that the table was moved by mechanical force impressed by the participants – yet they resolutely denied it. According to their testimony, the table appeared to be moving by itself.

In designing his experiment, Faraday was influenced by the very recent studies on involuntary movements carried out by the physician and physiologist William B. Carpenter (1813–1885). In the final paragraphs, Carpenter pointed out that "muscular movement" could be elicited and detected independently of the subject's conscious influence. This opened the possibility that unconscious action and expectation played a much greater role in everyday life than was usually thought: "the *anticipation* of a given result being the stimulus which directly and involuntary prompts the muscular movements that produce it [the result]" (Carpenter 1852, 153). Carpenter's findings resonated with Faraday's own intuitions as a self-educated experimentalist: developing a sound judgment in scientific and daily matters alike depends on one's commitment to a life-long education of one's faculties, just as reaching a "moderate facility" in playing a musical instrument depends on long hours of practice (Faraday [1854] 1859, 491). In time Carpenter would take up Faraday's experiment along with his insistence on education in the historical and scientific examination of mesmerism and spiritualism (Carpenter 1877).[6]

Faraday was convinced that the table moved by mechanical force, impressed upon it by unaware participants. He thus proceeded to devise an instrument (Fig. 6.1) to check whether the table moved first and the hands followed, as declared by the séance participants, or whether the opposite was true. The ingenuity and simplicity of Faraday's instrument capitalizes on his early years of apprenticeship as a book binder, as the description below illustrates.

The experiment consisted in interposing a "bundle" of four or five cardboard sheets between the table's surface and the hands of the participants. Pellets of wax and turpentine kept the bundle together. The "cementum" between the cardboard sheets was "strong enough to offer considerable resistance to mechanical motion,

---

[6] Carpenter's position on some psychical phenomena (e.g., telepathy) was, however, more nuanced than is possible to explore here; for more details, see Delorme (2014).

and also to retain the cards in any new position which they might acquire, – and yet weak enough to give way slowly to a continued force" (Faraday 1853b, 802). Faraday marked the position of each cardboard piece and arranged the pile in such a way that the marks would face down. He topped everything with a larger piece of cardboard to disguise the underlying structure, and put the pile on a piece of sandpaper to ensure it would not slide on the table's surface. After the table's first hop he examined the pile, and the way the cardboard pieces had been displaced showed the hands had exerted a force *first*, and the table then moved accordingly. He repeated the observations and found that the carboard showed signs of experiencing mechanical force even when the table did not move at all.

The first part of the experiment showed that the participants in the séance exerted mechanical force upon the table's surface. From their reports, however, they appeared utterly unaware of doing so, and believed that they felt pressure coming from the table to their hands. Interposing different materials between their hands and the table did not alter the participants' "effectiveness" in any way; would awareness of their involuntary action change the séance's outcome? To show participants whether they were applying a force and, if so, in what direction, Faraday built an indicator. It consisted in a piece of paper connected to the bundle of sliding cardboards under the participant's hands, lifted from the table's surface by a pin. The pin acted as the fulcrum of a lever, with the shorter arm near the participant's hands and the longer arm showing the direction of any force applied to the table by the hands, like an index. The party assembled around the table instantly became less "effective":

> The effect was never carried far enough to move the table, for the motion of the index corrected the judgment of the experimenter, who became aware that, inadvertently, a side force had been exerted. [...] now that the index was there, witnessing to the eye, and through it to the mind, of the table turner, not the slightest tendency to motion either of the card or of the table occurred. (Faraday 1853b, 803)

Without falling into the trap of ridiculing his audience for their beliefs, Faraday took this opportunity to make a more general point on the role of instruments in enabling reliable observation. He did not criticize the senses as a fallible source of knowledge; rather, he emphasized that they cannot be reliable if unaided by some instrument. In this case, the instrument acts as a *control* for the very actions séance participants believe they are performing (or not performing). Called "index," "indicator," and "instructor," Faraday's paper strip allowed participants to perform a task, such as pressing downwards, in a reliable manner. The task's apparent simplicity probably prevented people from thinking about the need for a control, but Faraday's experiments had shown that pressing "directly downward" was not an easy task at all. As he clearly stated in his initial remarks to the *Times*:

> I think the apparatus I have described may be useful to many who really wish to know the truth of nature [...]. Persons do not know how difficult it is to press directly downward – or in any given direction against a fixed obstacle: or even to *know only* whether they are doing so or not; unless they have some indicator, which, by visible motion or otherwise, shall instruct them[.] (Faraday 1853a, 390-1; emphasis in original)

Faraday checked the working of the indicator when participants were prevented from seeing it and observed that it would stop moving as soon as the participants were made aware of it. The indicator's role had a pedagogical as well as a scientific role for Faraday. As a self-educated scientist, he was always interested in methods to improve or educate the mind, as his famous 1854 lecture testifies. In his letter to the *Times,* Faraday insisted on the power of instruments in correcting observations while at the same time educating the observer:

> [...] *the most valuable effect of this test apparatus* (which was afterwards made more perfect and independent of the table) *is the corrective power it possesses over the mind of the table turner.* As soon as the index is placed before the most earnest, and they perceive – as in my presence they have always done – that it tells truly whether they are pressing downwards only or obliquely, then all effects of table turning cease, even though the parties persevere, earnestly desiring motion, till they become weary and worn out. No prompting or checking of the hands is needed – the power is gone; and this only because the parties are made conscious of what they are really doing mechanically, and so are unable unwittingly to deceive themselves. (Faraday 1853a, 385, my emphasis)

Faraday's intervention aimed at showing that, while table turning was a real phenomenon, its causes were not necessarily immaterial in nature. Moreover, it showed that even with the best intentions a mind unaided by some "index" may be deceived. His inquiry did not stop either popular or expert interest in the matter, however, and some researchers continued both to expose fraud and to legitimize psychical phenomena as genuine manifestations of spiritual entities or spiritual forces.

Finally, it is important to note that Faraday's own efforts to mitigate the "table moving craze that swept England around 1853" (James 1996, *xxx*) were grounded as much in religious concerns as in broader educational and methodological ones.[7] While the bulk of his correspondence on table-turning was destroyed (according to his wish), we can glimpse his attitude toward the phenomenon in a letter to Caroline Deacon of 23 July 1853 – which he wrote *after* publishing the results from his table-turning experiment:

> ...the world is running mad after the strangest imaginations that can enter the human mind. I have been shocked at the flood of impious & irrational matter which has rolled before me in one form or another since I wrote my [T]imes letter and am more than ever glad that as a natural philosopher I have borne my testimony to the cause of common sense & sobriety [...]. (Faraday [1853d] 1996, 538–39)

Deacon (née Reid, 1816–1890) was a niece of Faraday's wife, Sarah, and the Faradays corresponded with her often. Faraday's experimental refutation of table moving caused a great stir among "common sense" scientists and spiritualists alike, and it is a pity that his replies are now lost. On the one hand we must recognize, as Frank James certainly does, that Faraday's identification of table-moving phenomena as "impious" helped motivate him to examine them and try to dispel, with the help of "common sense and sobriety," the lure of such "imaginations." On the other hand, however, Faraday believed in the broader educational value of properly conducted experiments,

---

[7] For a discussion of the links between Faraday's scientific and religious commitments, see Cantor (1991).

which he thought trained sound judgment. Those epistemic values, while on a continuum with Faraday's religious values, can still be distinguished from them, and Faraday himself argued for them independently. In the end, his religious motivations were only partly responsible for the development of the specific technology of control that he employed to analyse the phenomenon of table turning.

6.3 Controlling the Medium and Communicating with Spirits: Robert Hare's Apparatuses

Among the scientists originally praising Faraday's 1853 experiment there was the American chemist Robert Hare (1781–1858). Dubbed "the foremost chemist of his generation" (Kneeland 2008, 245), Hare was famous in his day for inventing the "oxy-hydrogen blow pipe," a forerunner of today's welding torch. He also contributed to funding the Society for the Advancement of American Philosophy in 1848 (Tymn 2021). Hare had seen Faraday's public lecture in London in 1841, and he occasionally wrote about Faraday's scientific views, on subjects such as electricity and "the nature of matter."[8] In 1853 he published a letter in the *Philadelphia Inquirer*, denying the possibility of table-turning phenomena. He alluded to self-deception as the only possible source of the motion, and paraphrased Faraday's original conclusion:

> I am of the opinion that it is utterly impossible for six or eight, or any number of persons, seated around a table, to produce an electrical current. Moreover, I am confident that if by any adequate means an electrical current were created, however forcible, it could not be productive of table turning. [...] The only subject for inquiry, was *how people could so deceive themselves as to suppose that what they really moved, moved them.* (Hare [1853] 1855, 35; my emphasis)

Hare concluded his letter with an explicit endorsement of Faraday's experiment, to which he appended a "moral" of his own:

> I recommend to your attention [...] Faraday's observations and experiments [...]. *I entirely concur in the conclusions of that distinguished experimental expounder of Nature's riddles.* A moral may be drawn from this susceptibility of self-deception. In our moral conduct, as in our physical movements, we sometimes take the effect for the cause, and blame others for that which has originated in ourselves. (Hare 1855, 36; my emphasis)

In line with Faraday's own views, Hare ultimately blamed some inherent "susceptibility of self-deception" for the blunder; however, he quicky turned this psychological insight into a moral maxim offering no means to correct for our tendency to judge wrongly.

Two years later, in his *Experimental Investigations on Spirit Manifestation* (1855), Hare reprinted his 1853 *Inquirer* letter with a response that allegedly moved him to investigate the matter further. The writer of the response – most likely the

---

[8] See Robert Hare papers, 1764–1858, Boxes 7, 8, and 10.

telescope-maker Amasa Holcomb[9] – claimed to have witnessed table-turning and related phenomena and then titillated Hare's curiosity with the prospect of a great scientific discovery:

> I cannot in this case doubt the evidence of my senses. I have seen the tables move, and heard tunes beat on them, when no person was within several feet of them. This fact is proof positive that the force or power is not muscular. [...] If these things can be accounted for on scientific principles, would it not be a great acquisition to science, to discover what those principles are? (Letter to Hare, in Hare 1855, 37)

While Holcomb supported his claims with the authority of first-person testimony, Hare rose to the challenge of accounting for spirit manifestations "on scientific principles." In his 1855 *Experimental Investigations on Sprit Manifestation* (1855) he presents himself as staunch empiricist eventually converted to spiritualism, implicitly inviting the reader on a journey from incredulity to belief. Indeed, Hare had a strong interest in the possibility of spiritualism. At a personal level, he had lost many close family members, including his father, a brother, a sister, and numerous children. Further, he believed that something bigger than himself was at stake in the matter: namely, the possibility of a Christian faith in the modern, scientific world. Hare sincerely believed that demonstrating the existence of psychical forces and spirit communication was crucial to the belief in heaven and in the immortality of the soul. As he wrote in a letter to Silliman, "the single hope for Christianity lay[s] in the verification of spiritualistic evidence; arguments based on the 'internal evidence' of Christianity could not survive in the 19th Century" (Hare to Silliman, rough draft, APS; in Moore 1972, 494, footnote 68). Ironically, religion was also a primary concern for Faraday, who opposed spiritualism because he thought it was *dangerous* for the Christian faith and for thorough scientific thinking.

In the following I illustrate the control strategies, such as the general design and plan, and the control practices, such as concrete actions, that Hare devised to investigate communication with spirits. As he narrates, after his letter "corroborating the inferences of Faraday, [...] in obedience to solicitations already cited, [I] consented to visit circles in which spiritual manifestations were alleged to be made" (Hare 1833, 38). There he witnessed some supernatural phenomena of striking personal appeal – the medium gave voice to his own deceased father – but still he would not believe until properly testing the phenomena.

Hare's chief concern was to safeguard his experiments from the medium's potential fraud. In a retrospective remark he admitted:

> It must be manifest that the greatest difficulty which I had to overcome during the investigation of which the preceding pages give a history, arose from the necessity of making every observation under such circumstances as to show that I was not deceived by the media. (Hare 1855, 54)

Controlling for the possibility that media might deceive was therefore the chief concern behind the construction of his instruments:

---

[9] In Hare (1855), this letter is signed "Amasa Holcombe," but Kneeland (2008, 256) identifies the writer with Amasa Holcomb, a celebrated telescope maker and friend of Robert Hare.

> Subsequently, I contrived an apparatus which, if spirits were actually concerned in the phe-
> nomena, would enable them to manifest their physical and intellectual power *independently
> of control by any medium*. (Hare 1858, 40; my emphasis)

Hare actually devised many different apparatuses for the purpose, but they all fall in
one of two categories: they are either testing instruments, aiming to prove the reality
of communication with spirits by controlling the medium's actions (Figs. 6.2 and
6.3); or they are personal instruments, aiming to enable anyone to communicate
easily with the spirits without the need of a medium or of a spiritual circle (Fig. 6.4).
When Hare says that he is performing "tests" with the latter ones, he is already

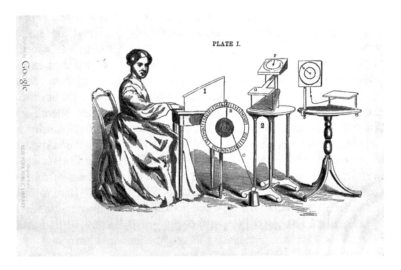

Fig. 6.2 The instrument that led to Hare's "conversion." [Plate I from Hare's book]

Fig. 6.3 Another testing instrument. Faraday's lesson is visible in the little board on casters interposed between the medium's hands and the table's surface. [Plate II from Hare's book]

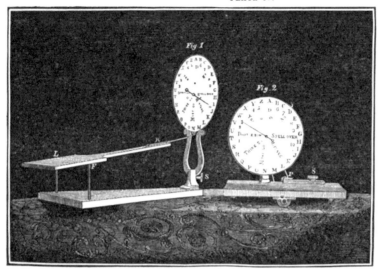

Fig. 6.4 Hare's spiritoscopes designed for "incipient mediumship," the one on the right (labelled "Fig. 2" in the original picture) being recommended for beginners. [Plate IV from Hare's book]

presupposing that communication with spirits is a genuine phenomenon. Although all instruments were indifferently called "spiritoscopes," the practices of "control" they instantiated were very different.

The relevance of apparatuses of the first type should not be underestimated. Although in the body of the text Hare seems to attribute at least part of his conversion to a persuasive manifestation of his father's spirit, the caption beneath Fig. 6.2 reads: "Engraving and description of the apparatus, which, being contrived for the purpose of determining whether the manifestations attributed to spirits could be made without mortal aid, by deciding the question affirmatively, led to the author's conversion" (Hare 1855, Plate 1). As seen above, the instrument's purpose was to control for the medium's possible manipulation of the manifestations during the séance. It is therefore also partially to the instrument and to its alleged controlling power that Hare attributed his change of feeling on the matter.

Hare described the instrument briefly next to the plate's reproduction, and then at length in the text:

Two weights were provided – one of about 8 pounds, the other about 2 pounds. These were attached one to each end of a cord wound about the pulley, and placed upon the floor immediately under it. Upon the table a screen of sheet zinc was fastened, behind which the medium was to be seated, so that she could not see the letters on the disk. A stationary vertical wire, attached to the axle, served for an index. On tilting the table, the cord would be unwound from the pulley on the side of the larger weight, being wound up simultaneously to an equivalent extent on the side of the small weight, causing the pully and disk to rotate about the axle. [...] Of course, any person actuating the table and seeing the letters, could cause the disk so to rotate as to bring any letter upon the index; but should the letters be concealed from the operator, no letter required could be brought under the index at will. (Hare 1855, 40)

As it appears, Hare took many precautions to prevent that the medium may tilt the table and thereby choose which letters would come under the disk's indicator; however, in spite of his awareness of studies on the ability of certain mediums to answer correctly to questions whenever *anyone* in the room happened to know the right answer, Hare did not engage in any practice aiming to control for the possibility of unconscious cueing.[10] Hare's strategy in addressing this and related objections was to report cases in which something completely unexpected happened. For example, the spirit of a friend would manifested himself when he was waiting for his sister's, or he would learn something that he did not know before. However, he kept devising more controls to prevent mediums form (voluntarily or involuntary) producing effects with physical rather than psychical means.

In designing his second testing instrument (Fig. 6.3), Hare took into account Faraday's concern about the motion of tables by mechanical force from unaware participants. To prevent such interferences, he had the medium rest her hands on a small wooden board, separated from the table's surface by brass balls. This way, involuntary pressure sideways would make only the upper board slide sideways on the ball, with no chance to take the table with it. With a few exceptions, Hare's instruments did not prevent tables from moving. In his eyes, this result was proof of the reality of spirits and of their ability to manifest themselves.

Once the spirits had passed the "test" of carrying their message to the rotating disk under the strictly controlled conditions provided by the instruments described (Figs. 6.2 and 6.3), they could have an easier time communicating with the living though instruments built for the purpose (Fig. 6.4). These were sorts of miniature tables on wheels, which could be operated by a single person alone. While the mechanism for one them (on the right in Fig. 6.4) was similar to that employed for the testing instrument of Plate 2 (Fig. 6.3), its functioning was allegedly much easier, so that even an "incipient" medium could achieve some effective communication. As Hare reported, "This instrument is preferred by the spirits, and is easier for the feeble medium to employ effectively. I cannot avail myself of Fig. 1 [on the left in Fig. 6.4]; through Fig. 2 [on the right in Fig. 6.4] I have had some interesting tests" (Hare 1855, Plate IV).

As Hare developed mediumship powers, the criteria for control underwent a subtle but essential change. Hare declared that he no longer needed the kind of control provided by his first two instruments to vouch for the phenomena's reliability; instead, the manifestations' trustworthiness depended only on his on his own good faith and "character":

> But having latterly acquired the powers of a medium in a sufficient degree to the interchange of ideas with my spirit friends, I am no longer under the necessity of defending media from the charge of falsehood and deception. It is now my own character only that can be in question. (Hare 1855, 54)

---

[10] Hare discusses at length the experiments of Bell on this topic, and concludes: "Dr Bell finds that in certain instances […] spirits could not communicate information nor ideas which did not exist in his mind or that of some mortals present" (Hare 1855, 66).

Hare rejected with various anecdotes the possibility that he may be deceiving himself, either though self-suggestion, involuntary movements, or the like. Perhaps the best anecdote in his own eyes – he recounted it three times in his 1855 essay (Hare 1855, 32–33, 52–54, 172) – is the following:

> The fact that my spirit sister undertook at one o'clock, on the 3rd of July, 1855, to convey from the Atlantic Hotel, Cape May Island, a message to Mrs. Gourlay, No. 278 North Tenth street, Philadelphia, requesting that she would induce Dr. Gourlay to go to the Philadelphia bank to ascertain the time when a note would be due, and report to me at half-past three o'clock; that she did report at the appointed time; and that on my return to Philadelphia, Mrs. Gourlay alleged herself to have received the message, and that her husband and brother went to the bank in consequence. With the idea received by the latter, my sister's report coincided agreeably to his statement to me. All this proves that a spirit must have officiated, as nothing else can explain the transaction. (Hare 1855, 54)

In this "spiritual messaging" between himself and a sensitive friend, his sister's spirit acted as messenger going back and forth between them. Here Hare shifted entirely from the domain of experimentation to the domain of testimony; the "controls" offered include (1) accurate descriptions of the event(s), including time, place, and names of the people involved, and (2) a profession of the witness(es)' integrity.

6.4 Controlling Telepathy Experimentally and Through Testimony: Edmund Gurney's Program

Experiment was the first but by no means the only method for establishing the reality of psychical phenomena. Such phenomena were indeed hard to observe under experimental conditions common to "any other physical investigation" (Faraday 1853b, 801), and even when they occurred, the risk that they would be exposed as fraud remained high. Eusapia Palladino (1854–1918), one of history's most celebrated mediums, cleverly admitted the allegation of fraud on psychical grounds: a medium naturally recurs to involuntary movements to "help" the manifestation of the spirit, which would otherwise require too much of her own psychical strength. She then put the ball back in the researchers' court by asking for stricter controls to prevent her from cheating (Blondel 2002, 157–58).

In this and the following section, I examine other strategies of control researchers used to capture psychical phenomena, focusing in particular on Edmund Gurney's (1847–1888). Before becoming interested in psychical phenomena, Gurney studied medicine and physiology at Cambridge and published a book on music theory (*On the Power of Music*, 1880).[11] He belonged to Sidgwick's circle and was a founding member of the Society for Psychical Research (SPR). Gurney chose both (1) to elaborate new forms of control for experiments in psychical topics (Sect. 6.4) and (2) to cast inquiry in psychical phenomena as a historical rather than experimental endeavour (Sect. 6.5).

---

[11] For a good introduction to this important figure, see Sommer (2015).

One of the main aims of writing a "natural history" – a supernatural history, in fact – of psychical phenomena was to reclaim them from the domain of experimental psychology, and therefore to rescue them from their dismissal as the product of involuntary motions by psychologists such as Carpenter. Indeed, Faraday's fame ensured a wide diffusion of his 1853 experiment; that fame had also sanctioned Carpenter's authority on the matter and, by extension, the authority of experimental psychology and of its methods of inquiry. Carpenter's further research on the topic eventually appeared as a book in 1877. This book was Gurney's point of departure for criticizing experimental psychologists' authority over psychical research.

Gurney explicitly challenged both Faraday and Carpenter's conclusions, claiming that (1) they were controlling ordinary experience, without attempting a genuine investigation of extraordinary phenomena, and (2) that, more generally, having acquired an expertise in a certain type of laboratory work did not make them experts in situations where the control strategies developed for the physical sciences did not apply. In the programmatic writing "On the Nature of Evidence in Matters Extraordinary" (1884), Gurney described Faraday's experiment as an ingenious study in the field of unconscious movements, which however did not investigate "extraordinary matters" such as mind-to-mind communication, or what was later to be called "thought-transference" or "telepathy" (Gurney 1884, 478). Moreover, Gurney dismissed Carpenter's account as based on "confused" notions of "scientific experts" and the "educated public." Rejecting these notions was part and parcel of Gurney's own claim to expertise and to the legitimization of his method of inquiry (Gurney 1884, 477).[12]

If psychical phenomena were not amenable to traditional forms of experiment, then experimental psychologists and physicists were not better qualified than any other honest investigator in adjudicating the claims of psychical research. In the 1884 essay, Gurney thus proposed a radical change of method, from laboratory work to "historical" research in the field:

> Where phenomena cannot be commanded at will (as is the case in some of the more striking departments of our research), the work of investigating them must consist, not in origination, but in the collecting, sifting, and bringing into due light and order, of experiments which Nature has from time to time given ready-made. And the due estimation of these depends, in the broadest sense, on the estimation of testimony; on what may be called historical, as opposed to experimental, methods of enquiry […]. (Gurney 1884, 482)

Indeed, if psychical phenomena could not be produced and re-produced at will, what was the advantage of a laboratory setting?[13] Especially given the many difficulties encountered in ensuring that manifestations from spirit circles were not artifacts of suggestion, involuntary movements, or trickery, collecting and analyzing testimonies of spontaneous apparitions could seem a safer and more productive

---

[12] See Noakes (2019, 239) for a recent account of Gurney's objections to Faraday and Carpenter.

[13] Lord Rayleigh made a similar comment in his presidential address at the Society of Psychical Research in 1919: psychical phenomena were challenging because "sporadic"; they could not be "reproduced at pleasure and submitted to systematic experimental control." See Christopoulou and Arabatzis, this volume.

alternative. Yet, history is not a straightforward matter, and "collecting, sifting, and bringing to light" those cases of spontaneous apparitions was no small challenge. It required a strategy to control testimonies – to check their reliability and to select among them accordingly. It also required familiarity with probability theory, since a testimony's truth was to be assessed on the prior probability that a given fact may be false. In addition, the approach required a particular understanding of the logic of induction so that, according to Gurney, the argument for psychical facts would not be considered "as weak as its weakest link." Instead, it must be allowed to accumulate every small piece of evidence into a collectively more robust construction (more on this in Sect. 6.5).

Eventually the SPR adopted a mixed-method approach, which integrated a revised experimental strategy with the historical research advocated by Gurney in 1884. In the two-volume work *Phantasms of the Living*, published in 1886 by Gurney, Frank Myers, and Frederic Podmore, the first chapters report the authors' *experiments* in thought-transference, which occurred with mediums in 1882 and 1883.[14] In the *Introduction*, the authors declare that "*Experiment* proves that telepathy […] is a fact in nature" (Gurney et al. 1886, lxv, emphasis in original). The bulk of the work, however, and the part most interesting to critics, consisted in testimonies of veridical apparitions and verified cases of mind-to-mind communication. The authors defined these phenomena as "the supersensory transference of thoughts and feelings from one mind to another," and defined "supersensory" as "independent of the recognised channels of sense" (Gurney, Myers, and Podmore 1886, e, note 1). The called them "telepathic" phenomena and added that, although they did not occur through ordinary sense channels, they could still be "analogous" to ordinary perception.

Neither Gurney's criticism for experimental methods nor his program of writing a sort of *super*natural history resulted in excluding experiment. In the *Phantasms* Gurney reintroduced experiments but redefined them around his subject matter: they should not be subjected to the same criteria as experiments in experimental psychology. Moreover, psychical experiments should be taken in conjunction with the historical research in spontaneous manifestations.[15] In the following, I examine the structure of experiment as reported in the first chapters of the *Phantasms* and the criteria introduced to control for the results. In Sect. 6.5 I look in greater detail at Gurney's implementation of the historical method and at Charles S. Peirce's criticism of it.

---

[14] The book was mostly written by Gurney, with the exception of the introduction (Myers); the fact-checking of the hundreds of testimonies of "supersensory" phenomena was Podmore's contribution. As the authors declared: "Mr. Myers is solely responsible for the Introduction, and for the 'Note on a Suggested Mode of Psychical Interaction,' which immediately precedes the Supplement; and Mr. Gurney is solely responsible for the remainder of the book" (Gurney, Myers, and Podmore 1886, v).

[15] For example, when talking of a specific experiment of "joined agency" in which two "agents" focus on two different pictures (a square and an X, respectively), and the "percipient" draws *both* – the X *inside* the square – Gurney comments that "the significance of this experimental proof of joined agency will be more fully realised in connection with some of the spontaneous cases" (1886, 50).

As the authors admitted, the structure and method for the telepathic experiment came to the SPR as a development of the "willing game," a Victorian parlor game that had become popular on both sides of the Atlantic. It consisted in a sort of telepathic hide and seek, where a "percipient" would wait outside a room while the remaining players hid an object or chose an action they might "will" the percipient to perform. Upon entering the room, the percipient had to find the object or perform the action, in each case following an inclination mysteriously communicated either by physical contact from the other "willers," or by sheer willpower (Gurney et al. 1886, 14). The game modeled the type of phenomenon the SPR wanted to reproduce, but much was needed for "controls." The first degree of control amounted simply to watching one's own actions and words:

> …when the experiments are carried on in a limited circle of persons known to each other, and amenable to *scientific control*, it is not hard for those engaged to set a watch on their own and each other's lips; and questions and comments can be entirely forbidden. (Gurney et al. 1886, 18; my emphasis)

The idea was that agents gathered together would be able to communicate telepathically, by *willing that the percipient receive a certain message*, with no exterior cues. At least in this phase the researchers did not entertain the possibility of unconscious cueing. The notion of "scientific control" remained vague, at best hinting at committed and intentional observation of certain behavior norms. Since so much of "control" was part of the observer's experience, Gurney recognized that "it would be rash […] to represent as crucial any apparent transferences of thought between persons not absolutely separated" (1886, 19). He therefore listed the "conditions of a crucial result" from the first-person observer's perspective:

> The conditions of a crucial result, for one's own mind, are either (1) that the agent or the percipient shall be oneself; or (2) that the agent or percipient shall be someone whose experience, as recorded by himself, is indistinguishable in certainty from one's own; or (3) that there shall be several agent or percipients, in the case of each of whom the improbability of deceit, or of such imbecility as would take the place of deceit, is so great that the combination of improbabilities amounts to a *moral impossibility*. The third mode of attaining conviction is the most practically important. (Gurney et al. 1886, 19; my emphasis)

The investigators conceived their inquiry as "experimental," because it involved the experimenter intervening on a "perceiving" subject. But the method for controlling results was much closer to that of a historical investigation than a physical one, because the experimenter's honest intention could only be measured against the honest intentions of his peers. The need to multiply such experiences, or the request that "several agents or percipients" be present, seemed the most important form of (implicit?) control for establishing the reality of telepathic phenomena. Even these conditions were not easy to achieve, however. Because the observers were both "controls" and crucial elements of the experiment's setup, and because "percipient" subjects were very sensitive to their environment, multiplying observations was likely to disrupt the phenomena. Reverend Creery, who, after some very successful sessions of the "willing game" with his daughters, had contacted the SPR and embarked on a series of "thought-transference" and telepathy experiments, mentioned the influence of the particular individuals involved on the success of the experiment:

> We soon found that a great deal depended on the steadiness with which the ideas were kept before the minds of "the thinkers," and upon the energy with which they willed the ideas to pass. Our worst experiments [...] have invariably been when the company was dull and undemonstrative... (Gurney et al. 1886, 21)

Gurney's own testimony echoed this statement regarding the SPR's experiments with the Creery family:

> Questions of mood, of goodwill, of familiarity, may hold the same place in psychical investigation as questions of temperature in a physical laboratory; and till this is fully realised, it will not be easy to multiply testimony to the extent that we should desire. (Gurney et al. 1886, 30)

In the case of the Creery family, the young age of the daughters further discouraged the experimenters from introducing stricter "scientific" controls. As a precaution against cueing, the girls would wait in another room while the object to be fetched was "willed" – and that was the extent of the external controls employed. Time and again, Gurney et al. return to the crucial tension between the desire to implement stricter degrees of "control" and the risk that such controls may contradict the very nature of the phenomena to be observed. For instance, in commenting on experiments with other subjects, they wrote:

> It is the "delicate psychological conditions" [...] that are in danger of being ignored, just because they cannot be measured and handled. The man who first hears of thought-transference very naturally imagines that, if it is a reality, it ought to be demonstrated to him at a moment's notice. *He forgets that the experiment being essentially a mental one, his own presence – so far as he has a mind – may be a factor in it;* that he is demanding that a delicate weighing operation shall be carried out, while he himself, a person of unknown weight, sits judicially in one of the scales. After a time he will learn to allow for the conditions of his instruments, and *will not expect in the operations of an obscure vital influence the rigorous certainty of a chemical reaction.* (Gurney et al. 1886, 51; my emphasis)

While apparently contrasting the reliability of experiments in chemistry to the unpredictable psychical phenomena, Gurney was in fact individuating in the investigation of the latter difficulties which would ring familiar to readers conversant with other better-established disciplines, such as physics. For example, in Lord Rayleigh's work on the determination of the Ohm (Rayleigh 1881, 1882) the experimenters had to adopt radical measures to control for interference, including conducting observations at night to prevent vibrations from traffic on nearby streets.[16] The many experiments on "Brownian motion" – i.e., the apparently spontaneous motion of matter particles suspended in a fluid – also illustrated the experimenters' awareness of the almost infinite ways in which disturbing and confounding factors could compromise an experiment (Coko, this volume). As Graeme J. N. Gooday notes, "Victorian instruments were not habitually used in anything like the disturbance-free convenience of the purpose-built late twentieth-century research

---

[16] See Christopoulou and Arabatzis, this volume.

laboratory" (Gooday 1997, 411). How could anyone, then, expect that experiments with psychical forces would be different? If "mood" and "temperament" in psychical research were comparable to the "temperature" in a physics laboratory, the "percipient" herself was then likened to a "delicate instrument." This analogy was common in the psychical literature of the time, and one which could be adopted by the medium herself.[17]

In short, while psychical researchers wished to distinguish their investigation from those in the physical sciences – and particularly from the limitations of Faraday and Carpenter – they still upheld these disciplines' ideal of a controlled experiment, which should ensure trustworthiness in the study of psychical phenomena. For one, the possibility of environmental disturbances in the environment did not automatically disqualify the observations reported. Any observation could be registered within an interval of confidence, and indeed part of Gurney et al.'s argument for telepathy was based on a mixture of statistical and probabilistic reasoning.[18] Moreover, because the manifestation of psychical phenomena was difficult to control, they directed their control efforts toward experimental design and setup. Thus Mr. Malcolm Guthrie, the owner of a textile factory in Liverpool, reported to the SPR his experiments on thought-transference with some of his employees. He emphasized his ability to control every aspect of the experiment, from its design to its realization:

> I have had the advantage of studying a series of experiments *ab ovo* [i.e., from the very beginning]. […] The experiments have all been devised and conducted by myself and Mr. Birchall, without any previous intimation of their nature, and could not possibly have been foreseen. In fact they have been to the young ladies a succession of surprises. (Gurney et al. 1886, 36–37, emphasis in original)

Guthrie concluded his description boasting that no series of thought-transference experiments had ever been as satisfactorily controlled as his: "No set of experiments of similar nature has ever been more completely known from its origin, or *more completely under the control* of the scientific observer" (Gurney et al. 1886, 37; my emphasis).

Guthrie's experiments consisted in having the "observer" fixate his attention on a diagram or simple picture, while the subject, sitting in a separate room, would attempt to draw it. Some of the results were reproduced in Gurney et al.'s *Phantasms* (Figs. 6.5 and 6.6). These experiments were soon repeated by Dr. Oliver J. Lodge,

---

[17] The celebrated medium Mrs. Eleanor Piper wrote that she considered her role within the SPR as "simply that of an automaton" (Bell ed., [1902] 1904, 101). This description echoes William Carpenter's comment on hypnotized subjects: "the individual [is] for the time (so to speak) a mere *thinking automaton*, the whole course of whose ideas is determinable by suggestions operating from without" (Carpenter 1852, 147; my emphasis).

[18] On the historical as well as conceptual relevance of telepathy for the development of probability, see Ian Hacking (1988). Charles S. Peirce will mockingly point out that the degree of probability assigned by Gurney et al. to psychical phenomena made them much more certain than any other physical phenomenon (see Sect. 6.5).

Fig. 6.5 Attempts at mental communication. (Selected from Phantasms)

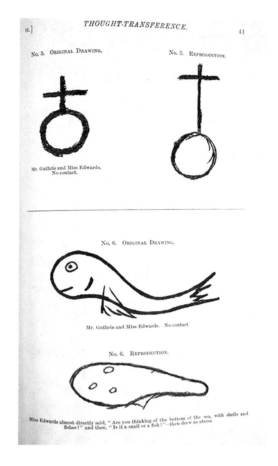

"professor of physics in University College, Liverpool." Lodge also spoke about the impact of control on one's confidence in the results of the experiments:

> ...to the best of my scientific belief no collusion or trickery was possible under the varied circumstances of the experiments. [...] *When one has the control of the circumstances, can change them at will and arrange one's own experiments, one gradually acquires a belief in the phenomena observed* quite comparable to that induced by the repetition of ordinary physical experiments. (Lodge, in Gurney et al. 1886, 49; my emphasis)

Indeed, while the kind of mind-to-mind communication investigated by the SPR was supposed to happen "outside the ken of senses," the requirements of control in the experimental setting strove to make the phenomena as ordinary as possible. This attitude rejected the thrills of ghost-stories to take a "cautious" and methodical approach to the matter, and extended from the experimental investigation to the collection and analysis of cases of "spontaneous" telepathy. As Shane McCorristine writes, the SPR embraced "quantitative, rather than qualitative methodology of reducing ghosts to statistics, as opposed to sensational experiences" (2010, 124). In the eyes of Gurney, Myers, and Podmore, experiment demonstrated the "fact" of

Fig. 6.6 Attempts at
mental communication.
(Selected from Phantasms)

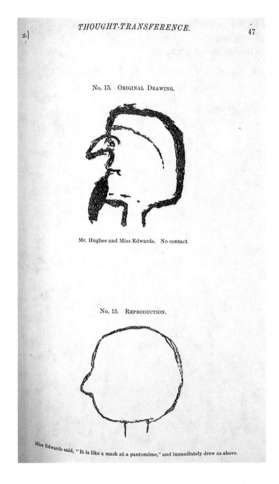

telepathy but this evidence had to be corroborated by a patient collection and verification of testimonies of ordinary apparitions, or "veridical hallucinations." These were manifestations whose content could be successfully cross-checked against historical facts. Such testimonies would become even more important after 1888, when Gurney published a *Note* admitting that the Creery sisters had been using a code to communicate in various tests (McCorristine 2010, 171).

6.5   Testimony and Its Tests: Gurney and Charles S. Peirce

If experiment, according to Gurney et al., proved the *reality* of telepathic phenomena, historical investigation into spontaneous telepathic manifestations played an important complementary role in establishing telepathy as a *natural* phenomenon, although somehow happening through other channels than those of ordinary sense-perception. Indeed, spontaneous and experimental cases of telepathy were considered two distinct manifestations of the same phenomenon: "The great point which connects many of the more inward impressions of spontaneous telepathy with the experimental cases is this – that what enters the percipient's mind is the exact reproduction of the agent's thought at the moment" (Gurney et al. 1886, 232). The recording of "spontaneous" telepathy relied on the collection of testimonies of apparitions, carefully cross-checked by Podmore. Testimony too had a probatory role, as Myers emphasized in the introduction: "*Testimony proves* that phantasms (impressions, voices, or figures) of persons undergoing some crisis, – especially death, – are perceived by their friends and relatives with a frequency which mere chance cannot explain" (Gurney et al. 1886, f; my emphasis).

In desiring to study psychical phenomena in a controlled manner, the SPR moved away from discourse on "spirits" and angelic intelligences to focus on mundane communication of thoughts or impressions – which ought to happen strictly between living persons. "Death" was reconceptualized as a "crisis," an episode (if perhaps the last one) of someone's life. "Our subject is phantasms *of the living*: we seek the conditions of the telepathic impulse on the hither side of the dividing line, in the closing passage of life; not in that huge negative fact – the apparent cessation or absence of life – on which the common idea of death and of its momentous importance is based" (Gurney et al. 1886, 230). The expression "of the living" was intentionally ambiguous, suggesting that the appearances both came from and represented (still) living people. This positivist setup left some room, however, for radically reconceptualizing the notion of the self as involving "a more fundamental unity, which finds in what we call life very imperfect conditions of manifestation" (1886, 231). Accordingly, the "individuality" that each of us regards as our "self" would be a "partial emergence" of this deeper "unity," and it is to this deeper level that "telepathic" manifestations may be tied:

> And this hypothesis [i.e., the hypothesis of a deeper self] would readily embrace and explain the special telepathic fact in question; while itself drawing from that fact a fresh support. By its aid we can at once picture to ourselves *how it should be that the near approach of death is a condition exceptionally favourable to telepathic action*, even though vital faculties seem all but withdrawn, and the familiar self has lapsed to the very threshold of consciousness. For to the hidden and completer self the imminence of the great change may be apparent in its full and unique impressiveness; nay, death itself may be recognised, for aught we can tell, not as a cessation but as a liberation of energy. (1886, 231, my emphasis)

The deep-self hypothesis might explain the perplexing fact that most of, if not all, the reported "phantasms" belonged to someone near death. However, the authors never explicitly argued for it, since the declared purpose of the *Phantasms* was to "bring the evidence to light" before arguing about possible explanations for it. As the authors declared, "we may contrast telepathy, not only with the comparatively modern superstition of witchcraft, but with phenomena of much older and wider acceptance – the alleged apparitions of the dead" (Gurney et al. 1886, 121). Yet the authors had to introduce further conditions for deciding when to accept a testimony. As they put it: "The evidence for telepathy has a certain type and structure of its own, and we must realise what this is, in order to know where to look for the weak points. What, then, are the essential elements of a typical telepathic phenomenon?" (Gurney et al. 1886, 131).

Let us examine these elements. First, Gurney et al. insisted on a "sober" and unimaginative nature of their witnesses: "a very large number of our first-hand witnesses are educated and intelligent persons, whose sobriety of judgment has never been called in question," 1886, 120–21). Second, they wished to ensure that news of a person's death did not precede their apparition: "one of the points to which we have, throughout our inquiry, attached the highest value, is the proof that evidence of the percipient's experience was in existence prior to the receipt of the news of the agent's condition" (1886, 134). However, information about times of death and apparitions was not easy to get. In order not to discard too many testimonies, the authors agreed on a protocol allowing a 12-h window between a person's death and their alleged apparition, to accommodate for uncertainty, memory lapses, and time zones. Finally, the percipient's condition must be taken into account, and testimony from sick, drunk, anxious, or habitually hallucinating subjects had to be rejected.[19] If enough phenomena were found to satisfy these conditions, then telepathy might stand a chance at being investigated as any other natural phenomenon:

> Amid all their differences, the cases present one general characteristic – an unusual affection of one person, having no apparent relation to anything outside him except the unusual condition, otherwise unknown to him, of another person. It is this characteristic that gives them the appearance, as I have just said, of a *true natural group*. (Gurney et al. 1886, 164; my emphasis)

According to Gurney et al., saying that telepathic phenomena constituted a "true natural group" was just another way of stating that they were real phenomena, worthy of further investigation. Their work, while deploring the inevitable loss of evidence on similar cases, aimed to build a repository of that little evidence that still existed and to be a model for further research:

> By far the greater part of the telepathic evidence, even of the last twenty years, has undoubtedly perished, for all scientific purposes; we want the account for the next twenty years to be different. But it is only by a decided change in the attitude of the public mind towards the subject that the passing phenomena can be caught and fixed; and it is only by a wider knowledge of what there already is to know that this change can come about. (Gurney et al. 1886, 169)

Phantasms of the Living was written to change attitudes on the matter of psychical research, which the authors hoped would become a respectable topic of inquiry. This change of "attitude of the public mind" hinged on proposing a new object of study – telepathic communication – on which to focus the scientific efforts, but it also offered a new conception of experiment and of the relevance of testimony to establish the reality of the phenomenon at stake. Like his father before him, Charles

---

[19] "Briefly, then, if the account of some alleged instance of telepathy is evidentially faulty, there must be misrepresentation as to one or more of the following items: (1) the state of the agent; (2) the experience of the percipient; (3) the time of (1); (4) the time of (2)" (Gurney et al. 1886, 132).

S. Peirce was asked to act as "control" for the case in question; his task was not to sit in a box with putative mediums, but to read and evaluate the *Phantasms*.

When Gurney et al.'s work launched, Peirce had just edited a collected volume on logic, to which he contributed the chapter "A Theory of Probable Inference," containing sections on both statistical and probabilistic reasoning (see Peirce 1883). With Joseph Jastrow he had also just co-authored a paper in experimental psychology that concluded there was no fixed threshold of sensation (Peirce and Jastrow, 1884). Interestingly, the conclusion of his experimental psychology paper linked his results on perception to the telepathy debate of the time:

> The general fact [that there is no "least perceptible difference" in sensation] has highly important practical bearings, since it gives new reason for believing that we gather what is passing in one another's mind in large measure from sensations so faint that we are not fairly aware of having them, and can give no account of how we reach our conclusions about such matters. The insight of females as well as *certain "telepathic" phenomena may be explained in this way*. Such faint sensations ought to be fully studied by the psychologist and assiduously cultivated by every man. (Peirce and Jastrow 1884, 83, W5, 135, my emphasis)

The scare quotes for "telepathic" show that Peirce in 1884 may not have been convinced of the phenomena's reality.[20] Following Faraday and Carpenter's example, he approached these enigmatic cases as signs of physiological or psychological processes that would have to be better understood by better knowledge of how our senses work. His experiment did not focus on psychical or telepathic manifestations, but its conclusions allowed Peirce to hypothesize that telepathy too may be understood as an extreme case of perceptual acuity without awareness.

Reading the *Phantasms* did not change Peirce's opinion on the nature of "telepathic" phenomena. He had two kinds of objections for Gurney. The first concerned the use of statistics to evaluate the incidence of the phenomena, and the use of probability to estimate the likelihood that a given apparition was indeed a case of telepathic communication and not a hallucination surrounding an untimely death. Peirce's verdict was blunt:

> ... these gentlemen, having addressed, as they estimate, a public of only 300,000 persons, claim to have found thirty-one indubitable cases of this kind of coincidence within twelve years. From this, they cipher out some very enormous odds in favor of the hypothesis of ghosts. I shall not cite these numbers, which captivate the ignorant, but which repel thinking men, who know well that no human certitude reaches such figures as trillions, or even billions to one. (Peirce 1887a, W6, 75)

The second objection was specific to the 31 cases the authors presented as "indubitable." Going case by case, Peirce argued against including each in the collection. He said that "every one of their thirty-one coincidences sins against one or more of the eighteen different conditions to which such an argument must conform to be valid" (Peirce 1887a, W6, 75). These "eighteen conditions" are all variations on the three proposed by Gurney et al.: (1) the status and trustworthiness of the witness, (2) the

---

[20] Luckhurst notes that "Peirce was among the first to cite the term telepathy in a scientific paper" (Luckhurst 2002, 71).

12-h window between a person's death and their apparition, (3) the good physical and mental health of the percipient at the moment of the apparition (and that the percipient was not habitually subject to hallucination). Among the variations on these conditions – introduced first by Gurney et al., although scattered throughout the book – are the requirements that the percipient be "wide awake" at the moment of the apparition; that the apparition be "clearly recognized"; that the subjects not have hallucinations "fortuitously"; that a "supposition of trickery," if available, trump the telepathic explanation. Witnesses are discounted on grounds of class ("No case should be admitted upon the unsupported and unverified statement of a superstitious, ignorant, and credulous person. And a common sailor or skipper may be assumed to be such a person," Peirce 1887a, W6, 79), of physical conditions ("intoxicated," "delirium of fever," W6, 79), and of sloppiness ("No case can be admitted which rests largely on the testimony of a loose or inaccurate witness," W6, 80). Moreover, Peirce protested that "where there is only a meagre story told in outline, we are not furnished with any means of judging of the reliability of the witness," and that in some cases "questions might have been asked which would have brought the matter to a test, and have not been asked" (ibid.). For him, an extended account was particularly necessary since the reader had no other way to examine the information than by the words of the authors. Moreover, if the authors did not provide the information necessary to evaluate each case, they would affect the reader's confidence in entire project:

> After all, the reader, who cannot cross-examine the witnesses, and search out new testimony, must necessarily rely upon Messers. Gurney, Myers, and Podmore having on the whole performed this task well; and we cannot accept any case at all at their hands, unless, as far as we can see, they have proved themselves cautious men, shrewd observers, and severe logicians. (Peirce 1887a, W6, 81)

Peirce circled back to his criticism of the quantitative aspects of the work at the end of his assessment, although again without entering in much detail; his concluding remarks feign some charity on the methodological side but hit at the heart of Gurney et al.'s epistemic aim, which was of establishing telepathic phenomena as scientific facts:

> The argument might, certainly, have been constructed more skillfully [sic.]; but I do not think that there is much prospect of establishing any scientific fact on the basis of such a collection as that of the *Phantasms of the Living*. (Peirce 1887a, W6, 82)

The discussion did not end after Peirce's report. Gurney responded to Peirce's assessment (Gurney 1887), Peirce provided a new (and more acrimonious) criticism of the evidence presented in the *Phantasms*, and this was followed by a last reply from Gurney (1889), who unfortunately died that very year. The rejoinders present more extensive sections dedicated to discussing the data's statistical significance and the antecedent probability of telepathic manifestations. Both parties rely strongly on some numerical estimates, which however are based on the same qualitative assessment of the received testimonies. Hence, the quantitative elements don't seem to add a meaningful layer of control over the case at hand, but rather to offer the opportunity for Peirce to definitely reject the possibility of establishing telepathic phenomena as "scientific truths":

... until telepathic theory of ghost-stories has been rendered far more antecedently probable
than it now is, it is useless to try to establish it as a scientific truth by any accumulation of
scientific observations. (Peirce 1887b, W6, 142)

Gurney's final words were a plea to continue the independent and honest examination of the matter, as well as a new argument over the kind of probability used to
evaluate the phenomena. Gurney claimed the *posterior* probability that a series of
telepathic hallucinations were all due to coincidence was low, and that consequently
the probability they were real phenomena was high. He also argued that Peirce con-
founded that probability with the *prior* probability[21] that a given hallucination would
be a case of true telepathic communication and not of mere coincidence or fraud.
Eventually, even if Peirce did not accept Gurney's arguments nor his evidence on
telepathy, Gurney won the methodological battle: an opponent of telepathic communication had engaged with the phenomenon in the terms Gurney, Myers, and
Podmore had set for it.

6.6   Conclusion

This chapter focused on nineteenth-century discussions of method, experiments,
and data collection for psychical phenomena. These cases were either based on
experiments or on "spontaneous" manifestations, which, by definition, may happen
at any time of day or night and in ordinary, non-controlled settings. Nonetheless, in
the context of spontaneous manifestations "controls" came back as standards for
respectable testimony and as "historical" cross-examinations of the information.
The account presented here is far from exhaustive; its aim is not to present general
conclusions on psychical research, but rather to show the value of this field of
inquiry for historians and philosophers of science interested in the notion of control.
A few points are of particular interest.

First, the word "control" appears in this context (specifically in the work of
Gurney, Myers, and Podmore) as an *actors' term*, a point missing from most discussions of experiment in this and earlier times (as other chapters in this volume illustrate). Second, Gurney et al. connect issues regarding controlled conditions (or the
lack thereof) in psychical research with parallel troubles in physical laboratories.
Since they cannot multiply or radically change their "instruments" (i.e., the
medium), they corroborate their findings with historical and observational studies
and argue for their statistical significance. Third, for both SPR researchers and for
Hare, "control" is about controlling the *instruments* or the experiment's *setting*;
there is no discussion of "comparison with a control." Faraday's experiment did
include this element, however, insofar as he observed table-turning with and without an "indicator" showing participants the direction they were actually pressing on

---

[21] Gurney uses the terms "à posteriori" and "à priori" (Gurney 1889, W6, 143), but, for the scopes
of this chapter, their meaning is that of posterior and prior probability.

the table. Only the table without the indicator could move, which led to Faraday's conclusion that it is pressure exercised *unconsciously* by the participants' hands – and not some spirit or new psychical force – that moves the table.[22]

The focus on experimental strategies and controls does not deny important connections between psychical research and spiritualism, and the religious interpretations for these manifestations. As shown in the introduction and throughout the chapter, psychical manifestations could be seen as confirmation for Christianity and for the theory of the soul's immortality, or as blasphemous and materialistic distortions of the Christian faith. Religion, personal circumstances, and methodological concerns all influenced how prominent scientists engaged the debate. Yet, the ability to justify methodological choices in methodological terms, and the attention paid to the instruments involved in the séances or to the collection and sifting of testimonies cannot be sufficiently explained with recourse to religious and biographic elements only. The very practices in which those scientists and investigators were involved testify to the notion of "science" at play in the second half of the nineteenth century, and looking at those practices more carefully sheds new light on the nineteenth-century equivalents of "control" in scientific research.

For example, both Faraday and Hare conceived methods and strategies to control for phenomena they had a strong religious interest in. But those strategies are no less interesting from an epistemological standpoint because the investigation's result lay close to the inquirers' hearts, nor can their differing conclusions be accounted for purely by their different starting points. Methodologically, Hare's first point of departure from Faraday concerns the phenomena he chooses to control for. For Faraday there was a single, crucial physical manifestation to control, namely the direction of the movement: did it go from the table to the participants' hands, or was the opposite true? For Hare, the physical manifestations to control for were many and complex ones, involving dials, disks with letters and short phrases written on their circumferences. Motion was no longer enough to prove a spiritual agency; the spirit(s) had to be able to move the table in a specific way in order to select the right letters and deliver a comprehensible message. Hare's attention to control is even more striking if compared with the apparent lack of it in Holcomb's letter, which prompted him to look closely at psychical phenomena in the first place. While Holcomb appealed to the testimony of the senses, Hare constructed complex instruments to control for the phenomenon he wished to observe – communication between spiritual entities and living beings.

Gurney and the other members of the SPR attempted to "control" the phenomena by renegotiating the scientific practices deemed adequate to capture them. Thus, Gurney criticized Faraday's and Carpenter's experimental approaches and proposed an alternative based on collecting and analyzing "spontaneous" telepathic phenomena, rather than on attempts at producing supernatural effects in the laboratory. Although he did not explicitly advocate for reforming the experimental method in such matters, *Phantasms of the Living* contains an experimental section, which the

---

[22] I am grateful to Jutta Schickore for her insightful comments on these points.

authors endorse as at least as authoritative as the historical data contained in the rest of the book. Analyzing these experimental cases shows how the notion of control is articulated in these new settings. The criteria for control are then examined in conjunction with Charles S. Peirce's criticisms. The controversy between Gurney and Peirce reveals how the battle for the "right" control practice is also an existential battle to affirm (or deny) the reality of telepathic phenomena.

Eventually, control over the setting is both a contingent and a constitutive feature of the knowledge gained in empirical inquiry. What is controlled (for) depends on how the experimenter conceptualizes what is going on (voluntary or involuntary movements, "willing" influences, or "spontaneous" manifestations); once the phenomenon is framed in a certain way, that determines what will need to be controlled. While the cases examined in this chapter do not offer one coherent methodology for investigating psychical phenomena, they offer many examples of designing and applying controls, and these examples help us understand how knowledge is construed in empirical inquiry.

References

Author Unknown, 1859. Spiritualism Shown as It Is. Boston Courier Report of the Proceedings of Professed Spiritual Agents and Mediums in the Presence of Professors Peirce, Agassiz, Horsford, Dr. B. A. Gould, *Boston Courier*, 3–24.

Bensaude-Vincent, Bernardette, and Christine Blondel, eds. 2002. *Des Savants face à l'occulte. 1870–1940*. Paris: Éditions La Découverte.

Blondel, Christine. 2002. Eusapia Palladino: la méthode expérimentale et la « diva des savants ». In *Des Savants face à l'occulte. 1870–1940*, ed. Bernardette Bensaude-Vincent and Christine Blondel, 143–172. Paris: Éditions La Découverte.

Bower, Brady M. 2010. *Unruly Spirits. The Science of Psychic Phenomena in Modern France*. Champaign: University of Illinois Press.

Cantor, Geoffrey. 1991. *Michael Faraday: Sandemanian and Scientist. A Study of Science and Religion in the Nineteenth Century*. Houndmills/London: The Macmillan Press.

Carpenter, William B. 1852. On the Influence of Suggestion in Modifying and Directing Muscular Movement, Independent of Volition. *Notices of the Proceedings of the Meetings of the Members of the Royal Institution* 1: 147–153.

———. 1877. *Mesmerism, Spiritualism, &c. Historically & Scientifically Considered: Being Two Lectures Delivered at the London Institution, with Preface And Appendix*. London: Longmans, Green, and Co.

Coon, Deborah J. 1992. Testing the Limits of Sense and Science. *American Psychologist* 47 (2): 143–151.

Delorme, Shannon. 2014. Physiology or Psychic Powers? William Carpenter and the Debate over Spiritualism in Victorian Britain. *Studies in History and Philosophy of Biological and Biomedical Sciences* 48: 57–66. https://doi.org/10.1016/j.shpsc.2014.07.004.

Faraday, Michael. 1853a. On Table Turning. *Times*, June 28, 382–386.

———. 1853b. Experimental Investigation of Table-Moving. *Athenaeum*, 2 July, 801–803.

———. 1853c. Table Turning. *Illustrated London News*, July 16.

———. 1853d. [1996]. Letter 2703. Faraday to Caroline Deacon, 23 July 1853. In *The Correspondence of Michael Faraday*. Vol. 4, 1849–1855, ed. Frank A. J. L. James. London: The Institution of Electrical Engineers.

———. [1854] 1859. Observations on Mental Education. In *Experimental Researches in Chemistry and Physics*, 463–491. London: Richard Taylor and William Francis.

Gooday, Graeme J. N. 1997. Instrumentation and Interpretation: Managing and Representing the Working Environments of Victorian experimental Science. In *Victorian Science in Context*, ed. Bernard Lightman, 409–437. Chicago: University of Chicago Press.

Gurney, Edmund. 1884. The Nature of Evidence in Matters Extraordinary. *The National Review* 4 (22): 472–491.

———. 1887. Remarks on Professor Peirce's Paper. *Proceedings of the American Society for Psychical Research*. Now in Peirce, Charles S. 2000. *Writings of Charles S. Peirce. A Chronological Edition, Volume 6, 1886–1890*, edited by the Peirce Edition Project, 83–101. Bloomington: Indiana University Press.

———. 1889. Remarks on Mr. Peirce's Rejoinder. *Proceedings of the American Society for Psychical Research*. Now in Peirce, Charles S. 2000. *Writings of Charles S. Peirce. A Chronological Edition, Volume 6, 1886–1890*, edited by the Peirce Edition Project, 143–155. Bloomington: Indiana University Press.

Gurney, Edmund, Frederic W.H. Myers, and F. Podmore. 1886. *Phantasms of the Living*. London: Trübner and Co., Ludgate Hill, E.C. https://doi.org/10.1007/BF01797824.

Hacking, Ian. 1988. Telepathy: Origins of Randomization in Experimental Design. *Isis* 79 (3): 427–451.

Hajek, Kim. 2015. Imperceptible Signs: Remnants of *Magnétisme* in Scientific Discourses on Hypnotism in Late Nineteenth-Century France. *Journal of the History of the Behavioral Sciences* 51 (4): 366–386.

Hare, Robert. 1853. Letter from Dr. Hare. In Reply to an Inquiry Respecting the Influence of Electricity in Table Turning. *Philadelphia Inquirer*, July 27, 1853. In Islandora Repository – Text Collection 973.C683 – Broadsides Collection, https://diglib.amphilsoc.org/islandora/object/philadelphia-inquirer-letter-dr-hare-reply-inquiry-respecting-influence-electricity, last accessed 22 Apr 2023.

———. 1855. Experimental Investigation of the Spirit Manifestations. New York: Partridge and Brittan.

James, Frank A. J. L., ed. 1996. *The Correspondence of Michael Faraday*. Vol. 4, 1849–1855. London: The Institution of Electrical Engineers. https://doi.org/10.1080/0003379 0.2014.949859.

Kneeland, Timothy W. 2008. Robert Hare: Politics, Science, and Spiritualism in the Early Republic. *Pennsylvania Magazine of History and Biography* 132 (3): 245–260.

Luckhurst, Roger. 2002. *The Invention of Telepathy, 1870–1901*. Oxford: Oxford University Press. https://doi.org/10.7208/chicago/9780226753416.001.0001.

McCorristine, Shane. 2010. *Spectres of the Self: Thinking about Ghosts and Ghost-Seeing in England, 1750–1920*. Cambridge: Cambridge University Press.

Moore, Laurence R. 1972. Spiritualism and Science: Reflections on the First Decade of Sprit Rappings. *American Quarterly* 24 (4): 474–500. https://doi.org/10.2307/2711685.

Noakes, Richard. 2019. *Physics and Psychics. The Occult and the Sciences in Modern Britain*. Cambridge: Cambridge University Press.

Owen, Alex. 2004. *The Place of Enchantment: British Occultism and the Culture of the Modern*. Chicago/London: University of Chicago Press.

Peirce, Charles S. 1883. *Studies in Logic*. Boston: Little, Brown, and Company.

———. (ed.), 1887a. Criticism on *Phantasms of the Living*: An Examination of an Argument of Messrs. Gurney, Myers, and Podmore. *Proceedings of the American Society for Psychical Research*; now in *Writings of Charles S. Peirce. A Chronological Edition, Volume 6, 1886–1890*, edited by the Peirce Edition Project, 2000, 75–82. Bloomington: Indiana University Press.

———. (ed.), 1887b. Mr. Peirce's Rejoinder. *Proceedings of the American Society for Psychical Research*; now in *Writings of Charles S. Peirce. A Chronological Edition, Volume 6, 1886–1890*, edited by the Peirce Edition Project, 2000, 102–142. Bloomington: Indiana University Press.

Peirce, Charles S., and Joseph Jastrow. 1884. On Small Differences of Sensation. *National Academy of Sciences* 3 (5): 75–85. now in *Writings of Charles S. Peirce. A Chronological Edition, Volume 6, 1884–1886*, edited by Christian J. W. Kloesel, Nathan Houser et al., 1993, 122–135. Bloomington: Indiana University Press.

Plas, Régine. 2012. Psychology and Psychical Research in France around the End of the 19th century. *History of the Human Sciences* 25 (2): 91–107. https://doi.org/10.1177/0952695111428554.

Schickore, Jutta. 2019. The Structure and Function of Experimental Control in the Life Sciences. *Philosophy of Science* 86 (2): 203–218. https://doi.org/10.1086/701952.

Sommer, Andreas. 2012. Psychical Research and the Origins of American Psychology: Hugo Münsterberg, William James and Eusapia Palladino. *History of the Human Sciences* 25: 23–44.

———. 2013. Normalizing the Supernormal: The Formation of the 'Gesellschaft für Psychologische Forschung' ('Society for Psychological Research'), c. 1886–1890. *Journal of the History of the Behavioral Sciences* 49: 18–44.

———. 2015. Edmund Gurney. *Psi Encyclopedia*. London: The Society for Psychical Research. https://psi-encyclopedia.spr.ac.uk/articles/edmund-gurney. Retrieved 15 June 2023.

———. 2018. Geisterglaube, Aufklärung und Wissenschaft – Historiographische Skizzen zu einem westlichen Fundamentaltabu. In *Jenseits des Vertrauten. Facetten Transzendenter Erfahrungen*, ed. H. Schwenke, 183–216. Freiburg i. Br: Verlag Karl Alber.

———. 2020. James and Psychical Research in Context. In *The Oxford Handbook of William James*, ed. by Alexander Klein. Oxford: Oxford University Press.

Taylor, Eugene. 1996. *William James on consciousness beyond the margin*. Princeton: Princeton University Press.

Tymn, Michael. 2021. Robert Hare. *Psi Encyclopedia*. London: The Society for Psychical Research. https://psi-encyclopedia.spr.ac.uk/articles/robert-hare. Retrieved 10 October 2022.

Claudia Cristalli is currently post-doctoral researcher at the University of Tilburg, working on the reception of logical empiricism in the United States. Previously, she was a Mellon-funded post-doc researcher at Indiana University, Bloomington, working on methodologies and notions of control, analysis, and synthesis in a broad historical perspective. Her PhD thesis is on Charles S. Peirce's theory of perception and his pragmatist philosophy more broadly.

Chapter 7
Controlling Nature in the Lab and Beyond: Methodological Predicaments in Nineteenth-Century Botany

Kärin Nickelsen

7.1 Introduction

In January 1872, the plant physiologist Julius Sachs gave a speech to the members of his university on "The State of Botany in Germany" (Sachs 1872). Sachs recounted how botany had undergone a radical transformation over the previous 40 years. The botanists of the 1870s, Sachs explained, were no longer obsessed with collecting plants and arranging them in systems, and they had liberated themselves from the harmful influence of the German *Naturphilosophie*. These days, botanists investigated the processes of cellular reproduction, the laws of growth and development, the effects of gravity and light, and the influence of climate on the distribution of plant species. Botany had become a scientific discipline—although, Sachs added, it could unfold its full potential only if there were more positions for academic botanists in Germany.

Sachs's speech clearly had a political agenda, and one should not take his rhetoric of awakening at face value. But the methodological shift that Sachs described was real. Botany changed dramatically in the nineteenth century, particularly in German-speaking countries. The rise of new microscopy techniques and other precision instruments opened new perspectives and prompted botanists to revisit almost

I am very grateful to all the organizers of the two workshops that gave rise to this volume. This goes especially to Jutta Schickore for insightful comments on earlier drafts of this paper, and for our productive and enjoyable conversations on the history of scientific methods and methodology over the past years. Comments by others were immensely helpful too, with special thanks to Klodian Coco, Claudia Cristalli, and Caterina Schürch. I also would like to thank Elizabeth Hughes, Berlin, and Louise R. Chapman, who carefully edited this essay.

K. Nickelsen (✉)
Ludwig Maximilian University Munich, Munich, Germany
e-mail: K.Nickelsen@lmu.de

© The Author(s) 2024
J. Schickore, W. R. Newman (eds.), *Elusive Phenomena, Unwieldy Things*,
Archimedes 71, https://doi.org/10.1007/978-3-031-52954-2_7

every area of the field, including the study of forms and functions, or morphology and physiology, where the influence of *Naturphilosophie* had been particularly strong (e.g. Mendelssohn 1964; Coleman 1971; Jahn 2000; Bowler and Morus 2005; Morange 2016).

An important part of this reform program was the development of new methodological principles for a "scientific botany," and this development is the focus of this essay. The new botany would no longer base itself in lofty imagination but in empirical facts, supporters of the agenda agreed. Hardly anybody used the term "control," or its equivalents in other languages, in this context (see the introduction to this volume). However, as this essay shows, many botanists were deeply concerned with making observations accurate, experimental design meaningful, inferences safe, speculations respectable, and interpretations reliable. In terms of this volume's categories, the chapter addresses general "methodological ideas," as well as "control strategies" and implemented "control practices."

Specifically, I shall examine the work of Julius Wiesner (1838–1916), acclaimed plant physiologist and protagonist of the new botany. Prior to his professorship in Vienna, Wiesner completed his doctoral dissertation in Jena with Matthias J. Schleiden (1804–1881), who was already famous for his *Principles of Scientific Botany* (first published in 1842). In this textbook, Schleiden called for a new beginning of botany as an "inductive science," with a set of rigorous control and validation strategies at its core. For many botanists, including Wiesner, this textbook became an important source of inspiration. I shall therefore begin this chapter by looking briefly at how methodology in general, and how different forms of "control" in particular, were discussed in Schleiden's *Principles*. I shall then note how this agenda unfolded in Wiesner's work, especially in his studies on the influence of light on plants. This influence occurred first in his botanical laboratory and then in the field, where he had to adapt his concepts and practices to entirely new conditions. The questions emerging from these studies led him far beyond plant physiology in its narrow sense, and demanded a different set of methodological principles and control strategies—so different, in fact, that Wiesner eventually helped to found a new sub-discipline: the "Biology of Plants."

7.2   Matthias J. Schleiden and *The Principles of Scientific Botany*

Matthias Jacob Schleiden (1804–1881) began his career as a lawyer but in 1835, after a personal crisis, dropped this profession and switched to botany (e.g., Möbius 1904; Jahn and Schmidt 2005). Four years later, Schleiden received a second PhD in this field, and in 1840 he became Associate Professor in Botany and Director of the Botanical Garden at the University of Jena.[1] In 1842, Schleiden published the first edition of a widely read (and celebrated) textbook, *Grundzüge der*

---

[1] In 1863, Schleiden became professor of botany at Dorpat (today's Tartu), which then was part of Russia; in 1864, he withdrew from this position and moved to Dresden as an independent scholar.

wissenschaftlichen Botanik nebst einer methodologischen Einleitung als Anleitung zum Studium der Pflanze ("Principles of scientific botany, with a methodological introduction as a guide to the study of plants"), which is the focus of the following sections.[2] Second and third revised editions were published in 1845 and 1849, the latter of which was reprinted in unaltered form in 1861.[3]

In this textbook, Schleiden introduced his groundbreaking theory of how plant cells developed and how they formed tissues and structures (first published as Schleiden 1838). In addition, he also addressed fundamental methodological questions of the field (e.g. Buchdahl 1973; Charpa 2003, 2010). Botany was in a deplorable state, Schleiden thought, especially in the German countries. Under the influence of Hegel, Schelling, and others, the field had degenerated into a "dogmatic science," Schleiden lamented, and "a widespread lack of orientation [prevailed] about the challenges to the human ability to gain knowledge, and about the means to meet them" (Schleiden 1861, 12).[4] For Schleiden, botany was urgently in need of a sound methodological and epistemological foundation, and this became the focus of his substantial, 100-page-plus introduction. He added methodological comments throughout the rest of the book.[5] The aim was to transform botany into an empirically based "inductive science," and Schleiden's textbook was to serve as an important step in this direction.[6] All of this is highly relevant for the question how the

---

[2] See Schleiden (1842, 1843) for the two volumes of the first edition. A second, revised edition was published as Schleiden (1845); this was again reworked for a third edition (Schleiden 1849a). The last of these was reprinted unaltered (albeit with a new preface) as Schleiden (1861). The English translation (Schleiden 1849b) was based on the 1845 edition. (This translation is not always reliable; if not stated otherwise, I am using my own translation for quotations from the German original.) In the English version of the text, the methodological introduction was omitted with a two-page "summary" in its place, while some of Schleiden's introduction to scientific microscopy was added as an appendix to the volume. The "Translator's Preface" explained that the methodological introduction was considered too long and also unnecessary, given that "two admirable works" on the principles of scientific inquiry had already been written in English by John Herschel and William Whewell.

[3] The introduction changed substantially from the first edition (1842) to the second (1845). Some of the personal attacks were dropped, while the discussion of general topics in epistemology was expanded. The subsequent third and fourth editions (1849, 1861) introduced only minor changes but they represent the most mature version of Schleiden's thoughts. In most cases, the 1861 edition is therefore used for quotations; relevant differences to earlier versions will be indicated. See, on these changes, e.g., Jost (1942), which also provides an overview of the textbook's reception.

[4] German original: "Es fehlt im Allgemeinen an einer richtigen Orientirung über die Aufgaben des menschlichen Erkenntnissvermögens und die Mittel zu ihrer Lösung." Schleiden vigorously promoted the post-Kantian philosophy of Jakob F. Fries, although in a slightly adapted version, and polemically criticized others, especially the protagonists of *Naturphilosophie*, with Hegel and Schelling as arch-villains.

[5] While the introduction grew over the years and through subsequent editions, the first version of 1842 already included 166 pages (and in smaller print format than later editions!).

[6] For a comprehensive treatment of how "induction" was understood by Schleiden (in the tradition of Fries), see, e.g., Apelt (1854). Apelt dealt primarily with the physical sciences but repeatedly cited Schleiden for topics related to biology *sensu largo*. In turn, Schleiden explicitly referred to Apelt in the 1861 edition and calls Apelt's contributions "the most important work of philosophy published in this century" (Schleiden 1861, Vorrede, VII).

notion of "control" gained a foothold in botany. The following sections, however, cannot do full justice to the program Schleiden pursued in this remarkable treatise. I restrict myself to a summary of Schleiden's comments on two areas that became so important for Wiesner and others: observation and experiment.

7.2.1 Observation

Schleiden treated the topic of observation in particular detail. This included a philosophical discussion of the faculty of "seeing," its weaknesses and its epistemological function, but also concrete recommendations of how to make observations reliable. In fact, large parts of the introduction were dedicated to the principles of scientific microscopy, where Schleiden was rightfully regarded an expert.[7] He believed that every botanist should master the techniques of microscopy, and he provided a comprehensive survey of how the microscope worked and how it should be used. But even without a microscope, Schleiden maintained, botanists ought to revisit their practices of observation, train their eyes, and refine their habits. "He who wishes to observe successfully," Schleiden lectured his readers, "must observe frequently and with the most profound attention, so that he gradually learns how to see, for *seeing* is a difficult art" (Schleiden 1861, 84).[8] Most importantly, botanists should observe the relevant phenomena themselves whenever possible, and always document them either in accurate sketches (which, according to Schleiden, helped to control and discipline one's observation) or, preferably, in durable preparations and slides that their colleagues would later be able to consult (e.g., Schleiden 1861, 85–91).[9] In cases where one had to rely on reports by others, Schleiden warned his readers to place their trust carefully.[10] There were so many pitfalls in the process, so many potential sources of error, and so many misguided minds, Schleiden thought, that one should always scrutinize another person's judgment carefully, even those made by alleged experts. He gleefully called out colleagues whose errors Schleiden found particularly outrageous.

---

[7] For an illuminating analysis of Schleiden's view of microscopy, see Schickore (2007).

[8] German original: "Wer mit Glück beobachten will, muss viel und mit angestrengter Aufmerksamkeit beobachten, damit er allmälig sehen lerne, denn *Sehen* ist eine schwere Kunst." Emphasis in original. The same phrase already appears in earlier versions of the introduction.

[9] Schleiden explains in detail how he thought observations should be documented in different forms and media, and control practices loom large in this context. However, as this essay focuses on control practices in experimental research, I do not discuss these interesting passages in more depth.

[10] On Schleiden's preference of "autopsia," see, e.g., Schleiden (1861, 54–55). On his warning against false authorities, and his plea to consider a person's character and scientific ethos in this context, cf. Schleiden (1861, 91–95).

Reliable observations and accurate descriptions were the foundations of everything. But Schleiden also reminded his audience that it was insufficient to observe a cellular configuration at one stage only. The risk for misinterpreting structures was high. The explanation of a plant's forms and functions had therefore to be rooted in observing the full process of cellular development, the "*Entwickelungsgeschichte*," as Schleiden called it (Schleiden 1861, 100–102).[11] This mantra was repeated many times over the course of the textbook. In fact, this "maxim of developmental history" was one of Schleiden's two guiding principles for inductive inferences in botany. (The other was the "maxim of the independence of the plant cell" which implied that cell physiology had to precede the physiology of the whole plant.)[12] Schleiden sharply criticized earlier traditions of morphology and physiology, which had violated this heuristic principle. They did so especially under the influence of *Naturphilosophie* and so had reached false conclusions. Diligent observation of the origin and development of a plant's cells and tissues were the only way to avoid such errors in the future.

In this context, Schleiden also called for a precise definition of the respective *explananda* and *observanda*. For him, conceptual clarity was as important as the accurate handling of the microscope. One of his most striking examples was the phenomenon of vegetable "growth," a central topic of investigation in nineteenth-century plant physiology. According to Schleiden, many of his colleagues failed to distinguish carefully between two processes: growth in its narrow sense, which was a division and multiplication of cells ("Zellvermehrung"), and growth in a more general sense, which was an effect of cellular elongation and the increase in cellular volume ("Zellstreckung"). These processes had similar effects but were very different in nature. Their widespread conceptual conflation, Schleiden argued, had led to flawed and nonsensical hypotheses on plant growth (Schleiden 1861, 574).[13]

Schleiden's message was clear: if botanists did not want to go amiss, they needed to control their observational practices. These included mastering the techniques of microscopy; taking into account the cells' developmental histories; documenting

---

[11] See also the textbook's concluding remarks (Schleiden 1861, 668): "Where should advice come from? From observing the external shapes, but not in the way it has been done up to now, without any principles and in a superficial manner. Instead, observation ought to be guided by the pursuit of morphology as a science which can only be founded on *developmental history*." German original: "Woher soll denn Rath kommen? Von der Betrachtung der äusseren Formen, aber nicht in der Weise, wie sie bisher principlos und oberflächlich getrieben, sondern von dem Erstreben einer Morphologie als Wissenschaft, deren Princip nur die Entwickelungsgeschichte sein kann." Emphasis in original. Apelt (1854, 53) also emphatically underlines this point.

[12] See Buchdahl (1873, esp. 36–39) for instructive and illuminating details. Some more "general maxims," Schleiden thought, were the same for all fields of inductive investigation; these included, e.g., parsimony, validity, unity, etc. Equally important were "specific maxims" for individual fields of study, such as botany. For Schleiden, induction and hypotheses without the guidance of these "maxims" would necessarily fail.

[13] The same point was already made in Schleiden (1842, §190, 458–59), although it was expanded and refined in subsequent editions of the textbook.

one's perception in accurate sketches and precise descriptions; double-checking the factual basis of claims made by others; and making sure that concepts and *explananda* were clearly defined. If botanists complied with these rules (or control strategies, as we may want to call them), Schleiden thought, botany might finally achieve something.

7.2.2 Experiment

Schleiden's remarks on experimental research were much shorter than his discussion of observation; in fact, the methodological introduction to the first edition of Schleiden's textbook does not even mention it. Experimentation appeared as a topic only in the second edition, that is, in 1845, and the respective passages remained unaltered in 1849 and 1861.[14] In contrast to observation, Schleiden found it difficult to formulate general guidelines for performing experiments, "because each one is modified differently according to the particular case" (Schleiden 1861, 85).[15] He also found experimentation more demanding. Anybody who was willing to receive adequate training and was ready to practice persistently could learn how to observe, but experiments required "innate talent." Probably as many as two out of three botanical experiments were inconclusive, Schleiden maintained, because their authors "did not have the gift to present questions to nature in an appropriate way, so that a clear yes-or-no response would be given" (Schleiden 1861, 85).[16] Besides talent, experimentation required comprehensive scientific and philosophical training, in order to develop one's power of judgment ("Urtheilskraft"). Only in rare cases could one rely on the felicitous instinct of a genius—Humboldt being the exception to the rule.

For Schleiden, the essence of experimentation was "placing natural bodies in a situation such that one can subject all aspects of their internal processes to measurement" (Schleiden 1861, 85).[17] Experiments, in other words, first and foremost required the design of a "controlled" setting—control in a broad sense (see the introduction to this volume)—which allowed for precise, quantitative examination. The measurements in question might entail the chemical analysis of the plant's organ but also the determination of the effects of physical forces on the plant's behavior, including temperature, light, gravity, magnetism, and electricity. To this end Schleiden distinguished two approaches:

---

[14] Cf. Schleiden (1845, 120–21) and Schleiden (1849a, b, 121–22).

[15] German original: "Für das Experiment dagegen lassen sich weniger allgemeine Vorschriften geben, weil jedes nach dem speciellen Fall sich verschieden modificirt."

[16] German original: "Es werden nur zu viele Experimente angestellt, die gar kein Resultat geben und geben können, weil ihre Urheber nicht die Gabe hatten, der Natur Fragen auf die zweckmässige Weise vorzulegen, so dass wirklich eine Antwort, Ja oder Nein, darauf folgen musste."

[17] German original: "Naturkörper in eine solche Lage zu versetzen, dass wir die an ihnen vorgehenden Processe in ihren einzelnen Elementen der Messung unterwerfen können."

(1) that the plant is deprived of the natural conditions needed for its growth as little as possible, and that it is made to grow in such a manner that the products of its vital processes, such as the emission of gas, the evaporation of water, and so forth, can be measured in terms of quantity and quality; (2) that one single, precisely defined condition of the plant's natural growth is excluded, or an alien condition is added, and the outcome is then compared with that state of a plant growing in natural conditions.[18]

These two types of experiment are intriguing, each in its own way. The first one (type 1) is unusual. Philosophically speaking, it might not be considered an experiment at all, because it involves as little intervention as possible, with procedures of measurement as the only exception. The botanist was to monitor the course of physiological processes under natural conditions and record carefully their manifestations, to better understand how these processes proceeded: a very valid intention, given the poor state of plant physiological knowledge at the time. The second type of experiment (type 2), in contrast, looks very familiar in its resemblance to Mill's "method of difference," where two specimens were compared, one of them as control (in a narrow sense), to test the effect of selected and possibly manipulated factors (Mill 1843).

Why Schleiden decided to include this paragraph in the introduction of 1845, as well as similar remarks in later chapters of his textbook, is not entirely clear. The timing is certainly suggestive—the passage was included only after John S. Mill's *System of Logic*, including its discussion of experimental methodology, was published in 1843. But there is no reference to Mill in Schleiden's textbook, neither in the introduction nor elsewhere, so that it is difficult to identify a specific connection. Given that the principal strategy was practiced long before Mill wrote his treatise (see, e.g., the chapters by Schürch and Coko in this volume), Schleiden's precise source of inspiration remains to be clarified elsewhere.

For Schleiden, a sufficiently "controlled" experimental setting included a thorough understanding of potential factors of influence and their reaction patterns:

These experiments can only bring us closer to our goal of understanding the phenomena of life if we at the same time subject all the individual substances and forces that might possibly affect the vital processes of plants, independently of the plant, to a careful examination and comprehensively investigate all their properties. (Schleiden 1861, 578).[19]

---

[18] German original: "(1) Dass man sie so wenig wie möglich den natürlichen Verhältnissen, unter denen sie wachsen, entzieht, dass man sie nur in denselben auf solche Weise wachsen lässt, dass man bestimmte Erfolge des Lebensprocesses, z.B. die Gasausscheidung, die Wasserausdünstung u.s.w. nach Quantität und Qualität dem Maass und Gewicht unterwerfen kann; (2) dass man eine einzelne genau bestimmbare Bedingung ihrer natürlichen Vegetation ausschliesst oder eine fremdartige hinzufügt, und den Erfolg dann quantitativ und qualitativ mit der unter natürlichen Bedingungen vegetirenden Pflanze vergleicht." See on this point also Schleiden (1861, 578), where it becomes clear that he really conceived of these as two separate approaches. (See, for the same remarks, Schleiden 1845, vol. 1, pp. 144–145 and Schleiden 1846, vol. 2, pp. 441.)

[19] German original: "Beide Arten von Versuchen können uns aber allein unserem Ziele, ein Verständniss der Lebenserscheinungen herbeizuführen, noch nicht näher rücken, wenn wir nicht gleichzeitig alle einzelnen, bei dem Pflanzenleben irgend in Frage kommenden Stoffe und Kräfte unabhängig von der Pflanze, für sich einer genauen Untersuchung unterworfen und in allen ihren Eigenschaften vollständig erforscht haben."

Schleiden demanded that, to assess the impact of certain substances and forces on life processes, botanists must first study these substances and forces *in vitro* and investigate their effects individually and in mutual interaction. Schleiden explicitly asked for preparatory trials to clarify reaction patterns outside the organism; only then the specificities of reactions in the living body could be identified. This was especially important if one wanted to minimize the role of so-called vital forces in the organism. Schleiden believed vital forces should be considered as explanatory factors only if it were utterly impossible to find a satisfactory physicochemical explanation. In most cases, Schleiden claimed, reference to vital forces was short-hand for "we do not know yet" (e.g. Schleiden 1861, 41–42).[20]

Plant nutrition was one such case. There were endless series of experiments about the alleged ability of plants to choose their nutrients, Schleiden lamented. "The theories based on them, the disputes about them fill a small library." All of this was pointless, he maintained, without prior investigation of reaction patterns outside the plant. If one knew the affinity of proteins, gums, sugars, and other elements of the plant toward minerals in solution, this might explain the mineral absorption of roots without any need to assume intentional action on plant's part. But experiments along these lines did not exist, and consequently almost nothing was known about the principles of plant nutrition (see Schleiden 1846, 442).[21]

Schleiden finally warned his readers that one should never jump to conclusions from only one set of experimental data, which may for arbitrary reasons not be entirely accurate. But one should also not be confused by persistent differences in the outcome. Individuals of the same species might very well behave differently and, therefore, yield different data in the same experimental set-up. These differences, however, were of no importance for the actual target of investigation, which for Schleiden was the general, characteristic, and lawful behavior of plant species.

This must suffice as a painfully brief survey of Schleiden's thoughts on the principles of observational and experimental methodology. His methodological introduction is a prime example of how control strategies and control practices (primarily in view of microscopy techniques *sensu largo*) were discussed at the time. Schleiden hardly ever used the term "control," but he had a clear concept of experiments in the sense of controlled intervention in otherwise stable settings (see on this point the

---

[20] The only exception was the phenomenon of development, which Schleiden explained as the effect of the "nisus formativus" in organic matter, that is, the "instinct" of development.

[21] German original: "So z.B. sind seit *De Saussure* eine endlose Reihe von Versuchen über das Vermögen der Pflanzen, ihren Nahrungsstoff zu wählen, angestellt worden und die darauf gebauten Theorien, die darüber geführten Streitigkeiten füllen eine kleine Bibliothek. Ich dächte, wenigstens seit *Dutrochet's* Entdeckung wäre es gar leicht einzusehen, dass alles Reden darüber leer ist, so lange wir nicht untersucht haben, ob den organischen oder unorganischen in der Pflanze vorkommenden Stoffen nicht auch ausser derselben, unabhängig vom Leben der Pflanze, ein Wahlvermögen zukommt und welches, und in wiefern dieses mit dem bei der Pflanze beobachteten übereinstimmt." On the controversy around vital forces as explanatory factors for natural processes in nineteenth-century science, see also the chapter by Coko in this volume on the different explanations for Brownian motion.

introduction to this volume). He was familiar with experiments according to Mill's "method of difference" (although it is unclear whether Schleiden had actually read Mill's treatise), but also supported an alternative approach, which entailed monitoring life processes quantitatively under natural conditions. Generally, Schleiden demonstrated that he was keenly aware of the challenges inherent in experimentation with living organisms, although he did few experiments himself. Schleiden emphasized that experiments would be meaningful only if they were appropriately designed. He warned his readers that individual differences in outcome should not be overrated. He called for a deeper understanding of causal factors and their effects outside the organism before claims about their effects inside the organism. Finally, he emphasized the need for extensive scientific and philosophical training: the aspiring scientific botanist had to bring far more to the table than curiosity and good will.

In the following sections, I shall trace how this agenda unfolded in the work of one of Schleiden's former doctoral students, the plant physiologist Julius Wiesner.

7.3 Plant Physiology and Its Control Practices in the Laboratory

Julius Wiesner (1838–1916) was born in Moravia, then part of the Habsburg Empire. He spent most of his childhood in Brünn (today's Brno), but moved to Vienna for his university studies.[22] Wiesner attended classes in botany with Eduard Fenzl (1808–1879) and Franz Unger (1800–1870). But he was also attracted to the group of physicists and physiologists around Ernst Brücke (1819–1892), with whom Wiesner received his initial training in the use of precision instruments. Wiesner then moved to Jena, where he completed his PhD with Schleiden in 1860, that is, shortly before the reprint of the third edition of Schleiden's textbook. Thereafter Wiesner returned to Vienna and, in 1873, after various positions at the Technical University, he was appointed chair for "Anatomy and Physiology of Plants" at the University of Vienna. Wiesner became known for his expertise in microscopy and experimentation, for his sophisticated methods and techniques, and for his success as a discipline-builder.

A long-standing research interest of Wiesner's was the influence of light on plants' forms and functions. The question was as important as it was complex. By the 1860s, it was beyond doubt that many, if not all, characters and vital processes of plants were strongly influenced by their exposure to light. This was obviously true for photosynthesis, or "carbon assimilation," as it was called at the time, which depended upon illumination and ceased in darkness. But it was equally true for the processes of growth and development, for the shape and outer appearance of plants,

---

[22] On Wiesner see, e.g., Wurzbach (1888), Molisch (1916), and Wininger (1933). On Wiesner's research in old paper, see Musil-Gutsch and Nickelsen (2020).

and for their internal cellular and subcellular constitution. In many cases, light of different wavelengths and intensities seemed to prompt different effects, but the details of which kind of illumination led to which plant characteristics, and why, were obscure.

One of the issues Wiesner addressed in this context was the influence of light on chlorophyll, the green pigment of plants (e.g., Wiesner 1877). Very little was known about this substance at the time. It was known to be involved in assimilating carbon, but nobody knew exactly what the pigment did.[23] Wiesner decided that chlorophyll deserved more attention, and he started with the basics. In line with Schleiden's idea of botanical experiments, Wiesner first analyzed the elementary composition of chlorophyll and its chemical behavior *in vitro*. Only then did he begin investigating its reactions *in vivo*, that is, within the cell. One of Wiesner's specific interests was how and under which conditions chlorophyll developed in plant cells. It was known that this process depended on light: plants grown in darkness or shoots covered with earth remained pale or, botanically speaking, "etiolated." But there was no consensus on how and why these etiolated plant organs turned green upon illumination. One of the open questions for Wiesner was which part of the incident light prompted this greening process. Sunlight, the usual light source in nature, encompassed the full spectrum of rays, from very short to very long wavelengths. It therefore exposed the plant to two very different physical factors at the same time, namely, light and heat.[24] Wiesner wondered which of these were effective for the formation of chlorophyll. Was it necessary for a plant to receive light rays in the narrow sense, that is, comparatively short wavelengths, from the visible part of the spectrum? Or was it sufficient to provide plants with heat rays, which were also part of the spectrum but invisible to the human eye? (Wiesner 1877, 39).

The standard assumption at the time was based on an 1857 study by one of Wiesner's French colleagues, Claude Marie Guillemin, who claimed that heat rays were, in fact, as effective as light rays in prompting chlorophyll formation (Guillemin 1857). Guillemin had passed sunlight through a prism, so that the light split into its different components, and observed what happened to seedlings that grew under different parts of this light spectrum. He found that all seedlings developed chlorophyll, including those illuminated by the "dark" part of the spectrum, that is, by the range of heat rays. These experiments and their interpretation were favorably received by most of Guillemin's colleagues. Wiesner, however, considered them methodologically flawed in almost every respect. He objected that the hypothesis was based on only two experimental runs of limited duration; that the spectral rays were not sufficiently separated from each other, so that overlapping illumination could not be reliably excluded; and, finally, that insufficient methods had been used to detect the formation of chlorophyll, namely, visual inspection and external

---

[23] It would take another 80 years before this issue was fully resolved (cf. Nickelsen 2015).

[24] The nature of the multitude of different rays in nature, including their chemical and physical properties and effects, were widely debated at the time. On this topic, see, e.g., Hentschel (2007).

appearance. There was no doubt, Wiesner thought, that the case had to be revisited in a more appropriate experimental set-up.

To this end, Wiesner created an entirely controlled environment. He ordered the construction of double-walled glass jars, and had the space between the glass walls (9 mm) filled with a solution of iodine in carbon disulfide. If illuminated, this liquid layer fully absorbed all visible light but was permeable to heat rays. Hence the test factor was fully isolated, and its effect was monitored with a precision thermometer recording temperature changes within the jar. Wiesner then carefully grew sets of etiolated seedlings, which had not yet formed chlorophyll, and he transferred them to the jar in full darkness. Finally, Wiesner used gas light instead of sunlight as a source of illumination. In contrast to the sun's rays, the spectral composition and effects of gas light were well known. Gas light was also much easier to control. In Wiesner's own words:

> Now, I had it in my power to manipulate the incident radiation within a wide range of possibilities, by combining different gas flames, and by varying the distance between flames and test plants. I was able to operate under conditions of constant radiation; and there was the great advantage that I was in full command of how long the experiments would last. (Wiesner 1877, 43)[25]

The last point was especially important: being in command of the course and duration of the experiment. When Wiesner first tried these experiments with sunlight, he had to stop early because of unexpected clouding in the sky. He was not able to draw reliable conclusions. The new experimental set-up, in contrast, yielded crystal-clear results: not a single seedling in the jar turned green, and a sensitive fluorescence test confirmed that not even traces of chlorophyll had formed in any of them. Wiesner corroborated these findings with two control experiments. First, some of the etiolated seedlings were not placed in the jar but exposed to full gaslight. Wiesner found that these seedlings formed chlorophyll without problems, indicating that there were no inherent problems with the seedlings and light. Second, Wiesner exposed the jar seedlings after their heat experience to the full spectrum of the gaslight, where most of them recovered and turned green. Wiesner was satisfied, and concluded that heat rays were insufficient to induce the formation of chlorophyll in plants.

This meticulous care in experimentation characterized Wiesner's work in general. The double-walled glass jar is only one example of a sophisticated apparatus he specifically constructed to meet his standards. Others included a so-called Clinostat, which neutralized the gravitational pull by slow rotation, thereby allowing it to distinguish the influence of light on plants from the influence of gravitation. He was also responsible for innovative applications of the Auxanometer, a self-registering instrument that continuously monitored a plant's growth. Besides isolating the test factor, one also had to control and measure its impact as precisely as

---

[25] German original: "Ich hatte es nunmehr in meiner Gewalt die Strahlung durch Combinirung von Gasflammen, Regulirung der Entfernung zwischen Gasflamme und Versuchspflanze innerhalb weiter Grenzen zu nuanciren, konnte bei constanter Strahlung operiren und hatte den grossen Vortheil, die Dauer der Versuche völlig zu beherrschen."

possible. In this latter respect, Wiesner thought, the then-current investigation of the influence of light was highly deficient:

> In physiology, one is not satisfied with the mere distinction between warm and cold but examines the surroundings of the plant with a thermometer, to the great benefit of the discipline. In a similar way, we must finally begin to measure the intensity of light that a plant receives in order to learn how much influence certain light intensities exert on the vital processes of plants. (Wiesner 1894, 1079)[26]

If light fundamentally affected a plant's growth and development, as everybody agreed that it did, it was high time to measure it in a way that allowed quantifying those effects. In other words, Wiesner was calling for a better-controlled method of recording.

In physics and chemistry these techniques already existed. In the 1860s, chemists Robert Bunsen (1811–1899) and Henry E. Roscoe (1833–1915) had developed a procedure to measure the intensity of the so-called "chemically active" rays, that is, rays of short wavelengths (blue-violet). They used standardized photographic paper and a color chart of blackness. The technique allowed the user to determine the intensity of these rays with high precision, but it was very demanding in practice. When Wiesner finally mastered the procedure, he used it in several experiments but almost immediately set out to develop a slightly adapted version (see Wiesner 1893). His version was less precise but easier to use and, as Wiesner emphasized, more reliable for high-light intensities. It was, therefore, more appropriate for measuring light conditions in nature, where Wiesner had taken his studies in the meantime.

7.4 New Concepts in the Field

After he had established the influence of light on the formation of chlorophyll, Wiesner also wanted to know how light affected the development of buds, the movement of tendrils, the shapes of leaves, the phenomena of differential growth, and other things. He first studied these questions in a series of greenhouse experiments. But when he started to investigate the same phenomena outside, Wiesner made two important observations. First, the intensity of chemically active light was dramatically higher outside than in the laboratory, so that the Bunsen–Roscoe method no longer worked. Apparently glass had a strong shielding effect, so that only a small part of the incident sunlight was actually effective behind windowpanes. One had to be careful, Wiesner concluded, in transferring the results of glasshouse experiments

---

[26] German original: "Aber so wie man sich in der Physiologie nicht mit der blossen Unterscheidung von warm und kalt begnügt, und die Medien, in welchen die Pflanzen sich ausbreiten, thermometrisch prüft, zu grossem Nutzen dieser Wissenschaft, so müssen wir endlich anfangen, die der Pflanze zu Gute kommenden Lichtstärken zu messen, um den Grad der Einwirkung der Lichtintensität auf die Lebensprocesse der Pflanzen kennen zu lernen."

to plants outside, living under natural conditions. Second, Wiesner found that in nature, plants received very different intensities of incident light, even if they grew almost side-by-side:

> The quantity of light that a plant receives is not only determined by the place on Earth where the plant grows but is also influenced by the specific characteristics of its location and, finally, by the form, number, and position of its organs. (Wiesner 1894, 1079)[27]

For Wiesner these differences were too important to be ignored, and had to be investigated through careful and precise measurements. This was an interesting move: inside the glasshouse, Wiesner might have tried to control this additional parameter (in the sense of Schleiden's type 2 experiments), but under natural conditions, in contrast, where full control was impossible, these differences became part of the research question and therefore had to be monitored (along the lines of Schleiden's type 1 experiments).

The relevant parameter under natural conditions, Wiesner concluded, was not the intensity of light incidence in general but the amount of light that a plant actually received ("factischer Lichtgenuss")—or, as Wiesner termed it, a plant's or a plant organ's *specific* light reception" ("specifischer Lichtgenuss").[28] He defined this parameter as the fraction of light that a plant received at its specific location compared to the full amount of light, the "full daylight" ("gesammtes Tageslicht"), that a hypothetical plant would receive in the same place fully in the open. Only under exceptional circumstances were the two parameters identical: if leaves were growing on the surface of a pond in full sunlight, for example, or if desert plants developed in full exposure. But these cases were very rare. Wiesner was greatly surprised by this finding: "The influence of the specific location on a plant's actual reception of light is, according to photometric investigation, far more significant than one would assume at first glance" (Wiesner 1894, 1081–82).[29] Even within the crown of one tree, the specific light reception of different organs varied enormously, from very high intensities at the tip of the branches to very low intensities near the trunk.

For Wiesner, these differences in actual or specific light reception were possibly the most import factor of influence for the development of plants and the shape of their organs. As he reminded his colleagues, light acted on plants as a double-edged sword: it was an indispensable catalyst of vital processes, but too much light was also harmful, as Wiesner himself had confirmed in his chlorophyll studies. Plants were creative in providing their organs with the optimal balance of light and shade.

---

[27] German original: "Das Lichtquantum, welches einer Pflanze zufliesst, ist nicht nur durch den Erdpunkt gegeben, auf welchem die Pflanze vorkommt, sondern wird auch mitbedingt durch die specifischen Eigenthümlichkeiten ihres Standortes, endlich durch die Form, Zahl und Lage ihrer Organe."

[28] This term is difficult to translate into English, as even Wiesner's colleagues from Anglo-Saxon countries acknowledged. "Specific light incidence" is an alternative term that was sometimes in use.

[29] German original: "Der Einfluss des Standortes auf die Grösse des Lichtgenusses der Pflanze ist, wie die photometrischen Untersuchungen lehren, viel beträchtlicher, als der Augenschein vermuthen liesse."

The morphology of leaves and stems, the patterns of arrangement and branching, the formation of buds, flowering periods, the growing of hairs and cuticle layers: Wiesner thought that all these phenomena, and many more, could be explained as reactions to a plant's specific light reception. And this might not only be true for individual specimens, but could also hold for the properties of plant species or for even larger patterns of vegetation.

But before these latter questions could be investigated, which all involved the long-term effect of certain kinds of illumination, Wiesner had to close a methodological gap. As mentioned earlier, his parameter of "specific light reception" was determined as a fraction of the hypothetical value of "full daylight" at the same location. The latter quantity, however, was not easy to determine. In some cases one could simply measure the light incidence nearby, in the open, but this was not always possible. Furthermore, given the daily and seasonal fluctuation of light incidence, one or two measurements were insufficient. A comprehensive investigation of "photochemical climates" was necessary, which became another new parameter Wiesner introduced (Wiesner 1897, 1907). It designated the average light conditions in a region, based on long-term data collected in one place under various conditions.

Wiesner emphasized that these investigations required commitment and persistence. He noted dismissively that some people had tried to extrapolate photochemical climates from just a few data and a set of equations. Wiesner found this approach not only careless but illegitimate and flawed: "With regard to the chemical intensity of light, as with temperature, the law of distribution on Earth can only be found by experiment" (Wiesner 1897, 75).[30] In collaboration with two assistants, Wiesner initiated more appropriate measurement series in Vienna, but he soon decided that he needed to investigate different climate zones to learn from their comparison. This was the main reason for Wiesner's extensive travel activities rather late in life: to the Botanical Station in Buitenzorg in the East Indies, Yellowstone National Park in the United States, Cairo in North Africa, and Tromsö in the Arctic.[31] Wiesner clearly had come a long way from his laboratory experiments on the formation of chlorophyll—geographically, methodologically, and intellectually. His new agenda was extremely innovative and ambitious. The question remains, however, whether it was still in line with the methodology so forcefully advocated by his famous teacher.

---

[30] German original: "Wie bezüglich der Temperatur wird also auch rücksichtlich der chemischen Intensität des Lichtes das Gesetz der Vertheilung auf der Erde erst durch das Experiment gefunden werden können."

[31] Fridolin Krasser and Ludwig Linsbauer contributed to the measurements in Vienna, Wilhelm Figdor worked with Wiesner on the climate in Buitenzorg, and Leopold Portheim travelled with Wiesner into Yellowstone National Park. See, e.g., Wiesner (1897, 75, 1898; 1907). On climate research in the Habsburg Empire (with a focus on the time before Wiesner), see Coen (2018).

7.5   A New Methodology for a New Field of Study: The Biology of Plants

The answer to this question is complex and requires a brief digression. For many scholars at the time, plant physiology, that is, Wiesner's discipline, was the embodiment of Schleiden's scientific botany, which inherently meant a strong commitment to empirical work and physicochemical explanations wherever possible. But there was no general consensus about what exactly this field entailed. In his keynote to the 1895 meeting of the American Association for the Advancement of Science (AAAS), the botanist Joseph C. Arthur circumscribed the field as follows:

> [V]egetable physiology […] is like a western or African domain, long inhabited at the more accessible points, more or less explored over the larger portion, but with undefined boundaries in some directions, and with rich and important regions for some time known to the explorer, but only now coming to the attention of the general public. In fact, our domain of vegetable physiology is found to be a diversified one, in some parts by the application of chemical and physical methods yielding rich gold and gems, in other parts coming nearer to every man's daily interest with its fruits and grains. (Arthur 1895, 360)

For Arthur, plant physiology covered a wide range of subjects and approaches; the boundaries with other fields of study were blurred and exciting discoveries still lay ahead, just as exciting as the discoveries waiting in Africa for the Europeans or in the Wild West for the Americans. Wiesner shared this broad understanding of the discipline, albeit without the dubious colonial metaphors. He thought that plant physiology "encompasses the study of everything regarding the plant's structure, development and life" (Wiesner 1898, 106). In his textbook of botany, first published from 1881 to 1884, he distinguished four divisions of plant physiology. They were: *first*, Anatomy and Physiology in the narrow sense, that is, the physicochemical explanation of vital processes (similar to Wiesner's own investigation of the formation of chlorophyll); *second*, Organography, the investigation of shape, development, and changeability of plant organs; *third*, a modernized Taxonomy and Systematics that also considered physiological and chemical properties of plants; and *fourth*, the "Biology of Plants" (Wiesner 1881b, 1884). This final division expanded substantially over time, and starting from 1899, it became a full separate volume of the textbook (Wiesner 1889).[32]

How are we to understand this "Biology"? The term is ambiguous and its history complex (see, e.g., Toepfer 2011). According to a still-popular narrative, it was in 1800 that the French naturalist Jean-Baptiste de Lamarck (1744–1829) and the German naturalist Gottfried Reinhold Treviranus (1776–1837) allegedly invented biology independently from each other, as the science of life. Over the course of the nineteenth century, the field then developed "from natural history to biology," to borrow a widely used expression, that is, in linear progression from a descriptive, old-fashioned enterprise into the scientific discipline we know today. Joseph Caron criticized this narrative already in 1988 and argued that the invention of a name

---

[32] For Wiesner's concept of "Biologie," see also Nickelsen (2023).

must not be confused with the founding of a discipline. He thought that the first real attempts at the discipline were made by Thomas H. Huxley, who in 1858 attempted to institutionalize a class in "Principles of Biology" in Cambridge, albeit with moderate success (Caron 1988). Kai T. Kanz then pointed out that we cannot extrapolate to other countries from this episode in England. With several examples Kanz showed how, from the eighteenth century, the term "biology" was used as either an umbrella term, a subordinate term, or a term synonymous with various others, and often the term was not used at all (see, e.g., Kanz 2002, 2006, 2007). Even more complicated is the combination of biology with other terms. Highly illuminating in this context is Eugene Cittadino's (1990) discussion of "Biologie der Pflanzen" as precursor of evolutionary ecology for plants in the German-speaking countries.[33] This was exactly how Wiesner used the term in his pioneering textbook, as the following passage demonstrates:

> The word biology has very different meanings. Huxley, and probably most British naturalists ("*Naturforscher*") with him, use the word in its broadest sense, as the study of organisms. Other naturalists have significantly limited the concept and regard biology as that part of science that deals with the way of life of plants and animals.
>
> The majority of today's naturalists fall somewhere between these two extremes and see biology as the science of the habits, heredity, variability, adaptation, origin and natural distribution of organic beings. In this last sense, the word biology will be understood in the present book. (Wiesner 1889, 1)[34]

Wiesner obviously was aware of the terminological difficulties. He therefore tried to clarify his own usage in reference to the emergent field of study, which he introduced as part of "plant physiology"—understood widely but differently from "plant physiology" in a narrow sense (although he had to admit that the boundaries were blurred). Wiesner's colleagues similarly struggled with defining this new field, including the Munich-based botanist Karl Goebel (1855–1932) in a paper of 1898:

---

[33] See Cittadino (1990), esp. 149. Lynn K. Nyhart made a similar observation of a new "biological perspective" on animal life, albeit mostly beyond the circles of academic zoology; see Nyhart (2009). Already Arthur (1895) pointed to this German peculiarity and specifically cited Wiesner's book as the first to have been published on the theme (the only other book that Arthur cited was the one by Friedrich Ludwig, see below; Arthur also declared that there was so far no analogous publication in English). Arthur acknowledged that the name "biology" was justified, yet given that Huxley had already used "biology" differently, he favored the alternative designation of this area as "ecology." See Arthur (1895, 365).

[34] German original: "Man bezeichnet mit dem Wort *Biologie* sehr Verschiedenes. Huxley und mit ihm wohl die meisten britischen Naturforscher gebrauchen dieses Wort in seinem weitesten Sinne, als die Lehre von den Organismen. Andere Naturforscher schränken diesen Begriff wieder sehr stark ein und betrachten die Biologie als jenen Theil der Naturwissenschaft, welcher sich mit der Lebensweise der Pflanzen und Thiere beschäftigt. Die Mehrzahl der heutigen Naturforscher bewegt sich in der Mitte zwischen diesen beiden Extremen und begreift unter Biologie die Lehre von der Lebensweise, Erblichkeit, Veränderlichkeit, Anpassung, Entstehung und natürlichen Verbreitung der organischen Wesen. In dem zuletzt bezeichneten Sinne soll auch in diesem Buch das Wort Biologie verstanden sein."

We can compare the relationship between physiology and biology to that of two maps, one of which displays only the mountain ranges and rivers, the other also the political borders and settlements. How a country is populated clearly depends on its physical nature but also on the characteristic properties of its inhabitants and their varied history. Similarly, experimental physiology shows us a broad outline of the relationship of plants to their environment, but it does not reveal how the vital processes take place according to the plants' characteristic properties and history. On a general level, for example, the role of water is the same for all plant species. However, the ways in which plants go about meeting their demand for water, depending on their level of organization and the conditions of the environment, is infinitely different. (Goebel 1898, 4)[35]

Goebel, as we see here, drew the line between the two fields, physiology and biology, in terms of the questions being asked and the level of particularity being studied. Whereas physiology investigated the water balance of plants in general, biology studied the multiple adaptations of plant species, that is, their "manifold relationships to the outside world."[36] Like Wiesner, Goebel was in favor of the new field. He thought that the progress of physiology had come to a halt, while biology was on the rise, for two main reasons. First, the ongoing "exploration of tropical areas": botanists were no longer satisfied with lists of new species but had started to investigate the multitude and variability of vital processes on display in tropical climates. Second, the new approach of "Darwinism," which pointed to the interplay of an organism's morphology with its natural environment. Both had not only raised important questions but also opened paths to answer them (on these points, see Goebel 1898, 4–5).

For Goebel, "Darwinism" did not primarily refer to the transformation of species by means of natural selection: "In fact, if we look at today's botanical literature, we find that the actual Darwinism, that is, the theory in which natural selection is the main factor that causes adaptations, is hardly represented anymore, at least in Germany" (Goebel 1898, 10–11).[37] Goebel explained that even Darwin himself had increasingly downgraded the importance of natural selection. It might well be that it contributed to the transformation of species, but for Goebel, direct adaptation was

---

[35] German Original: "Das Verhältniss zwischen Physiologie und Biologie können wir etwa dem zweier Landkarten vergleichen, von denen die eine uns nur die Gebirgszüge und Flüsse, die andere auch die politischen Grenzen und Ortschaften gibt. Wie nun die Besiedelung eines Landes zwar abhängig ist von seiner physischen Natur, aber ausserdem auch von den charakteristischen Eigenschaften seiner Bewohner und ihrer wechselnden Geschichte, so zeigt uns auch die Experimentalphysiologie nur in grossen Zügen die Beziehungen der Pflanzen zur Aussenwelt, nicht aber, wie je nach der besonderen Eigenthümlichkeit und nach der Geschichte einer Pflanzenform ihre Lebensvorgänge sich abspielen. So ist die Bedeutung des Wassers im Wesentlichen für alle Pflanzenformen dieselbe, unendlich verschieden aber die Art, wie je nach der Organisationshöhe oder den äusseren Lebensbedingungen der Wasserbedarf gedeckt wird."

[36] German original: "mannigfaltige Beziehungen zur Aussenwelt."

[37] German original: "In der That, sehen wir uns in der heutigen *botanischen* Literatur um, so finden wir, dass der eigentliche Darwinismus, d.h. die Richtung, welche der natürlichen Zuchtwahl die *Haupt*rolle bei dem Zustandekommen der Anpassungen zuschreibt, in Deutschland wenigstens fast keine Vertreter mehr hat."

clearly the most significant factor. Its effects were also apparently transmitted to the next generation, which should remove the last lingering doubt about its significance, Goebel maintained. He predicted that biology would gain important insights along these lines in the near future, and concluded on a lyrical note: "The young biological science resembles the man which the poet sings about: 'There he goes without hesitation/His soul filled with dreams of harvest/And he sows and hopes.'" (Goebel 1898, 21).[38]

A slightly different relationship between physiology and biology was suggested by the botanist Friedrich Ludwig in his textbook on *Biologie der Pflanzen* (Ludwig 1895). In accordance with his Italian colleague Federico Delpino, Ludwig defined biology as "the doctrine of the external relationships of plants," whereas physiology was the "doctrine of the internal processes of plants."[39] To this Ludwig added a methodological observation: "While the latter amounts to physicochemical transformations, the former sneers at all attempts of mechanical explanation, as will always be the case with the mechanical explanation of life in general" (Ludwig 1895, V).[40]

For Ludwig, the difference between investigating (and explaining) the inner processes of plants, and investigating (and explaining) their relationship with the external world, was correlated with different types of explanation. Whereas physiology aimed at a "mechanical explanation," to be understood as causal explanation based on physicochemical factors, biology strove for non-mechanical, primarily teleological explanations. Ludwig left no doubt that he, like Delpino, considered "mechanical" approaches insufficient, and so supported a vitalist perspective on the manifestations of life. "Biology," thus understood, came dangerously close to *Naturphilosophie*, which many botanists at the time regarded as the epitome of "unscientific," and from which they had only just emancipated themselves. Wiesner was clearly getting into troubled methodological waters with his new area of interest.

---

[38] German original: "Da geht er ohne Säumen / Die Seele voll von Ernteträumen / Und sät und hofft". Goebel cites these verses, without any explicit reference, from a poem written by J. W. Goethe, "Ein zärtlich jugendlicher Kummer" (which approximately translates to "A tender adolescent sorrow").

[39] German original: „die Lehre von den äußeren Lebensbeziehungen der Pflanze" vs. „die Lehre von den Vorgängen des inneren Pflanzenlebens."

[40] German original: "Während die letzteren auf physikalisch-chemische Umwandlungen hinauslaufen, spotten die ersteren aller mechanischen Erklärungsversuche in dem Maße, wie dies mit der mechanischen Erklärung des Lebens überhaupt immer der Fall sein wird." Federico Delpino (1833–1905) pioneered the study of how floral morphology related to pollination. He also investigated the topic of "plant intelligence" and supported a teleological, spiritual interpretation of the processes of evolution.

7.6   A New Role for Speculation?

The introduction to Wiesner's textbook on the biology of plants is highly instructive in this respect. "Physiology" (in a narrow sense) and "biology" differed in their subject matter, he explained, and therefore necessarily also differed in methodology. Physiology focused on specific processes, such as transpiration and respiration, and sought to spell out the effects of isolated factors in this context. To this end, physiologists used the "*inductive* method" of chemistry and physics, which Wiesner understood as drawing inferences about causal links from experimentation. Biology, in contrast, focused on so-called "vitalistic" problems, "which we cannot yet resolve with exact scientific methods", and it aimed to understand the effect of all factors combined. It therefore "mostly arrives at the desired outcome by way of *speculation*" (Wiesner 1889, 2; both emphases in original).[41] As an example, Wiesner pointed to the complex relationship between insect behavior and flower morphology (which incidentally was an important research area of Delpino's, one of the authorities with a vitalistic inclination referred to by Ludwig). If one wished to illuminate these phenomena, Wiesner explained, it was not only practically impossible to separate the different factors from each other; it was also nonsensical, because the investigation aimed at the interplay of factors.

Given the ill repute of speculation at the time, in the wake of Schleiden's critical campaign against "speculative botany," this was dangerous ground. But the methodological schism between physiology and biology was not as radical as it might appear, Wiesner hastened to add: "For physiology too, like every other natural science, has to draw on speculation from time to time, to quickly open up new ways of induction, or to accelerate its often sluggish pace. And biology will only gain a sufficient basis for its speculation from the facts that have actually been ascertained" (Wiesner 1889, 2).[42] This commitment to an empirical basis implied that not all speculation was legitimate. In line with Schleiden, Wiesner insisted that vital forces or instincts were unacceptable as explanatory factors in biology as well:

> Overall, the assumption of a special vital force is only justified insofar as we have not yet succeeded in tracing all manifestations of life back to the effects of mechanical forces. However, since the assumption of a specific vital force loses its justification in proportion to the advances of the natural sciences, and since the assumption itself has turned out to

---

[41] German original: "*inductive* Methode"; "[vitalistische Probleme], welchen wir mit exacten naturwissenschaftlichen Methoden noch nicht beizukommen vermögen"; Biology arrives "vornehmlich auf dem Wege der *Speculation* zu den erstrebten Resultaten" (emphases in original).

[42] German original: "Freilich zeigt sich auch hier wieder die Zusammengehörigkeit beider; denn auch die Physiologie muss, gleich jeder anderen Naturwissenschaft, zeitweilig die Speculation heranziehen, um rasch neue Wege der Induction zu erschliessen, oder um den oft schleppenden Gang der Induction abzukürzen, und auch die Biologie wird nur aus dem thatsächlich Erhobenen eine zureichende Basis für ihre Speculation gewinnen."

be absolutely unfruitful, [...], one must approve the point of view delineated at the beginning of this paragraph: that the existence of a specific vital force cannot be accepted (Wiesner 1889, 14).[43]

And to make his position perfectly clear, Wiesner added, "The peculiarity of the life processes is not to be found in a principle independent of matter, or in a specific vital force, but in the combination of mechanical forces" (Wiesner 1889, 14).[44]

For Wiesner, biology was not an invitation to revitalize elusive forces. It was the attempt to include complexity and long-term effects into the realm of science. But translating this ideal of biological investigation into methodologically sound research practice remained a challenge. Wiesner's own research shows how he dealt with this dilemma, with examples as early as the 1870s. In his chlorophyll studies, Wiesner found that this pigment, which was essential for a plant's survival, was extremely sensitive to light and easily harmed in direct illumination. One should therefore expect to find in the organs and tissues of plants "special means of protection to preserve this substance," Wiesner explained, and this is exactly what he then identified (Wiesner 1875, 22).[45] Wiesner described in detail the striking differences between a plant's morphology in the sun and in the shade, including the shape and structure of stems, leaves, cuticles, and hairs, and also between patterns of vegetation, periodic movements, and other factors. On this basis, Wiesner explained the emergence of these characters in nature as a protection strategy against too much light—that is, with reference to their purposes and not their causes.[46] Wiesner conceded that the approach entailed methodological risk, but assured his readers that the risk was limited: "Since biology builds its speculations upon a broad factual basis, its hypotheses—notably Darwin's important doctrine, which in a way inaugurated the age of biological research—gain strength and support"

---

[43] German original: "Alles in allem genommen hat die Annahme einer besonderen Lebenskraft nur insofern eine Berechtigung, als es bisher noch nicht gelungen ist, alle Lebensäusserungen auf die Wirksamkeit mechanischer Kräfte zurückzuführen. Da aber die Annahme einer specifischen Lebenskraft desto mehr an Berechtigung verliert, je weiter die exacte Naturforschung vorwärtsschreitet, und da diese Annahme sich durchaus als unfruchtbar herausgestellt hat [...], so wird man den im Eingange dieses Paragraphen markirten Standpunkt, von welchem aus eine besondere Lebenskraft nicht zugestanden werden kann, nur billigen müssen."

[44] German original: "Das Eigenartige der Lebensprocesse ist also nicht in einem von der Materie unabhängigen Principe oder in einer specifischen Lebenskraft, sondern in der Combination mechanischer Kräfte zu suchen."

[45] German original: "besondere Schutzmittel zur Erhaltung dieser Substanz [waren] schon von vornherein [zu] erwarten."

[46] One anonymous reviewer of Wiesner's book on the "Lichtgenuss der Pflanzen," however, felt that Wiesner's research had yielded important insight but was methodologically problematic: "The book is by no means free from doubtful generalizations and generous assumptions; indeed, it seems that everyone who deals with adaptations must allow his imagination a rather loose rein. Withal there is in the work an important nucleus of no little value, and even an occasional flight of fancy may be permitted, if it stimulates interest" (C.R.B. 1908, 343).

(Wiesner 1889, 3).[47] While the speculation that *Naturphilosophie* employed had been unfounded and fruitless, speculation in biology was based on facts and therefore legitimate, Wiesner wanted to persuade his readers. It was, in a way, a "controlled" form of speculation.

The reference to Darwin in this context is significant.[48] Wiesner's biology included all the phenomena Darwin had wanted to explain, such as the "habits, heredity, variability, adaptation, origin and natural distribution of organic beings" (see Wiesner's definition of the field, quoted above). Darwin had likewise been accused of speculation: of presenting hypotheses insufficiently based on empirical evidence. In turn, Darwin had rejected this critique as unfounded and justified his speculation, like Wiesner, with reference to the explanatory power of his hypothesis: it simply had to be true because so many phenomena could be explained with his theory that otherwise would remain mysterious. This argument was common practice in other areas of nineteenth-century science, where it was impossible to provide experimental proof. The argument was always contested and certainly fallible but not illegitimate (see the chapter by Coko in this volume, in particular the discussion of how Gouy tried to explain Brownian motion). Wiesner now pointed to the same principle for the biology of plants.

However, there were limits to the speculation that Wiesner was prepared to accept. In 1880, Darwin published a comprehensive treatise on the movements of plants, in which he presented the result of studies (undertaken with his son Francis) on the question of how plants responded to external stimuli (Darwin 1880). Like others of Darwin's major publications after 1859, the book provided further evidence for his theory of transmutation and common descent. In a letter to his colleague Alphonse P. de Candolle, Darwin described his main finding with glee: "I think that I have succeeded in showing that all the more important great classes of movements are due to the modification of a kind of movement common to all parts of all plants from their earliest youth."[49] A second claim of Darwin's was that environmental factors, such as light, gravity, etc., acted like stimuli on certain areas of the plant with their effects transmitted to others, similar to transmission processes in the sensory and nervous systems of lower animals. Wiesner greatly admired Darwin's work, but in this particular case, he was unimpressed and found much to criticize:

> I soon recognized that Darwin had entered an area in which the methodology is just as powerful, and perhaps I do not exaggerate when I say, more powerful than the genius, namely the area of experimental plant physiology. In this field, no step forward can reliably be taken unless accurate physical or chemical methods are used to solve the problems,

---

[47] German original: "Da aber die Biologie ihren Speculationen eine möglichst breite thatsächliche Unterlage gibt, gewinnen ihre Hypothesen—namentlich die bedeutungsvolle Lehre Darwins, welche die Epoche der biologischen Forschung geradezu inaugurirte—Halt und Stütze."

[48] Arthur agreed with Wiesner on this point: "We may call Darwin the father of vegetable ecology, for had he not written, the field would have lain largely uncultivated and uninteresting" (1895, 368).

[49] Darwin to DeCandolle, 28 May 1880, in F. Darwin (2009 [1887]), 333.

even if the question is precisely formulated. Darwin has not conducted his experiments with the required rigor, which is why many of his results are uncertain, even doubtful. (Wiesner 1881a, 3)[50]

Darwin, apparently, failed to meet Wiesner's methodological standards in experimental work. With this assessment, Wiesner fully agreed with his colleague from Würzburg, Julius Sachs, who not only rejected Darwin's conclusions as inaccurate but also ridiculed his experiments as unskillful and meaningless (e.g., Sachs 1882, 843). Soraya de Chadarevian (1996) has convincingly interpreted this strong reaction by Sachs as an attempt to maintain authority on the right way to do experiments, namely under fully controlled conditions in the laboratory. From this perspective, Darwin's naturalist approach could not possibly produce useful results, because it failed to meet the requirements of a scientific botany—which Sachs publicly championed as the only legitimate approach to plant science. De Chadarevian's claim is probably correct for Wiesner too, because he shared Sachs's methodological standards and worked in the same project of discipline-building. However, it is worthwhile to look at the specific targets of Wiesner's critique, in order to see where Wiesner tried to draw a line between the legitimate methodological approach of plant biology, which necessarily violated some plant physiological conventions, and illegitimate work, which yielded unreliable data and untenable conclusions.

Given the time and effort he invested, Wiesner clearly considered the issue important. To demonstrate where Darwin went amiss, Wiesner carefully replicated many of Darwin's experiments, compared the findings, and in most cases challenged Darwin's interpretation (and in many cases the experimental design as well). The result was devastating for Darwin. Wiesner presented his critique respectfully and with nuance, and he acknowledged that the work presented many interesting and valuable observations. Nevertheless, he fundamentally disagreed with its claims. "No man was ever vivisected in so sweet a manner before, as I am in this book," Darwin maintained in a letter to his friend and colleague Joseph D. Hooker (Chadarevian 1996, 38).[51]

---

[50] German original: "*Darwin's* Buch enthält, wie ich mich alsbald überzeugte, wieder eine Fülle neuer interessanter Beobachtungen und geistreicher biologischer Bemerkungen über den Zweck der Bewegung für das Leben der Pflanze. Allein, ich musste bald erkennen, dass *Darwin* hier ein Gebiet betreten, in welchem die Methode ebenso mächtig, und vielleicht ist es keine Uebertreibung, wenn ich sage: mächtiger ist als das Genie, das Gebiet der experimentellen Pflanzenphysiologie, in welcher bei aller Schärfe der Fragestellung kein sicherer Schritt nach vorne gemacht werden kann, wenn nicht genaue physikalische oder chemische Methoden zur Lösung der Probleme in Anwendung gebracht werden. *Darwin* hat nun seinem Experiment nicht die erforderliche Strenge gegeben, wesshalb viele seiner Ergebnisse unsicher, ja zweifelhaft werden."

[51] De Chadarevian cites a letter by Darwin to Hooker of 22 October 1881. Darwin clearly did not take Wiesner's critique lightly but discussed the matter in a series of letters with his son Francis Darwin and also with Wiesner himself (de Chadarevian 1996, 38; see also the online edition of letters provided by the Darwin Correspondence Project at: https://www.darwinproject.ac.uk/letters). While Darwin conceded that Wiesner's critique was convincing in many points, he was not prepared to change his mind on the subject entirely.

Wiesner's main critique was that Darwin introduced hypotheses that were impossible to substantiate empirically—in other words, they were unfounded speculations. A prime exemplar, Wiesner thought, was Darwin's claim that all plant movements were derived from growth in the form of circumnutation. Wiesner admitted that there was something "tremendously appealing" about this idea (Wiesner 1881a, 23). However, in the absence of any conclusive evidence, it was just as likely that the exact opposite was true, namely, that all plant movements were derived from straight growth. Wiesner concluded that there was little value in Darwin's claim "because it is entirely based on speculation" (Wiesner 1881a, 23).[52] Wiesner, in contrast, tried to explain the same phenomena as the effect of a number of well-known factors combined, which for him made any additional hypothesis unnecessary.

The second target of Wiesner's critique was Darwin's claim that environmental stimuli were transmitted through the plant—from the sites of perception to adjoining tissues or organs, where then the reactions took place. This transmission would be similar to the functioning of sensory organs in animals. In particular, Wiesner set out to demonstrate that "the tip of the radicle did not have the peculiar and apparently mysterious properties, which Darwin attributed to it and which prompted him to claim that this part of the root directed all its movements and worked similarly as the brain of a lower animal" (Wiesner 1881a, 12–13). Wiesner firmly rejected the claim and was particularly critical of the comparative approach. The analogy between plant and animal characteristics was not illuminating at all, Wiesner maintained; even worse, he found it dangerous.[53] Wiesner acknowledged that Darwin had vital interest in drawing this analogy, because the unity of plants and animals was inherent to Darwin's theory of common descent. However, in this case Darwin went further than was compatible with a "rational investigation of nature" (Wiesner 1881a, 15). This was unfortunate and set a bad example:

> Darwin's comparison of plants and animals is always spirited and original, and it gives us intellectual pleasure even if we must disagree. However, these digressions raise concern since they encourage less talented students of nature to emulate this approach and

---

[52] As Wiesner wrote: "Allein man wird zugeben müssen, dass man auch den umgekehrten Fall setzen kann, d.h. dass man alle diese Bewegungen auch aus der einfachsten Form, dem geraden Wachsthum, ableiten könnte. So annehmbar dies klingt, so gering ist einstweilen der Werth dieser Anschauung, da sie doch nur auf Speculation beruht. Will man eine Grundlage für den Zusammenhang der Formen finden, so muss man den Weg der Beobachtung einschlagen. Es ist dies auch der Weg, den Darwin verfolgte, auf dem er aber zu Resultaten kam, die ich in der von ihm ausgesprochenen Allgemeinheit nicht bestätigen kann."

[53] This part of Darwin's work was also the main target of Sachs's criticism, as Soraya de Chadarevian (1996) has shown. Sachs even prompted one of his assistants, Emil Detlefsen, to replicate Darwin's experiments in order to refute their conclusion (de Chadarevian 1996, 29). Detlefsen notes that his experiments are in full agreement with Wiesner's findings, which he, however, only saw after he had already completed his studies (Detlefsen 1882, 627). German original: "Die kritische Studie von Wiesner […] erhielt ich leider erst, als meine Arbeit schon vollendet war, und ich konnte dieselbe daher nicht berücksichtigen. Es freut mich, constatiren zu können, dass ich in manchen wesentlichen Punkten zu Resultaten gelangt bin, die mit denen Wiesners übereinstimmen."

steer them in a speculative direction, which turns away from strict investigation and proved to be a veritable impediment for the science of organisms not so long ago. (Wiesner 1880, 16)[54]

For Wiesner, drawing analogies between plants and animals was not only methodologically questionable but also implied a relapse into the aberrations of *Naturphilosophie*. Schleiden had specifically castigated Hegel, Schelling, and others for drawing analogies of this kind between the different realms of nature (e.g., Schleiden 1842, 46–47; see also Jahn 2006). In line with this assessment, Wiesner strictly rejected this type of reasoning, even when it came from Darwin. Speculation was only legitimate if it was principally possible to test the resulting claim empirically. Furthermore, no additional speculative factors or hypotheses were admissible if the phenomenon in question could be sufficiently explained by the effect of well-established factors. For Wiesner, this was where Darwin had failed, even in those cases where his experimental set-up was fine and the measurements beyond reproach.

7.7   Wiesner's Legacy

Wiesner's search for a methodologically sound experimental biology provided critical inspiration for the founding of one of the most interesting research institutions at the time: the *Biologische Forschungsanstalt* in Vienna, also known as the *Vivarium*. This remarkable institution was founded in 1903 by three scientists of Jewish origin: the zoologist Hans Leo Przibram (1874–1944) and the two plant physiologists Leopold von Portheim (1869–1947) and Wilhelm Figdor (1866–1938), who had been Wiesner's students.[55] All three had been unsuccessful in their attempts to gain academic positions in the anti-Semitic atmosphere of the Habsburg Empire at the time and, therefore, used private capital to set up their own research institution. Many people believe that this institution was called "biological" because it investigated questions from both botany and zoology, but a different interpretation is more convincing. The institute was called *Biologische Versuchsanstalt*, I propose, because it engaged in biological research in the sense of Wiesner and others.

---

[54] German original: "Darwin's Vergleich der Pflanze mit dem Thiere zeichnet sich stets durch Geist und Originalität aus und gewährt uns auch dann einen geistigen Genuss, wenn wir ihm unsere Zustimmung versagen müssen. Allein diese Excurse haben auch ihre bedenkliche Seite, indem sie weniger begabte Naturforscher zur Nacheiferung anspornen und zu einer speculativen, von der strengen Forschung abgekehrten Richtung hinleiten, welche vor nicht allzu langer Zeit als ein wahrer Hemmschuh für die Wissenschaft von den Organismen sich gezeigt hat."

[55] Portheim had accompanied Wiesner on his journey to Yellowstone National Park. Figdor had travelled with Wiesner to Buitenzorg and Ceylon, and pursued Wiesner's research on the influence of light on leaf arrangements at the *Vivarium*. On the history of this institution, see, e.g., (Reiter 1999; Taschwer et al. 2016; Müller 2017). On the history of plant sciences in the *Vivarium*, see, (Nickelsen 2017).

This interpretation is supported by a remark in the first report of the institution written by Hans Przibram, one of the three founders:

> While it may be enough for the physiologist to keep his research organisms alive for as long as he wants to monitor a particular function, and then take fresh specimens for his further observations, the biologist is usually concerned with tracing the changes of form over a longer experimental period. (Przibram 1908, 235)[56]

This juxtaposition makes sense only if we assume that Przibram shared Wiesner's concept of biology. Biologists need more time than physiologists to complete their experiments, because they are doing biological research: they investigate, as Wiesner had detailed, the habits, heredity, variability, and adaptation of organisms, in interaction with their environment. The institute's particular focus was experimental morphology, the study of the causes of forms and functions of living organisms. This project the founders and their colleagues tried to establish in quantitative terms, but without necessarily aiming for mechanistic explanations. The institute was equipped with sophisticated light and dark chambers and hosted precision instruments of all kinds. Its members were trained in botany, zoology, physiology, chemistry, physics, and mathematics; and they investigated a wide range of vital processes, as well as their morphological basis, in long-term studies. From a certain perspective, the *Vivarium* had turned Wiesner's ambitious vision of a biological research program into reality. Its life span, however, was brief. It survived World War I and became highly successful thereafter, but the annexation of Austria by Nazi Germany in 1938 ended it. All three of its founders were murdered, and today almost nothing is left of this remarkable institution.

7.8   Concluding Remarks

Methodological considerations, statements, and critiques—in other words, "methods discourse" (Schickore 2017)—loomed large in nineteenth-century botanical research. The question of adequate control strategies and practices, the central focus of this volume, was an important part of this discourse, although the term itself was hardly used at the time. Control was even part of a programmatic change: Schleiden and Wiesner were both important protagonists in unfolding a "scientific" botany that ventured beyond descriptive taxonomy and set a comprehensive group of new methods, techniques, and approaches at its core. Wiesner was also instrumental in promoting the standards of a different field, plant biology. For both disciplines, plant physiology and plant biology, textbooks served as highly instructive sources for a reconstruction of methodological attitudes.

---

[56] German original: "Während es dem Physiologen genügen mag, seine Versuchsobjekte so lange am Leben zu erhalten, als er eine bestimmte Funktion verfolgen will, und dann zu weiterer Beobachtung frische Exemplare zu nehmen, kommt es dem Biologen meist auf Durchverfolgung der Formänderungen während einer längeren Versuchszeit an."

This essay began with Matthias Schleiden's agenda of a fundamental reform of botany, which he thought had been badly damaged by the dogmata of *Naturphilosophie*. To this end he provided his fellow botanists with clear guidelines. Schleiden touched on general issues, such as the principles of empirical work based on *autopsia* and reliable induction, but also gave detailed, hands-on introduction in how to use a scientific microscope. Schleiden reminded his colleagues that conceptual confusion would necessarily lead to unsatisfactory interpretations, and that explanations based on vital forces indicated factual ignorance. Interestingly, experimentation only started to appear in Schleiden's introduction from the second edition onwards, with two types he found permissible: type 1 was a sophisticated monitoring of life processes, and type 2 a difference test according to Mill's method. They required the full range of control dimensions for both physical and cognitive activities, and they were not for everybody. In contrast to observation, performing experiments required philosophical training and the talent to ask the right questions.

Schleiden's influence on subsequent generations of botanists was enormous, especially in the German-speaking countries. Even Julius Sachs, cited at the beginning of this essay, was full of praise: "The difference between this and all previous textbooks is like the difference between day and night," Sachs wrote in his otherwise hypercritical survey of the history of botany (Sachs 1875, 203). Julius Wiesner, one of Schleiden's former students, served as a case in point for this chapter. Wiesner was widely known as an excellent experimenter, and his research in the formation of chlorophyll is a model of Schleiden's type 2 experiments (and Mill's method of difference). Wiesner made painstaking efforts to create experimental set-ups that allowed reliable causal inferences. The careful description of how Wiesner separated potential factors of influence in the lab also demonstrated that, in botany, these measures were far from self-evident at the time. The influence of light on plants was a widely debated topic but few of his colleagues were as successful in studying it as Wiesner. In his research papers or in his textbook Wiesner never used terms such as "control" or "confounding factors," which we might expect in this context, nor did he refer to methodological treatises. But he clearly tried to exert control on all experimental circumstances.

This became impossible, however, when Wiesner moved these studies into the field, where his work started to resemble Schleiden's type 1 experiments in requiring the quantitative monitoring of vital processes of plants in reaction to their environment. Wiesner encountered difficulties of both a practical and conceptual nature. He responded by developing new techniques, such as an adequate procedure to measure light intensities in the field, and by defining new parameters, such as the new unit of specific light reception. But Wiesner increasingly became interested in questions that were impossible to answer in controlled experimental set-ups. He eventually decided that these questions required a subdisciplinary field of their own—the "biology of plants"—with its own methodological principles that deviated from established control strategies and practices. But Wiesner's discomfort in doing so was palpable. He was deeply worried that this approach would lead botany down on a slippery slope into the realm of wild speculation about vital forces; the attitude manifested by his colleagues Ludwig and Delpino confirmed that his worries were not unfounded. "Speculation" was necessary in biology, Wiesner argued, but only within boundaries: if it was based on facts and observation, if it was parsimonious, and if it had the potential to be tested empirically. In other words, if it was *controlled* speculation. The most ambitious attempt to put this program into effect was made by some of Wiesner's former students in the *Biologische Versuchsanstalt* in Vienna, but their institute was short-lived. The difficult transition from fully controlled physiological experimentation via field studies under limited control, to the challenges of methodologically sound research in the ecology and evolution of plants, would therefore be completed elsewhere.

References

Allen, Garland. 1975. *Life Sciences in the Twentieth Century*. New York: Wiley.
Apelt, Ernst F. 1854. *Die Theorie der Induction*. Leipzig: Engelmann.
Arthur, Joseph C. 1895. Development of Vegetable Physiology. *Science* 2 (38): 360–373.
Bowler, Peter, and Iwan Rhys Morus. 2005. The New Biology. In *Making Modern Science. A Historical Survey*, 165–188. Chicago: University of Chicago Press.
Buchdahl, Gerd. 1973. Leading Principles and Induction: The Methodology of Matthias Schleiden. In *Foundations of Scientific Method: The Nineteenth Century*, ed. Ronald N. Giere and Richard S. Westfall, 23–52. Bloomington: Indiana University Press.
C. R. B. 1908. The Lighting of Plants, review of *Der Lichtgenuss der Pflanzen*, by J. Wiesner. *Botanical Gazette* 45, no. 5: 342–343.
Caron, Joseph A. 1988. 'Biology' in the Life Sciences: A Historiographical Contribution. *History of Science* 26: 223–268.
Charpa, Ulrich. 2003. Matthias Jakob Schleiden (1804–1881): The History of Jewish Interest in Science and Methodology of Microscopic Botany. *Aleph* 3: 213–245.
———. 2010. Darwin, Schleiden, Whewell, and the 'London Doctors': Evolutionism and Microscopical Research in the Nineteenth Century. *Journal for General Philosophy of Science/ Zeitschrift Für Allgemeine Wissenschaftstheorie* 41 (1): 61–84.
Cittadino, Eugene. 1990. *Nature as the Laboratory: Darwinian Plant Ecology in the German Empire, 1880–1900*. New York: Cambridge University Press.
Coen, Deborah R. 2018. *Climate in Motion: Science, Empire, and the Problem of Scale*. Chicago: University of Chicago Press.
Coleman, William. 1971. *Biology in the Nineteenth Century*. New York: Wiley.
Darwin, Charles. 1880. *The Power of Movement in Plants*. London: Murray.
Darwin, Francis. 1882. Das Bewegungsvermögen der Pflanzen: eine kritische Studie über das gleichnamige Werk, von Charles Darwin, nebst neuen Untersuchungen. *Nature* 25: 597–601.
Darwin, Charles. (1887) 2009. *The Life and Letters of Charles Darwin*, edited by Francis Darwin. *Vol. 3*. Cambridge: Cambridge University Press.
de Chadarevian, Soraya. 1996. Laboratory Science versus Country-House Experiments. The Controversy between Julius Sachs and Charles Darwin. *The British Journal for the History of Science* 29 (1): 17–41.
Detlefsen, Emil. 1882. Ueber die von Charles Darwin behauptete Gehirnfunktion der Wurzelspitzen. *Arbeiten des Botanischen Instituts in Würzburg* 2: 627–647.
Goebel, Karl. 1898. Ueber Studium und Auffassung der Anpassungserscheinungen bei Pflanzen. Festrede gehalten in der öffentlichen Sitzung der k.b. Akademie der Wissenschaften zu München, zur Feier ihres 139. Stiftungstages am 15. März 1898. Munich: Verlag der k.b. Akademie.

Guillemin, Claude Marie. 1857. Production de la Chlorophylle et direction des tiges sous l'influence des rayons ultra-violets, calorifiques et lumineux du spectre solitaire. *Annales des Sciences Naturelles: Botanique* 4th ser., 7: 154–172.

Hentschel, Klaus. 2007. *Unsichtbares Licht? Dunkle Wärme? Chemische Strahlen? Eine wissenschaftshistorische und -theoretische Analyse von Argumenten für das Klassifizieren von Strahlungssorten 1650–1925 mit Schwerpunkt auf den Jahren 1770–1850.* Stuttgart: GNT-Verlag.

Jahn, Ilse. 2000. "Die Spezifik der romantischen Naturphilosophie." In Geschichte der Biologie. *Theorien, Methoden, Institutionen, Kurzbiografien.* Reprint of the third edition, 290–301. Heidelberg: Spektrum Akademischer Verlag.

———. 2006. Christian Gottfried Nees von Esenbeck und Matthias Jacob Schleiden: Zwei konträre Entwicklungsvorstellungen in der Botanik. In *Christian Gottfried Nees von Esenbeck*, ed. Daniela Feistauer, Uta Monecke, Irmgard Müller, and Bastian Röther , 123–138. Stuttgart: Wissenschaftliche. Verlagsgesellschaft.Acta Historica Leopoldina, nr. 47

Jahn, Ilse, and Isolde Schmidt. 2005. *Matthias Jacob Schleiden (1804–1881). Sein Leben in Selbstzeugnissen*, Acta Historica Leopoldina, nr. 44. Stuttgart: Wissenschaftliche Verlagsgesellschaft.

Jost, Ludwig. 1942. Matthias J. Schleidens 'Grundzüge der Wissenschaftlichen Botanik' (1842). *Sudhoffs Archiv für Geschichte der Medizin und der Naturwissenschaften* 35 (3/4): 206–237.

Kanz, Kai Torsten. 2002. Von der *Biologia* zur Biologie. Zur Begriffsentwicklung und Disziplingenese vom 17. bis zum 20. Jahrhundert. In *Die Entstehung Biologischer Disziplinen II*, ed. Uwe Hossfeld and Thomas Junker, 9–30. Berlin: VWB-Verlag.

———. 2006. '… die Biologie als die Krone oder der höchste Strebepunct aller Wissenschaften.' Zur Rezeption des Biologiebegriffs in der romantischen Naturforschung. *NTM* 14: 77–92.

———. 2007. Biologie: Die Wissenschaft vom Leben? Vom Ursprung des Begriffs zum System biologischer Disziplinen (17. bis 20. Jahrhundert). In *Lebenswissen. Eine Einführung in die Geschichte der Biologie*, ed. Ekkehard Höxtermann and Hartmut H. Hilger, 100–121. Rangsdorf: Natur & Text.

Ludwig, Friedrich. 1895. *Lehrbuch der Biologie der Pflanzen.* Stuttgart: Enke.

Mendelsohn, Everett. 1964. The Biological Sciences in the Nineteenth Century: Some Problems and Sources. *History of Science* 3 (1): 39–59.

Mill, John S. 1843. *A System of Logic, Ratiocinative and Inductive.* London: Parker.

Möbius, Martin. 1904. *Matthias Jakob Schleiden zu seinem 100. Geburtstag.* Leipzig: Engelmann.

Molisch, Hans. 1916. Obituary of Julius von Wiesner. *Berichte der Deutschen Botanischen Gesellschaft* 34: 71–99.

Morange, Michel. 2016. *A History of Biology.* Princeton/Oxford: Princeton University Press.

Müller, Gerd, ed. 2017. *Vivarium. Experimental, Quantitative, and Theoretical Biology at Vienna's Biologische Versuchsanstalt.* Cambridge, MA: MIT Press.

Musil-Gutsch, Josephine, and Kärin Nickelsen. 2020. Ein Botaniker in der Papiergeschichte: Offene und geschlossene Kooperationen in den Wissenschaften um 1900. *NTM* 28: 1–33.

Nickelsen, Kärin. 2015. *Explaining Photosynthesis: Models of Biochemical Mechanisms, 1840–1960.* Dordrecht: Springer.

———. 2017. Growth, Development, and Regeneration: Plant Biology in Vienna around 1900. In *Vivarium*, ed. Gerd Müller, 165–187. Cambridge, MA: MIT Press.

———. 2023. Julius Wiesner und die Biologie der Pflanzen. In *Von der Wissenschaft des Lebens zu den Lebenswissenschaften in Zentraleuropa*, ed. Mitchell Ash and Juliane Mikoletzky, 85–102. Berlin: LIT Verlag.

Nyhart, Lynn K. 2009. *Modern Nature: The Rise of the Biological Perspective in Germany.* Chicago: University of Chicago Press.

Przibram, Hans Leo. 1908. Die Biologische Versuchsanstalt in Wien. Zweck, Einrichtung und Tätigkeit während der ersten fünf Jahre ihres Bestandes (1902–1907). Bericht der Zoologischen, Botanischen und Physikalisch-Chemischen Abteilung. *Zeitschrift für Biologische Technik und Methodik* 1: 234–264.

Reiter, Wolfgang L. 1999. Zerstört und vergessen: Die Biologische Versuchsanstalt und ihre Wissenschaftler/innen. *Österreichische Zeitschrift für Geschichtswissenschaften* 4: 585–614.

Sachs, Julius. 1872. *Über den gegenwärtigen Zustand der Botanik in Deutschland. Rede zur Feier des 290. Stiftungstages der Julius-Maximilians-Universität.* Würzburg: Thein'sche Druckerei.

———. 1875. *Geschichte der Botanik vom 16. Jahrhundert bis 1860.* Munich: Oldenbourg.

Schickore, Jutta. 2007. *The Microscope and the Eye. A History of Reflections, 1740–1870.* Chicago: University of Chicago Press.

———. 2017. *About Method. Experimenters, Snake Venom, and the History of Writing Scientifically.* Chicago: University of Chicago Press.

———. 2019. The Structure and Function of Experimental Control in the Life Sciences. *Philosophy of Science* 86 (2): 203–218.

Schleiden, Matthias J. 1838. Beiträge zur Phytogenesis. In *Archiv für Anatomie, Physiologie und Wissenschaftliche Medicin,* ed. J. Müller, 136–176. Berlin.

———. 1842. *Grundzüge der wissenschaftlichen Botanik nebst einer methodologischen Einleitung als Anleitung zum Studium der Pflanze.* Vol. 1. Leipzig: Engelmann.

———. 1843. *Grundzüge der wissenschaftlichen Botanik nebst einer methodologischen Einleitung als Anleitung zum Studium der Pflanze.* Vol. 2. Leipzig: Engelmann.

———. 1845. *Grundzüge der wissenschaftlichen Botanik nebst einer methodologischen Einleitung als Anleitung zum Studium der Pflanze.* Second revised edition. Leipzig: Engelmann.

———. 1849a. *Die Botanik als inductive Wissenschaft.* Third revised edition. Leipzig: Engelmann.

———. 1849b. Principles of Botany, or: Botany as an Inductive Science. Translated by Edwin Lankester. London: Longman, Brown, Green & Longmans.

———. 1861. *Die Botanik als inductive Wissenschaft.* Reprint of the third edition. Leipzig: Engelmann.

Taschwer, Klaus, Johannes Feichtinger, Stefan Sienell, and Heidemarie Uhl, eds. 2016. *Experimentalbiologie im Prater. Zur Geschichte der biologischen Versuchsanstalt 1902 bis 1945.* Vienna: Österreichische Akademie der Wissenschaften Verlag.

Toepfer, Georg. 2011. Biologie. In *Historisches Wörterbuch der Biologie. Geschichte und Theorie der biologischen Grundbegriffe,* vol. 1, 254–295. Stuttgart: Metzler.

von Wurzbach, Constantin. 1888. Wiesner, Julius. In *Biographisches Lexikon des Kaiserthums Oesterreich,* vol. 56, 88–92. Vienna: k. k. Hof- und Staatsdruckerei.

Wiesner, Julius. 1875. Die natürlichen Einrichtungen zum Schutze des Chlorophylls der lebenden Pflanze. In *Verhandlungen der Zoologisch-Botanischen Gesellschaft in Wien, Festschrift 25 Jahre,* 19–49. Vienna: Braumüller.

———. 1877. *Zur Entstehung des Chlorophylls in der Pflanze.* Vienna: Hölder.

———. 1881a. *Das Bewegungsvermögen der Pflanzen: Eine kritische Studie über das gleichnamige Werk, von Charles Darwin nebst neuen Untersuchungen.* Vienna: Hölder.

———. 1881b. *Anatomie und Physiologie der Pflanzen. Vol. 1 of Elemente der wissenschaftlichen Botanik.* Vienna: Hölder.

———. 1884. *Organographie und Systematik der Pflanzen. Vol. 2 of Elemente der wissenschaftlichen Botanik.* Vienna: Hölder.

———. 1889. *Biologie der Pflanzen. Mit einem Anhang: Die historische Entwicklung der Botanik. Vol. 3 of Elemente der wissenschaftlichen Botanik.* Vienna: Hölder.

———. 1893. Photometrische Untersuchungen auf pflanzenphysiologischem Gebiete. Erste Abhandlung. Orientirende Versuche über den Einfluss der sogenannten chemischen Lichtintensität auf den Gestaltungsprocesss der Pflanzenorgane. *Sitzungsberichte der Akademie der Wissenschaften mathematisch-naturwissenschaftliche Klasse* 102: 291–350.

———. 1894. Bemerkungen über den factischen Lichtgenuss der Pflanzen. *Berichte der Deutschen Botanischen Gesellschaft* 12: 1078–1089.

———. 1897. Untersuchungen über das photochemische Klima von Wien, Cairo und Buitenzorg (Java). *Denkschriften der Kaiserlichen Akademie der Wissenschaften mathematisch-naturwissenschaftliche Klasse* 64: 73–166.

———. 1898. Beiträge zur Kenntnis des photochemischen Klimas im arktischen Gebiete. In *Offprint from the Denkschriften der Kaiserlichen Akademie der Wissenschaften mathematisch-naturwissenschaftliche Klasse*, vol. 67. Vienna: Gerolds Sohn.

———. 1907. Beiträge zur Kenntnis des photochemischen Klimas des Yellowstone-Gebietes und einiger anderer Gegenden Nordamerikas. *Denkschriften der Kaiserlichen Akademie der Wissenschaften mathematisch-naturwissenschaftliche Klasse* 80: 1–14.

———. 1910. Die Beziehungen der Pflanzenphysiologie zu den anderen Wissenschaften. In *Natur—Geist—Technik. Ausgewählte Reden, Vorträge und Essays*, 103–138. Leipzig: Engelmann.

Wininger, Salomon. 1933. Wiesner, Julius. In *Große jüdische National-Biographie*, vol. 6, 282–283. Cernăuți: Arta.

Kärin Nickelsen is Professor of History of Science at Ludwig-Maximilians-Universität, Munich. Her research focuses on the history and philosophy of biology since 1800, with particular interest in the epistemology of plant sciences, from natural history to experimental research.

Chapter 8
Controlling the Unobservable: Experimental Strategies and Hypotheses in Discovering the Causal Origin of Brownian Movement

Klodian Coko

8.1 Introduction

Brownian movement is the seemingly irregular movement of microscopic particles—of a diameter less than approximately 10^{-3} mm—of solid matter when suspended in liquids.[1] Although experimentally investigated in the nineteenth century, it was only at the end of that century that the phenomenon's importance was recognized for the kinetic-molecular theory of matter, i.e., the theory that matter is composed of atoms and molecules in incessant motion. Historians of science have expressed both surprise and lament that Brownian movement played no role in the early development and justification of the kinetic theory of gases. Today, we know that the movement is an observable effect of the molecules' motions constituting the liquid state of matter. If molecular motion had been identified from the beginning as the cause of the phenomenon, some of the most important philosophical and scientific objections raised against the early kinetic theory could have been answered. For example, the molecular explanation of Brownian movement could have resolved the nineteenth-century philosophical debates over the empirical status of molecular hypotheses, which centered on the question of whether the existence of unobservable entities such as atoms and molecules could be resolved by observation and

---

[1] *Brownian movement, mouvement Brownien, moto Browniano, Molecularbewegungen* were the terms used in the nineteenth century to refer to the movement of microscopic particles suspended in liquids. In this chapter, I use these same terms to describe the nineteenth-century investigations of this phenomenon. I avoid the term *Brownian motion,* which is more recent, and which already includes the randomness of the motions; it therefore has wider connotations. According to Encyclopedia Britannica, for example, "Brownian motion" concerns "various physical phenomena in which some quantity is constantly undergoing small, random fluctuations" (Britannica, March 21, 2023).

K. Coko (✉)
Philosophy Department, Ben-Gurion University of the Negev, Beersheba, Israel
e-mail: coko@post.bgu.ac.il

© The Author(s) 2024
J. Schickore, W. R. Newman (eds.), *Elusive Phenomena, Unwieldy Things,*
Archimedes 71. https://doi.org/10.1007/978-3-031-52954-2_8

experiment. In addition, Brownian movement could have provided independent empirical evidence for one of the theory's controversial claims: that at a molecular level, the Second Law of Thermodynamics had only statistical as opposed to absolute validity. Relatedly, it is often claimed that most nineteenth-century experiments on the nature and cause(s) of Brownian movement were less rigorous than later experiments, which successfully established molecular motion as the proper and unique cause (Brush 1968, 1; Nye 1972, 9; Maiocchi 1990).[2]

In this chapter, I focus on the experimental practices and the reasoning strategies used by nineteenth-century investigators of Brownian movement, in their quest to determine the phenomenon's causal origin. By focusing on these practices and strategies, we may better appreciate the century's investigative efforts in and of themselves, and not only insofar as they relate to later scientific and methodological developments. Nevertheless, this account presents some of the practical and conceptual complexities of the investigations on the cause of Brownian movement, which help to make sense of its delayed connection with the kinetic-molecular theory of matter. I argue that there was extensive and sophisticated experimental work done on the phenomenon of Brownian movement throughout the nineteenth century. Most investigators were aware of the methodological standards of their time and tried to align their work with them. The main methodological strategies they employed were two.

The first was the traditional strategy of varying the experimental parameters to discover causal relations. In the nineteenth century, this strategy was codified into explicit methodological rules by John Herschel (1830) and then, perhaps more famously, by John Stuart Mill ([1843] 1974). In nineteenth-century investigations of Brownian movement, we find that the reasoning underlying this strategy was already embedded in experimental practices prior to this codification, and independently of Herschel and Mill (see also the chapters by Schürch and Nickelsen, Chaps. 3 and 7 in this volume). More specifically, the basic rationale underlying these investigations was that: (a) all the circumstances and factors that could be introduced, varied, or entirely excluded without influencing Brownian movement, were not causes of the phenomenon; (b) all the circumstances and factors whose introduction, variation, or exclusion influenced the phenomenon were considered to play a causal role in its production. As mentioned in the introduction to this volume, employing this strategy required (implicitly or explicitly) at least three notions of control: (1) control over the introduction, variation, or exclusion of the circumstance or factor whose causal influence was to be examined; (2) control over the rest of the circumstances or factors, which ought to be kept as much as possible the same; and (3) control in the more familiar sense, of comparing the experimental situation after the intervention (i.e., the introduction, variation, or exclusion of the factor whose causal influence was being investigated) or with it, with the (control) situation before the intervention or without it (see also Boring 1954; Schickore 2019).

---

[2] These sentiments echo those of the historical actors who played important roles in connecting Brownian movement with the molecular theory of matter. See, for example, Perrin (1910) and Poincaré (1905).

The strategy of varying the circumstances succeeded more in excluding various suspected causal factors than in establishing a positive causal explanation. Even when some causal influence was detected, not all investigators shared the conclusion. Disagreements over the influence of various causal factors led to the recognition of the importance of a different notion of "control": that of the independent confirmation of experimental results by other researchers. Despite the difficulties surrounding its implementation, the strategy of varying the circumstances, by showing the insufficiency of the various causal explanations of Brownian movement, enhanced the importance of the fact that the newly developed kinetic-molecular conception of matter seemed to provide a plausible explanation of the phenomenon.

The second strategy was similar to what at the time was called the *method of the hypothesis*. This method, at least according to some scholars, re-emerged in the nineteenth century as the proper strategy for validating explanatory hypotheses about unobservable entities, processes, and phenomena (Laudan 1981). Amid all the criteria for evaluating explanatory hypotheses, the ability of a hypothesis to explain, successfully predict, and/or be supported by a variety of facts—especially facts playing no role in the hypothesis' initial formulation—was considered to be the most important criterion for its validity. Proponents of this strategy appealed to the ability of the kinetic-molecular hypothesis to offer a natural explanation of Brownian movement. What was remarkable about this explanation, they argued, was the fact that the elements of the hypothesis invoked to explain the phenomenon were developed independently of it. The ability of the kinetic-molecular hypothesis to explain a variety of unrelated phenomena and experimental evidence was offered, by some investigators, as an important "control" for the validity of the kinetic-molecular explanation of Brownian movement.

Neither methodological strategy could, on its own, establish molecular motion as the cause of Brownian movement. Their combination and their accompanying notions and practices of control, at the end of the nineteenth century, to the recognition of molecular motion as the most probable cause. From then on, the goal of experimental practices and reasoning strategies shifted to that of probing and evaluating the kinetic-molecular explanation of Brownian movement.

8.2   First Observations of the Curious Phenomenon

The phenomenon of Brownian movement owes its name to the Scottish botanist Robert Brown (1773–1858), who experimentally investigated it beginning in the summer of 1827 (Brown 1828). An already eminent botanist, Brown was not the first to observe the phenomenon. All earlier investigators, however, seem to have connected it with the motion of infusory animalculæ, and had attributed it to some sort of vitality possessed by the moving particles (Brown 1829, 164; Brush 1968). Brown's main contribution, and his claim to priority, lies in establishing that the movement of microscopic particles when suspended in liquids was a general phenomenon exhibited by all microscopic particles, independently of their chemical nature. We start, therefore, by examining the methodological ideas and practices Brown used to establish this claim.

Brown offered an account of his initial investigations in a pamphlet he originally circulated privately among his friends, but which aroused enough interest to appear, in 1828, in the *Edinburgh New Philosophical Journal*. It appeared soon afterwards in numerous other journals (Mabberley 1985). The pamphlet provides an interesting step-by-step account of his investigations. Brown was investigating the mechanism of fertilization in the plant *Clarckia pulchella*, whose grains of pollen were filled with microscopic particles of different sizes that were easy to observe with a simple microscope. "While examining the form of these particles immersed in water, I observed many of them evidently in motion; …These motions were such as to satisfy me, after frequently repeated observations that *they arose neither from currents in the fluid, nor from its gradual evaporation, but belonged to the particle itself* (Brown 1828, 162–63, my emphasis).

Brown extended his observations to particles derived from the pollen of plants belonging to different families, and found similar spontaneous movements when they were suspended in water. Having found these movements in the particles of pollen of all the living plants he examined, Brown inquired whether they continued after the death of the plant and for how long they were retained (Brown 1828, 164). Unexpectedly, he found that specimens of dead plants, some of which were preserved in an herbarium for no less than one hundred years, produced similar moving particles. Soon he discovered that the moving particles—or *active molecules*, as he began to call the smallest particles of apparently spherical shape not exceeding 1/15000 of an inch—were not limited to the grains of pollen, for they could also be produced from other parts of the plant as well. Even more surprisingly, however, Brown found that these molecules were not limited to organic matter but could be equally acquired in inorganic matter. He found that fragments of window glass, various minerals,

> [r]ocks of all ages, including those in which organic remains have never been found, yielded the molecules in abundance. Their existence was ascertained in each of the constituent minerals of granite, a fragment of the Sphinx being one of the specimens examined…In a word, in every mineral which I could reduce to a powder, sufficiently fine to be temporarily suspended in water, I found these molecules more or less copiously. (Brown 1828, 167)

The next step for Brown was to investigate whether the movement of the molecules derived from organic substances was affected by the application of intense heat on the substance from which they were derived. A comparative experiment was conducted. Small portions of wood (both living and dead), linen, paper, cotton, wool, silk, and hair were heated, and immediately quenched in water. In all cases molecules could be derived, and they were found to be as evidently in motion as those obtained from the same substances before burning (Brown 1828, 168).

To sum up, during these initial investigations, Brown used the seeming invariance of the suspended particles' movements to the variation or change of the suspected causal factors—namely, currents and evaporation in the suspending liquid, the chemical nature of the suspended particles, the application of heat on the particles' originating material—to conclude the causal independence of these move-ments from the varied factors.[3] As already mentioned, this strategy of varying the circumstances to discover causal dependencies involves at least three notions of control: (1) control over the variation of the suspected causal factor, (2) control over the remaining circumstances that should remain the same as much as possible,[4] and (3) control in the sense of comparing the experimental situation with the variation or after it to the situation without the variation or before it. Brown did not use the term "control," and these three notions of control are only implied in the description of his observations and experiments. In the rest of this chapter, we shall see that these and other forms of control became more explicit when the validity of the ini-tial observations was challenged.

The invariance of the movements to the variation of some of the suspected causal factors led Brown to exclude these factors as causes of the surprising phenomenon. But they could not help him identify a positive cause. His conclusions regarding the cause of the movements of the "active molecules" were cautious: "I shall not at present enter in any additional details, nor shall I hazard any conjecture whatever respecting these molecules, which appear to be of such general existence in organic as well as inorganic bodies" (Brown 1828, 169).

In the pamphlet presenting the results of his early research, Brown stated that he knew close to nothing about the phenomenon before beginning, and that he was only acquainted with the abstract of a memoir that the French botanist Adolphe Brongniart (1801–1876) had read before *l'Académie des Sciences* in Paris, in December 1826. The abstract was later published in the *Annales des Sciences Naturelles* (Brown 1828, 171–72; Brongniart 1827). Brongniart was also studying the process of fertilization in plants. Using an Amici microscope, which provided a magnification of up to 1050 times, Brongniart found that the microscopic granules contained in the pollen grains of numerous plants, or *granules spermatiques*, as he called them, performed clearly distinguishable spontaneous movements when sus-pended in water. The granules formed *la poussiere fecondant* (i.e., the most essen-tial part of the pollen fertilizing the ovum). These movements seemed impossible to attribute to an external cause (Brongniart 1827, 45). These observations corrobo-rated, according to Brongniart, his initial hypothesis that the spermatic granules found in the pollen of plants were analogous to the *spermatic animalculæ* found "swimming" in the sperm of animals (Brongniart 1827, 48).

As they were published in prestigious scientific journals, Brown's and Brongniart's observations drew great attention and elicited a strong reaction against the claim that the moving microscopic particles were self-animated.[5] The most influential critique came from the French physiologist François Raspail (1794–1878),

---

[3] This early use of the varying-the-circumstances strategy seems to be a case of what Steinle (2002, 2016) has identified as *exploratory experimentation*.

[4] Brown explicitly stated that, to give greater consistency to his statements, and to bring the subject as much as possible to the reach of general observation, he continued to use the same microscope with one and the same lens throughout his initial investigations (Brown 1828, 161).

[5] Brush (1968) provides an extended bibliography of these reactions.

who claimed that his conclusions on the subject were the result of many repeated and varied experiments (Raspail 1829a, b). First, Raspail attacked Brongniart's claim that the granules discharged in the explosion of grains of pollen were analogous to the spermatic animalculæ. His numerous experiments, argued Raspail, showed that the granules derived from the explosion of the grains of pollen, even those of the same plant, varied in shape, diameter, size, and other characteristics (Raspail 1829a, 97). This result challenged the claim that these granules were of an organized nature and that they belonged to a distinguishable category of entities. Second, Raspail rejected the claim that the movements of the particles suspended in water belonged to the particles themselves. He argued that the movements were easily distinguishable from the spontaneous movements of the infusory animalculæ, and that they could be attributed to the influence of various mechanical causes (Raspail 1829a, b). Raspail listed several such causes that, based on "a great number of consecutive observations" (Raspail 1829a, 97), could communicate even to the most inactive particles the appearance of spontaneous motion. The list included the motion communicated to the granules from the explosion of pollen discharging them, capillarity, the evaporation of the suspending water, the evaporation of the volatile substances with which the granules issuing from pollen may be impregnated, the ordinary motions of great towns, the motions caused by the air's agitation, the motions caused by the observer's hands, the inclination of the object plate, and the electricity communicated to particles of metallic origin by friction (Raspail 1829a, 97; b, 106–7).

Raspail's list proved to be influential. For the greater part of the nineteenth century it constituted the essential list of causes that, singly or in combination, were invoked to explain the movements of microscopic solid particles suspended in liquids. The list is also important because it reveals the difficulties surrounding the ascertainment of the concrete cause(s) of the observed movements by means of the experimental strategy of varying the circumstances. Such an experimental effort would require rigorous control over the many suspected causes and possible confounding factors.[6] Regarding his own methodological efforts, and faced with claims about the existence of spontaneous motion, Raspail maintained that, although his numerous earlier experiments on the subject had made him aware of the various contributing causal factors, he felt it incumbent on himself "to repeat all my experiments, and to vary them in every way, as if I had doubted the accuracy of my former ones" (Raspail 1829a, 99).

Replying to this criticism, Brongniart defended his original observations on both methodological and experimental grounds. Besides claiming that his conclusions were the result of repeated experiments performed on pollen from different kinds of plants, Brongniart appealed to another kind of experimental control: that of

---

[6] Schickore (2022) and Schürch (Chap. 3, this volume) provide detailed accounts of the difficulties surrounding the concrete applications of the varying-the-circumstances strategy in establishing causal claims.

independent confirmation by other researchers.[7] Independent confirmation, Brongniart asserted, was essential for the verification of claims concerning phenomena that were not readily observable and that contradicted in certain respects widely established theories.[8] Brongniart emphasized especially the fact that some of this confirmation came from research done without prior knowledge of his conclusions (Brongniart 1828, 392–93).[9] This specific kind of independent confirmation was important because it precluded the possibility that the other researchers had simply adjusted their conclusions to achieve consensus.[10] Among the claims that Brongniart maintained had been independently confirmed by other researchers were that the granules contained in the pollen of the same plant were of a well-determined form, that they had exactly measurable dimensions, and that each one performed extremely small motions which, because of their irregularities, seemed to be independent of any external cause (Brongniart 1828, 382). To these independently confirmed observations Brongniart added new ones conducted on twenty-four species of plants from different families. He also discussed new experiments that, he claimed, established without any doubt that the "spermatic granules" were different from the irregularly shaped particles of non-organized matter also found in the pollen of plants (Brongniart 1828, 386–88).

Regarding the movement of the "spermatic granules," Brongniart cited the irregular way they changed their positions relative to one another in order to argue that the movement was not caused by any external influences. It was instead dependent, he said, on a cause existing in the granules themselves (Brongniart 1828, 389). He too used the strategy of varying the circumstances to show that the movement continued without the smallest difference, even when some of the mechanical causes in Raspail's list—like the agitation of the liquid caused from evaporation, the trembling of ground or air, or the influence of sunlight—were either excluded or varied. More specifically, Brongniart burst the grains of pollen in very small glass capsules filled with a drop of water. He then covered the capsules with a thin film of mica to stop evaporation and the agitation of the water's surface. He conducted microscopy

---

[7] This kind of experimental control is discussed in detail in the chapters by Schürch, and Christopoulou and Arabatzis, Chaps. 3 and 9 in this volume.

[8] "Les phénomènes de la nature, qui s'éloignent de ceux qui frappent habituellement nos yeux, qui contredisent à quelques égards les systèmes fondés sur des observations anciennes et généralement reconnues; qui, par cette raison, sont d'ordinaire plus difficiles à saisir, exigent, pour être admis au nombre des vérités non contestées, des recherches souvent répétées, présentées avec ces détails qui éloignent toute espèce de doute, et vérifiées par de observateurs différens; car le concours des opinions d'hommes indépendans les uns des autres, est la seule preuve de la vérité pour ceux qui ne peuvent pas la rechercher eux-mêmes" (Brongniart 1828, 381–82).

[9] "Cette observation est d'autant plus curieuse qu'elle a été faite par une botaniste qui ne pouvait avoir à cette époque aucune connaissance des résultats auxquels l'examen du pollen des plants phanérogames m'avait amené; qui n'y était conduit par aucune théorie, et qui même, par ces raisons n'a pas pu sentir la liaison de ces phénomènes avec d'autres analogues" (Brongniart 1828, 393).

[10] For a discussion of this notion of (genetically) independent confirmation and its differences from other notions of independent confirmation see Soler (2012) and Coko (2020b).

observations of this preparation under the lamp light but also during cloudy days. Despite the measures taken to control (i.e., to exclude or lessen the influence of the suspected mechanical causes), the movements of the suspended granules continued without any difference. In contrast, when he replaced water with alcohol in the same experimental setting, the movements ceased completely instead of becoming livelier, as one would expect if they were caused by the liquid's evaporation (Brongniart 1828, 389–90).

Of special interest is the *note additionelle* to the paper which Brongniart wrote after learning about Brown's observations of the irregular movement of suspended particles derived from inorganic matter (Brongniart 1828, 393–98). Brongniart stated that Brown's observations prompted him to conduct new ones on suspended inorganic particles. These observations generally agreed with Brown's.[11] Because Brongniart initially claimed that the "spermatic granules" in pollen were analogues of the spermatic animalcules in the sperm of animals, and that they were clearly distinguishable in both their form and movement from the (irregularly shaped) microscopic agglomerations of matter also found in pollen, asserting agreement with Brown's observations was an exaggeration. In fact, even in the *note*, Brongniart continued to distinguish between the movement of the "spermatic granules" in pollen from that of inorganic particles. The movements of the inorganic particles seemed to him less constant and more dependent on the nature of the inorganic substance from which they were derived. In general, the movements were more evident in inorganic particles derived from substances that were better conductors of electricity. Despite the differences between his observations and Brown's, and in line with his previous assertion about the importance of independent confirmation, Brongniart was eager to emphasize the points of agreement. The most important one was the claim that the movements of both the spermatic granules and inorganic particles seemed to be caused by a force inherent in the particles and not by any external factors.[12] The crucial point, he continued, was to determine whether they were attributable to the same cause(s). In particular he wished to determine whether they were caused by the particles' vitality or by some hitherto unaccounted for internal factor or external influence (Brongniart 1828, 394–96).

---

[11] "Quant aux molécules des corps inorganiques, on observe en effet assez souvent, dans plusieurs substances broyées dans l'eau de très-petits corpuscules arrondis semblables aux plus petites molécules du pollen, et doués de mouvemens analogues en apparence à ceux des granules du pollen" (Brongniart 1828, 394).

[12] "La seule chose sur laquelle je ne puis conserver aucun doute, et sur laquelle j'ai le bonheur de voir mon opinion entièrement confirmée par celle des commissaires de l'Académie et de M. Brown, c'est l'indépendance complète de ce mouvement de toutes les causes extérieures influant sur le liquide ambiant. Il me paraît bien certain que la cause du mouvement, quelle quelle soit, réside dans une force physique ou organique inhérente aux corpuscules mêmes qui se mouvent. C'était la seule chose que j'avais avancée dans mes premières observations sur ce sujet, puisqu'en disant que ce mouvement était spontané, j'avais observé que j'entendais seulement exprimer par ce mot que ce mouvement était inhérent aux granules eux-mêmes" (Brongniart 1828, 396).

Brown rejected too the charge that his original memoir had implied that the moving suspended particles were animated (Brown 1829, 161–62). He also claimed to have conducted additional research on the subject, this time using different microscopes and different kinds of particles suspended in various liquids (Brown 1829, 162). The additional research, Brown asserted, confirmed the main results he had advanced in his 1828 pamphlet:

> that extremely minute particles of matter, whether obtained from organic or inorganic substances, when suspended in pure water, or in some other aqueous fluids, exhibit motions for which I am unable to account, and which from their irregularity and seeming independence resemble in a remarkable degree the less rapid motions of some of the simplest animalcules of infusions…I have formerly stated my belief that these motions of the particles neither arose from currents in the fluid containing them, nor depended on that intestine motion which may be supposed to accompany its evaporation. (Brown 1829, 162)

Brown cited the complete irregularity of the movements—i.e., the seemingly total independence in the movements of every two particles—to reject the various mechanical explanations of the phenomenon. In addition, he described two experiments demonstrating that the particles continued to move with their usual degree of activity even when the principal mechanical causes suspected of their motion were either reduced or completely excluded.

In the first experiment, Brown was able to isolate minute drops of water, some of them containing few or only one microscopic particle, in almond oil. In this manner, the drops, which if exposed to air would dissipate in less than a minute, were retained for more than an hour. But in all the drops, the motion of the suspended particles continued with undiminished activity. This was true even though the mechanical causes suspected for their motion, namely evaporation and the particles' mutual attractions and repulsions, were either reduced or entirely excluded.

In the second experiment, Brown was able to show that the motion of the particles was not produced by causes acting on the surface of the water-drop—e.g., currents in the surrounding liquid. Inverting his first experiment, he mixed a very small proportion of almond oil with the water drops containing the particles and was able to produce almond oil drops of extreme minuteness, some of them not exceeding the size of the particles themselves, attached to the surface of the water drops. The oil drops remained nearly or altogether at rest while the material particles isolated in the water drops continued to move with their usual degree of activity (Brown 1829, 163–64).

Brown and Brongniart's observations and experiments seemed to have aroused much interest over the cause of this curious phenomenon. Because many researchers at the time considered vitalist explanations questionable, the idea of the particles' vitality was rejected and, despite Brown and Brongniart's experimental efforts, various mechanical causes, singly or in combination, were proposed as explanations. In 1829 Georg Wilhelm Muncke from Heidelberg cited experimental research on the phenomenon to conclude that: "The movement certainly bears some resemblance to the one observed in Infusoria, yet the latter shows more voluntary action. Vitality, like many possibly have believed, is out of the consideration [as an explanation]. I rather consider the motion to be purely mechanical and caused by the uneven temperatures in strongly illuminated water, evaporation, air and heat currents, etc."[13]

These mechanical explanations of the phenomenon persisted, despite Brown's and Brongniart's experiments showing the phenomenon's invariance even when explicit measures were taken to control and/or exclude the influence of the relevant mechanical causes. It seems that one important factor was the impression that rejecting those causes would leave the particles' vitality as the only plausible explanation. For example, the renowned Scottish physicist David Brewster, then-editor of the *Edinburgh Journal of Science*, referred to the sufficiency of Raspail's mechanical causes to explain the motions of the suspended particles. He remarked that "even if they did not afford a sufficient explanation of the motions in question;—nay, if these motions resisted every method of explanation, it is the last supposition in philosophy that they are owing to animal life" (Brewster 1829, 219). For Brewster, an explanation showing that the motions of the suspended particles obeyed physical laws like the ones governing the motions of larger bodies would always take precedence over any hypothesis claiming the particles to be in some way animated (Brewster 1829, 219–20).

8.3   Experimental Investigations of Brownian Movement: 1830–1860

Despite the disagreement regarding the causal origin of the curious phenomenon, Brownian movement was not neglected during the period 1830–1860, as is sometimes claimed. In fact, what was neglected was rather the study of some of the investigators of the phenomenon by subsequent historiography of science. One of these neglected figures was Giuseppe Domenico Botto (1791–1865), professor of experimental physics at the University of Torino, who conducted experimental investigations into Brownian movement in the late 1830s (Guareschi 1913). Knowing the disagreements about the characteristics and causes of the phenomenon, Botto called for a cautious, purely experimental approach, and for a multiplication of experiments.[14]

In his own investigations, Botto found that the movement of suspended particles derived from organic matter had different characteristics from that of inorganic particles. Using an Amici horizontal microscope, Botto conducted extensive

---

[13] "Die Bewegung hat allerdings einige Aehnlichkeit mit der bei Infusorien wahrgenommenen, jedoch zeigt letztere mehr Willkühr. An Vitalität, wie vielleicht Einige geglaubt haben, ist dabei gar nicht zu denken, vielmehr halte ich die Bewegung für rein mechanisch, und zwar durch ungleiche Temperatur des stark erleuchteten Wassers, durch Verdampfung desselben, durch Luftzug und Wärmeströmung u. s. w. Erzeugt" (Muncke 1829, 161).

[14] "Au milieu de ces contradictions, et dans un sujet aussi important et complexe, ce qu'il y a de mieux à faire, est de multiplier les expériences, sans franchir trop à la légère les limites de l'observation" (Botto 1840, 459).

microscopic observations on suspended microscopic globules derived from different plants, vegetable products, and inorganic substances. In all his observations of suspended microscopic globules derived from vegetable matter, Botto found the phenomenon exhibited in the manner described by Brown: "one sees them changing their relative positions every moment, approaching one another, receding from one another, spinning, as if these movements originated on their own."[15] The lively oscillatory movement was invariantly found on suspended globules derived from all the parts of the individual plant: the grains of pollen, the ovary before and after fertilization, the pistil, the stamen, the anther, the buds, the tubers, the seeds, and so on (Botto 1840, 465). However, Botto argued, the globules derived from pollen had a vivacity of motion not encountered in globules derived from other parts of the plant. Such lively motion, he claimed, qualified as the effect of a spontaneity peculiar to animal nature. Botto proceeded to investigate the influence of various chemical substances and physical agents on the movement of organic globules suspended in water. He found that a small quantity of ammonia ceased almost all movement. Sulfuric, nitric, and hydrochloric acids, as well as opium, produced similar deadening effects. The application of strong heat and electricity on the suspending liquid also immobilized the moving globules (Botto 1840, 462).

Contrary to Brown, Botto claimed that the movement of suspended inorganic particles had different features from that of organic globules: "Neither powdered glass, neither quartz, nor the granite of our Alps, nor the pebbles of our rivers, nor rocks of any kind, offered particles endowed with movements analogous to those of organic globules. I could not either certify their presence anymore in the organic substances after carbonization or incineration."[16] The explanation for the movement of the inorganic particles by familiar mechanical causes seemed to him to be "neither impossible nor difficult" (Botto 1840, 467). On the other hand, the movement of the organic globules could not be explained by known physical causes. It must, therefore, be considered a proper quality of the globules themselves, and of their organic and vital nature (Botto 1840, 468). Botto's research shows that vitalist claims, although distrusted by most researchers, remained a viable option, at least for the movement of organic particles. Although these observations did not seem to have much influence on subsequent Brownian movement research, they are important from a historiographical point of view. Once again, they reveal the difficulties in applying the varying-the-circumstances strategy for reaching consensus on the causal influence various factors had on the phenomenon.

One of the most widely accepted explanations of Brownian movement during this period was offered by Felix Dujardin (1801–1861). Although he used similar methodological reasoning, Dujardin reached entirely different conclusions

---

[15] "On les voit changer à chaque instant de position relative, s'approcher, s'éloigner, tournoyer, comme ci ces mouvements venaient de leur propre fait" (Botto 1840, 459).

[16] "Ni le verre pilé, ni le quartz, ni le granit de nos Alpes, ni les cailloux de nos rivières, ni les roches de toute espèce ne m'ont offert de globules doués de mouvements analogues à ceux des globules végétaux. Je n'ai pas pu en constater non plus la présence dans les substances végétales après la carbonisation ou l'incinération" (Botto 1840, 466–467).

from Botto regarding the generality of the phenomenon, the influence of physical agents such as heat and electricity, and the cause. Dujardin gave his view in his influential treatise on microscopy, in a chapter titled "Du Mouvement Brownien ou Mouvement Moléculaire" (Dujardin 1843, 58–60). This chapter followed one that expounded some of the main causes of illusions and errors in microscopy observations. It seems that disagreements regarding the basic features and causes of the phenomenon invited reflection about possible sources of error. Dujardin cited the phenomenon's invariance amid the influence of various physical and chemical agents—light, electricity, magnetism, chemical reagents—to argue that the movement was a purely physical phenomenon, belonging to all particles of solid matter sufficiently small to be suspended in liquids. In fact, he wished to warn the uninitiated observer who might perceive in it the manifestation of life and other kinds of organic activity (Dujardin 1843, 59–60). Studying oil globules suspended in milk, Dujardin found that the vivacity of the movements depended on the particles' size. The smallest particles, of a radius of less than 1/600 mm, moved the most vigorously, those of radii of between 1/400 and 1/300 mm showed movement noticeable only if one observed carefully, whereas those of larger size remained motionless. He also found the movement to be livelier as the density of the material from which the suspended particles were derived was less than that of water (Dujardin 1843, 59). Dujardin claimed heat as the only physical agent affecting the phenomenon: it caused the movements to become more rapid. Reflecting on these results, he concluded that the movements of the suspended particles could be attributed to the various impulses that each particle receives from the radiant heat emitted by the particles adjacent to it.[17]

Dujardin's views on the cause of Brownian movement were shared by Griffith and Henfrey in Britain and were included in their *Micrographic Dictionary* (Griffith and Henfrey 1856). Like Dujardin's treatise, the *Dictionary* too began with a methodological introduction concerning the proper use of microscopes and the main sources of errors in their employment. The remarks on Brownian motion were included in the entry *Molecular Motion*—where the term "molecule" refers to extremely minute particles of any substance. Although the entry suggests it was based on original experimental work, it was in fact a summary of Dujardin's text, with the part referring to the probable causes of motion being simply the English translation of Dujardin's words.[18]

---

[17] "si l'on chauffe le liquide, le mouvement devient notablement plus vif, et comme tout autre agent physique ou chimique, la lumière, l'électricité, le magnétisme, le contact des réactifs chimiques ou des divers solides est sans influence sur le mouvement Brownien, on est conduit à penser que c'est le résultat des impulsions variées que chaque particule reçoit de la part du calorique rayonnant émis par tous les corps voisins" (Dujardin 1843, 59–60).

[18] "Heat is the only agent which affects it [molecular motion]; this causes the motion to become more rapid. Hence it may be attributed to the various impulses which each particle receives from the radiant heat emitted by those adjacent" (Griffith and Henfrey 1856, 429).

In 1858, Jules Regnauld (1822–1895), physics professor at the École de Pharmacie in Paris, cited extensive experimental work on the phenomenon to conclude that Brownian movement was caused by the solar heat absorbed in suspended particles. When transferred to the surrounding liquid, this heat created very small currents responsible for the observed motions.[19]

Those investigating the phenomenon during these earlier phases of experimental research failed to agree on its essential characteristics and the influence of the various suspected causal factors. To clarify the disagreements, it would be useful to distinguish between *causal claims* and *causal explanations* made regarding the causal origin of Brownian movement.[20] A causal claim asserted the identification of a "difference-maker," i.e., the causal influence of a suspected factor—evaporation, heat, electricity, and so on—on the movement of the suspended particles. By changing or varying the suspected causal factors, the experimental strategy of varying circumstances tried to identify a difference-maker and thus make a causal claim. A causal explanation of Brownian movement, on the other hand, aimed at providing a more or less detailed account of a concrete mechanism linking a causal factor with the effect, i.e., the observed Brownian movements. A causal explanation was more speculative than a causal claim because its details could not be established by varying the circumstances. Causal claims, however, could identify the difference-maker, which could then be used to offer a probable causal explanation of the observed movements.

Early experimental investigations of Brownian movement failed to reach consensus in identifying a difference-maker. This was to be expected, given the difficulties with the varying-the-circumstances strategy in such a complex phenomenon. Even when reaching agreement on the influence of some (macroscopic) agent, like heat, on the movement of the suspended particles, researchers still disagreed about the exact mechanism by which this agent, at the microscopic level, produced the observed movements. In the rest of this chapter, we see various permutations of the relationship between causal claims and causal explanations in the nineteenth-century investigations of Brownian movement.

---

[19] "M. J. Regnauld est porté à conclure que les oscillations des corps très-divisés nageant au sein d'un liquide diathermane sont dues à leur échauffement par la portion de la radiation solaire que, absorbée par eux, les rend visibles. Cette faible quantité de chaleur se transmettant par voie de de conductibilité au liquide en contact avec les particules semblé la cause de petits courants rendus manifestes par les changements de position relative des substances tenues en suspension" (Chatin 1858, 141).

[20] In making this distinction, we are following Russo and Williamson (2007), who claim that a causal connection can be established only if it can be shown (a) that there is a difference-making relationship between the cause and the effect, and (b) that there is a mechanism linking the cause and the effect responsible for the difference-making relationship.

8.4 Non-molecular Causal Claims and Explanations of Brownian Movement: 1860–1880

The explanation of Brownian movement by the absorption and radiation of heat turned out to be quite popular. In Britain, a prominent defender of the view was John Benjamin Dancer (1812–1877), a microscopist from Manchester. Dancer claimed to base his conclusions on experiments performed over 30 years with various substances and solutions (Dancer 1868, 162). He asserted that the intensity of the movements depended on the size and shape of the particles as well as on the nature of the solutions. The particles approaching a spherical shape usually exhibited a more marked movement. To further support his claim, Dancer excluded chemical and electrical influences as causes. This he did by demonstrating that the particles showed no marked alterations in their movements when exposed to electric and chemical influences (Dancer 1868, 164; Jevons 1870, 83).

Dancer's claim went against another popular view in Britain regarding the causal origin of the movement, which presented it mainly as an electric phenomenon. The most prominent defender of this claim was William Stanley Jevons (1835–1882), the British philosopher and polymath. Jevons coined the name *pedesis* from the Greek πήδησις (meaning "leaping" or "bounding"), and the adjective *pedetic* from πηδητικός, as more appropriate for describing the dancing movement of the suspended particles. The term *molecular movement* used by Brown was inadequate because the moving particles were not molecules in the new chemical sense, whereas the term *Brownian movement* was an inconvenient two-word expression which, in addition, concealed the fact that Brown was not the first to observe the phenomenon (Jevons 1878, 171). Jevons too claimed that his conclusion was the result of extended experimental investigations (Jevons 1870).

In looking for its cause, Jevons conducted observations and experiments to test the validity of the various available causal claims (i.e., claims in the sense of identifying a difference-maker). First to be tested and disproved was the claim, by Dancer and others, that the movement was caused or excited by light or heat falling on the liquid. Working with particles derived from substances such as kaolin (or China clay, as it was known at the time), road dust, and red oxide of iron suspended in distilled water, Jevons found that their vibratory movements were the same both in relative darkness and in intense sunlight. The movements showed no apparent change even when differently colored glass screens were interposed between the liquid and the sunlight (Jevons 1878, 172). He reached the same conclusion by means of a comparative experiment. Two suspensions of China clay in water were taken, with one placed in a dark environment and the other exposed to the sun's direct rays for 3 hours. He saw no difference in the rapidity of subsidence of the particles (Jevons 1878, 172). Regarding the influence of heat in particular, Jevons' conclusions were surprisingly opposite to those of previous researchers. He thought that the increase of temperature decreased the motion. Jevons perceived no difference in the movements of the suspended particles when he warmed the microscope plate. He then tried a comparative experiment. A mixture of charcoal-powder and boiled water was surrounded with ice, while a similar mixture in boiling water was maintained at 100 °C. At the end of the hour the heated mixture had deposited nearly all the charcoal, whereas the ice-cold water had as much in suspension after 8 hours. A similar experiment with suspensions of China clay gave similar results. Trying to explain these surprising results, Jevons surmised that they were produced by the increase of electrical conductivity of liquids caused by the rising temperature (Jevons 1878, 173).

Jevons called these comparative experiments "indirect," but not because he sensed a difference in their epistemic import compared with traditional experimental intervention, where the comparison is between the situation before and the situation after an intervention or variation of circumstance. He called them indirect because, rather than investigating the effect of light and heat on the vibratory movements, the comparative experiments looked at how these agents affected the particles' rate of subsidence. In other words, Jevons ascertained the association of pedesis with the suspension of particles in water and then performed comparative experiments investigating the influence of various factors on the particles' suspension, rather than on their movement.

The comparative experiments, however, differed from traditional experimental interventions (or variations) with respect to their epistemic role.[21] Jevons used the comparative experiments to investigate the longer-term effects of the change or variation of the suspected cause, as opposed to its instantaneous or immediate effects. This difference in epistemic role manifests in another (indirect) comparative experiment, which convinced Jevons that no causes external to the suspending liquid were involved in the production of pedesis. Trying to test the effect of light and heat, Jevons took a suspension of China clay in water and frequently heated it in fire for 2 days, allowing it to cool at various intervals. A similar suspension was sunk in sawdust that had been undisturbed for several years in a wine-cellar. After remaining for 52 hours in complete darkness at a constant temperature of 9 °C, the second preparation was found to contain more clay in suspension than the first, which had been moved and heated many times. Even after 7 days the buried preparation "showed a slight cloudiness" (Jevons 1878, 173).

Another time-sensitive question was whether pedetic motion exhausted itself rapidly or was retained for a long time. Jevons found that ink many months or even years old exhibited the motions clearly. A slow, distinct motion of suspended particles was observed in a drop of lees from a wine bottle that had been undisturbed in a wine-cellar for several years. The drop was placed under the thin glass cover of the microscope with the least exposure to air. The motion did not increase when some of the dregs were shaken in a bottle with air. The most surprising and conclusive fact of this investigation, however, came from a comparative experiment. Old mixtures of China clay and water were compared with fresh ones. Two glass tubes containing China clay and distilled water were laid in a drawer for a long period of time.

---

[21] In her contribution to this volume, Schürch also discusses how eighteenth-century researchers investigating the influence of electricity on plant growth perceived the difference between comparative and intervention-based experimentation (see also Bernard 1856, 80–82).

The drawer was usually opened several times in a day, so the tubes would be shaken every now and then. Frequently the two tubes were shaken by hand. At long intervals the old tubes were opened and drops of the milky liquid were examined. Comparing the motion of the suspended China clay particles in the old mixtures with the motion of newly mixed particles found that "*no diminution of motion was apparent*; on the contrary, *the motion seemed to be even more remarkable than in a fresh mixture*" (Jevons 1878, 174; emphasis in original). This comparative trial lasted for 9 years and led Jevons to declare pedetic motion "the best approach yet discovered to perpetual motion" (Jevons 1878, 174).[22]

To investigate the relation of the movement with the shape of the particles, Jevons compared under a microscope "the fine needle-shaped particles of asbestos dust with the spherical globules of milk, the minute spheres of gamboge, the flat particles of talc, the small cubes of galena, and the wholly irregular fragments of glass." Given that all the differently shaped particles exhibited pedesis, he concluded that no particular shape was essential to its production. Contrary to Dancer, however, Jevons found that, *ceteris paribus*, sharp-pointed and irregularly shaped particles oscillated more quickly than spherically shaped particles (Jevons 1878, 173–74).

Jevons considered inconclusive all experiments rejecting the relevance of electricity for pedetic motion because external electrical currents applied to the liquid had no effect on the movements of the suspended particles. His conclusion that pedesis was caused by electricity was based on experiments that placed more weight on the variations of suspending liquid's chemical nature. He did not learn much by varying the nature of the suspended particles, finding that particles from substances of the most different chemical character exhibited similar pedetic motion (Jevons 1870, 78; 1878, 176). In varying the chemical nature of the liquid by dissolving various substances therein, however, he discovered that only the purest distilled water showed the movements in their highest perfection. With a few exceptions, all acids, alkalis, or salts tended to diminish the movement, but in a manner that was wholly independent of their peculiar chemical qualities and dependent only on their electric properties (Jevons 1870, 79; 1878, 179). More specifically, what convinced Jevons that pedesis was caused by electric action was the close analogy between his

---

[22] In *Against Method* ([1975] 1993), Paul Feyerabend used the example of "Brownian motion" to support the claim that empirical facts are not simply "given" but that the description of every single fact depends on *some* theory; in addition, some empirical facts cannot be unearthed except with the help of alternative theories to the one being tested. More specifically, Feyerabend claimed that *without the introduction of the kinetic theory*: (a) it is not clear whether the relevance of Brownian motion for the phenomenological second law of thermodynamics could have been discovered, and (b) it is certain that it could not have been demonstrated that Brownian motion actually *refutes* the phenomenological second law (Feyerabend ([1975] 1993, 27). Jevons' longer-term comparative experiments show that the relevance of Brownian movement for the phenomenological second law could be perceived without considering the kinetic theory. In addition, as we show in this chapter, the nineteenth-century investigations of Brownian movement, which ended up demonstrating the persistence of the phenomenon despite the variation of the factors external to the suspending liquid, make it less certain that an experimental investigation of Brownian movement could not, by itself, pose a challenge to the phenomenological second law.

findings when varying the chemical nature of the liquid, and the circumstances in which electricity was produced by the hydro-electric machine. Only pure water produced the greatest amount of electricity in the hydro-electric machine, and almost any salt, acid, or alkali prevented production by rendering the water a conductor (Jevons 1870, 79–80).[23] Pure caustic ammonia, a substance that, remarkably, did not render water a good conductor and did not prevent the hydro-electric machine from giving electricity, was used in a crucial experiment. Jevons dissolved ammonia in water in different amounts and found that it had no effect on the movement of the microscopic suspended particles (1870, 79–80). He emphasized that his conclusions were based on a great number of experiments done with suspended particles from different substances, and they involved a great number of substances dissolved in the suspending water in various amounts. All the variations in the chemical nature of the suspending liquid, with only few "doubtful exceptions," showed that dissolved substances turning the water into a conductor also inhibited pedetic motion. Jevons distinguished his causal claims regarding the relevance of electricity for the phenomenon—which he regarded as more or less certain, because they were based on a large number of observations and experiments[24]—from his more speculative explanations regarding the mechanism of electric action on the suspended particles. More specifically, regarding the exact *modus operandi* of the electric action, Jevons speculated that it was probably connected with the phenomenon of electric osmose (Jevons 1878, 183).

In later experiments, Jevons used a solution of common soap to decide between the causal claim of electric action and the newly proposed claim that asserted that pedesis was caused by surface tension in water (Jevons 1878, 175; 1879). Soap could serve as a crucial substance for deciding between the two alternative claims because it reduces the surface tension of water in which it is dissolved without affecting its electric conductibility. If pedesis was caused by surface tension, reasoned Jevons, then the motion of the suspended particles would be destroyed or diminished when soap was dissolved in the suspending water. He tried the experiment with particles derived from China clay, red oxide of iron, chalk, barium carbonate, etc., and it gave the opposite result: the pedetic motion of the suspended particles appeared to increase. For Jevons the experiment constituted further proof that pedesis was a phenomenon of electric origin, appearing only in liquids of high electric resistance (Jevons 1879, 435).

---

[23] "The analogy of these circumstances to those of pedesis is so remarkable that little doubt can be entertained that the same explanation applies. *It is perfectly pure water which produces electricity and pedesis.* Almost all soluble substances prevent both one and the other; but ammonia is one of a few exceptions—it allows both electric excitation and pedesis. Boracic acid is another exception, and gum a third one" (Jevons 1878, 182; emphasis in original).

[24] "My recorded observations amount to nearly eight hundred, and the solutions named were tried not only in different strengths, varying according to circumstances, from one part in ten to one part in a million, but they were tried with various suspended powders, such as charcoal, red oxide of iron, amorphous phosphorous, precipitated carbonate of lime, red oxide of lead, black oxide of manganese, and occasionally with other substances. I don't think, then, that I can be much mistaken in my chief conclusions" (Jevons 1878, 180).

Jevons' conclusions regarding the cause of pedesis were challenged, in turn, by William Ord. Ord preferred retaining the term "Brownian movement," because "everyone knows at once knows what is meant when Brownian movements are spoken of, and, what is of no little importance, the term is extensively used in the continent" (Ord 1879, 656). Although not aware of Jevons' experimental work before its publication, Ord claimed to have independently repeated and confirmed some of his experimental findings, such as the hindering action of acids on the movement of the suspended particles (Ord 1879, 658–60). While he admitted that heat, electricity, capillary action, water's surface tension, and chemical and other forces may each or all play a part in producing Brownian movements, Ord claimed its main cause to be "vibrations or intestinal disturbances in the colloid suspending fluid, such as attend its decomposition, or its metamorphosis or its resolution into a crystalloid" (Ord 1879, 658).[25]

This conclusion was based on reasoning similar to Jevons'. Ord found that the Brownian movements were more active and persistent under conditions that favored the activity of chemical changes in the suspending fluid; conversely, the movements were diminished or altogether stopped by introducing conditions that hindered such chemical reactions. Ord explicitly stated that he used, what Mill had recently named as, *the method of concomitant variations* and *the method of difference* to support his induction. Regarding the first, he found that "the concomitant variations set forth" showed "that the movement of particles is more or less active according to the presence in the surrounding fluid of conditions favouring or hindering chemical changes in the colloid" (Ord 1879, 660). Ord claimed he used the method of difference in studying mixtures of India-ink with distilled water.[26] When the solid ink was rubbed gently with water, a mixture of suitable thickness was obtained, consisting of particles of solid black matter suspended in water that was now dissolving the colloid matter binding the ink particles. On the other hand, when a large quantity of ink was rubbed with water, and the mixture left in a tall vessel to allow the subsidence of particles, the colloid matter was gradually washed away, leaving a mixture of particles with nearly pure water. When compared with particles of the same size and number in the first mixture, particles in the second showed less active and persistent movement (Ord 1879, 660).

Finally, Ord reinterpreted Jevons's experiments with solutions of soap in a way that supported his own conclusion. Whereas for Jevons introducing soap into the suspending fluid increased the movements of the suspended particles because soap retained or did not conduct electricity, for Ord it was a colloid that kept up the movements by revolutionary perturbations (Ord 1879, 660–61).

---

[25] "To sum up…I claim the intestine vibration of colloids as in many cases an agent in the process, and more especially in the fluid and semi-fluid parts of animal and vegetable organisms" (Ord 1879, 662).

[26] "I may cite an experiment in which the method of difference gives results in the same direction" (Ord 1879, 660).

8.5 Brownian Movement and Atomic-Molecular Theories of Matter: Early Investigations

According to historian of science Mary Jo Nye, a major reason for the delayed connection of Brownian movement with a molecular conception of matter was that, until the second half of the nineteenth century, there was no atomic theory of matter capable of offering a suitable mechanism to causally connect the atomic-molecular structure of liquids with a phenomenon having the characteristics of Brownian movement. Atomic theories prior to the middle of the century offered a static conception of atoms that interacted with one another primarily through acting-at-a-distance attractive and repulsive forces (Nye 1972, 46; Gouy 1895, 5).

Nye is right to observe that the explanation of Brownian movement in terms of the molecular motions constituting the liquid state of matter required a molecular theory capable of offering a suitable mechanism explaining how the cause (molecular motions) produced the effect (observed Brownian movements). We should acknowledge, however, the complexity of the nineteenth-century relationship between the ability to make a causal claim regarding Brownian movement, and the ability to provide a causal explanation of it, as noted at the end of Sect. 3. So far, we have seen that most nineteenth-century investigators of Brownian movement began with the experimental strategy of varying the circumstances aiming to identify a difference-maker (i.e., a causal circumstance influencing the phenomenon). In a second step, some of them speculated about a (more or less) concrete mechanism that, by linking the difference-making circumstance with the observed Brownian movements, was responsible for the experimentally detected difference-making relationship. In the rest of the chapter, I examine some of the permutations of the relationship between causal claims and causal explanations emerging in the efforts to connect the observed Brownian movements with an atomic-molecular theory of matter during the second half of the nineteenth century.

The first to explicitly connect Brownian movement with an atomic theory of matter was Christian Wiener (1826–1896), professor of descriptive geometry and geodesy at the University of Karlsruhe. In fact, Wiener used the phenomenon of Brownian movement (*Molecularbewegungen*) to provide support for his atomic theory of matter (Wiener 1863). Wiener's atomic theory was a hybrid between the older static conception of atoms and the newer kinetic conceptions, which were beginning to emerge at the time. According to Wiener, matter is composed of matter atoms, which attract one another, and aether atoms, which repel one another. The aether atoms are found in the empty spaces between the mutually attracting matter atoms, with aether and matter atoms repelling each other (Wiener 1863, 79). The network of forces exerted between matter and aether atoms meant that matter was in a state of permanent vibration. *Molecularbewegungen*—the trembling motion of microscopic particles suspended in liquids—was then the result of the constant vibrational atomic motions constituting the liquid state of matter (Wiener 1863, 85). Wiener supported his causal explanation of Brownian movement not by providing independent (empirical) evidence for it, but by rejecting other alternative claims about the causal origin of *Molecularbewegungen*.

Lacking positive evidence for his atomic explanation, Wiener used the strategy of varying the circumstances to experimentally disprove, one by one, (all) other alternative causal claims (Wiener 1863, 86). First, Wiener argued, the motion could not be that of infusoria or caused by the vitality of the particles, because he could observe it in finely divided suspended particles derived from inorganic matter. To reject the possibility that the moving particles derived from inorganic substances were actually organic particles trapped in inorganic matter, Wiener annealed quartz particles and found that this had no effect on their movements when suspended in liquids. This same possibility was also excluded by the fact that all the suspended particles exhibited the movements, as opposed to just a few (Wiener 1863, 86). Second, the movement was not caused by mechanical or any other external influences communicated to the suspending liquid. The movements of the suspended particles were more like vibrations, and no one had ever observed such irregular, tremulous movements being caused by external influences. In addition, if the movements were caused by external influences, they ought to change or decrease with time. But Wiener's microscopy observations, made over many days, revealed an incessant movement showing no signs of decrease (Wiener 1863, 86). Third, the movement could not be caused by attractive or repulsive forces, electric or otherwise, between the suspended particles. This was because it was independent both of the number of particles present in the liquid and of the distances between them. Suspended particles in a dilute emulsion and in relatively large distances from one another exhibited the same trembling motion as that of many particles close together (Wiener 1863, 87). Fourth, the movement could not be caused from temperature differences between the different parts of the liquid. These temperatures differences would offset or decrease with time, whereas the main characteristic of the particles' trembling motion was its invariance through time. In addition, the temperature differences would produce currents from the surface to the interior of the liquid and could not explain the trembling motion of the particles, which constantly changed direction even in very small volumes. If the temperature differences were the cause of the trembling motion, the motion would have to increase its liveliness when the environment temperature was changed abruptly. But no changes in the movement were observed despite sudden temperature changes in the surrounding environment (Wiener 1863, 87–89). Fifth, the movement was not caused by evaporation, because evaporation usually takes place near the surface of the liquid, whereas Wiener's microscopy observations revealed that the movement of the suspended particles occurred at all levels of the liquid, and it continued in the same manner even when measures to preclude any evaporation were taken (Wiener 1863, 89–90).

In short, Wiener excluded all the plausible causal claims that could provide the empirical basis for an alternative causal explanation of Brownian movement. He did this by showing that the phenomenon remained invariant when each of the suspected causal factors was either varied or entirely excluded from influencing the phenomenon. He concluded that the exclusion of all these suspected difference-makers left no other explanation besides the one attributing Brownian movement to the vibration of the atoms constituting the liquid state of matter: "It remains nothing left but for us to seek the cause [of the phenomenon] in the liquid, and *to ascribe it to the movements constituting the liquid state.*"[27]

Another investigator who connected Brownian movement with a mechanical theory of heat was Giovanni Cantoni, professor of experimental physics at the University of Pavia (Cantoni 1867). Cantoni's investigations on the phenomenon, like those of Botto, were ignored by his contemporaries and rediscovered only by the efforts of the historian Icilio Guareschi in the beginning of the twentieth century (Guareschi 1913).[28] Cantoni saw in the phenomenon of Brownian movement (*moto Browniano*) the confirmation of a mechanical theory of heat.

For Cantoni, the heat of a body consists in the vibratory movements of its constituent molecules. Every chemical substance, at a given temperature, has a characteristic vibratory motion of its constituent molecules. This was macroscopically indicated by the fact that different amounts of heat are required to increase by the same degree of temperature the same weight of different substances (i.e., by the existence of the different substances' specific heats).

According to Cantoni's proposed explanation, Brownian movement was caused by the different molecular velocities that must exist at the same temperature between the molecules constituting the solid suspended particles, on the one hand, and the molecules of the suspending liquid hitting the suspended particles from every direction, on the other.[29] Cantoni argued that this explanation could be experimentally tested and positively confirmed: *ceteris paribus*, Brownian movements ought to be livelier the greater was the difference between the velocities of the molecules constituting the solid particles from the velocities of the molecules constituting the suspending liquid. At the macroscopic level, the difference between the molecular velocities of different substances was simply the difference between their specific heats (Cantoni 1867, 163). If the difference between molecular velocities was the real cause of Brownian movements, then varying the difference between the specific heat of the suspended particles and the specific heat of the suspending liquid ought to bring a corresponding variation in the intensity of Brownian movements. Cantoni claimed that his numerous experiments, performed with various suspended particles and suspending liquids, showed that this was indeed the case. For example, particles derived from the same substance moved far more intensely in water than in alcohol. Because alcohol has a lower specific heat than water, there was a smaller difference between the specific heat of the suspending liquid and that of the suspended

---

[27] "[E]s bleibt uns daher Nichts übrig, als die Ursache in der Flüssigkeit an und für sich zu suchen, und sie *inneren dem Flüssigkeitszustande eigenthümlichen Bewegungen zuzuschreiben*" (Wiener 1863, 90, emphasis in original).

[28] According to Guareschi (1913, 50), Cantoni was the first to clearly discover the true cause of the phenomenon.

[29] "Ebenne, io penso che il moto di danza delle particelle solide estremamente minute entro un liquido, possa attribuirsi alle differenti velocità che esser devono ad una medesima temperatura, sia in codeste particelle solide, sia nelle molecole del liquido che le urtano d'ogni banda" (Cantoni 1867, 163).

particles. Following similar reasoning, one could explain why the Brownian movement of identical particles was even less marked in gasoline and ether than in water
(Cantoni 1867, 163–67). All this evidence, according to Cantoni, led to the conclusion that the cause of the phenomenon resided in the different velocities the molecules of different substances have at the same temperature. From here Cantoni
inferred that the existence of Brownian movement provided one of the most beautiful and direct experimental demonstrations of the fundamental principles of the
mechanical theory of heat, manifesting the assiduous vibratory state that must exist
both in liquids and solids, even when their temperature does not change.[30]

Wiener's atomic explanation of Brownian movement was based on the rejection of all other alternative causal claims. For Wiener, the rejection of all possible macroscopic difference-makers left no other explanation than the one
attributing the movement of suspended particles to the vibratory movements of
aether and matter atoms. These movements, according to Wiener's atomic theory, constituted the liquid state of matter. Embedded as it was in an idiosyn-
cratic theory of matter that had no independent empirical evidence in its favor,
Wiener's explanation was deemed inadequate. Cantoni, on the other hand,
explained Brownian movement in terms of the different molecular velocities
that, according to his molecular theory of heat, must exist at the same temperature between the molecules of the suspended particles and the molecules of the
suspending liquid. In contrast with Wiener's, Cantoni's explanation manifested
itself in a macroscopic difference-making relationship that could be experimentally manipulated to provide empirical support. Cantoni's work, however, did
not receive any attention and thus had no influence on subsequent research
(Guareschi 1913). To my knowledge, even the difference-making relationship
detected by Cantoni was not replicated by anyone else. One possible reason for
the neglect of Cantoni's explanation may have been his peculiar mechanical
theory of heat, which contradicted some of the basic tenets of the newly developed and more successful kinetic-molecular theory (see next section). The main
obstacle facing all (kinetic-) molecular explanation of Brownian movement during this period, however, was the emergence of arguments challenging the adequacy of the hypothesized molecular motions to cause a phenomenon with the
observable characteristics of Brownian movement (Nye 1972, 23; Nägeli 1879;
Ramsay 1882).

---

[30] "Ora tutti gli esposti particolari concorrano alla deduzione, che la condizione fisica del moto
browniano stia nella diversa velocità che hanno le molecole dei corpi differenti sotto una stessa
temperatura. E di tal modo il moto browniano, così dichiarato, ci fornisce una delle più belle e
dirette dimostrazioni sperimentali dei fondamentali principii della teoria meccanica del calore,
manifestando quell' assiduo stato vibratorio che esser deve e nei liquidi e nei solidi ancor quando
non si muta in essi la temperatura" (Cantoni 1867, 167).

8.6 Brownian Movement and the Kinetic-Molecular Theory of Matter

In this section, I examine the reasoning of the researchers who first explicitly connected the phenomenon of Brownian movement to the thermo-dynamic motion of molecules, as proposed in the recently developed kinetic theory of gases. These were a group of Jesuit scholars associated with the journal *Revue des questions scientifiques*, published by the Scientific Society of Brussels (Nye 1976). These proponents of the kinetic-molecular explanation did not start by varying the circumstances to exclude alternative causal claims and/or identify difference-makers. They tried to show that that the tenets of the kinetic-molecular conception of matter, which were developed independently to explain a different range of observable phenomena—namely the macroscopic behavior of gases and liquids—could give a causal explanation for the altogether different phenomenon of Brownian movement. The ability of the kinetic-molecular theory to account for a range of unrelated phenomena and experimental evidence was used to "control" its validity as well as the validity of the offered explanations.[31]

The first explicit connection of Brownian movement with the kinetic theory of gases was made by Father Joseph Delsaulx, a Brussels-born Jesuit, in a paper whose aim was to show "that all the Brownian motions of small masses of gas and of vapour in suspension in liquids, as well as the motions with which viscous granulations and solid particles are animated in the same circumstances, proceed necessarily from the molecular heat motions, universally admitted, in gases and liquids by the best authorized promoters of the mechanical theory of heat" (Delsaulx 1877, 2).

Delsaulx gave a detailed account of how the invisible molecular motions, postulated by the kinetic theory of heat to explain the macroscopic behavior of gases, would cause the dancing movement of microscopic particles suspended in liquids. More specifically, it followed from the principles of the mechanical theory of heat that a favorable concourse of the movements of oscillation, rotation, and translation of the molecules of the suspending liquid would, by necessity, produce a pressure of an exceptional intensity at isolated points on the surface of a suspended particle. These pressures were averaged out in particles of larger dimensions, but not in the microscopic dimensions of Brownian particles. They were thus the real cause of the particles' continuous oscillatory motions (Delsaulx 1877, 3–6). "All these [Brownian] movements," Delsaulx concluded, "result from the interior dynamic state that the mechanical theory of heat attributes to liquids, and are a remarkable confirmation of it" (Delsaulx 1877, 5).

The kinetic-molecular explanation of Brownian movement could make sense of the phenomenon's observed features: Brownian movement is more active in heated liquids than in those of a low temperature; supposing equal diameters, the oscillatory displacement is more rapid and more extended in fatty granulations than in

---

[31] This way of reasoning is similar to that which we encounter in William Whewell's (1847, 1858) notion of the *consilience of inductions*. See also Coko (Forthcoming).

metallic granulations, whose density is very great; and the duration of the phenomenon may be said to be without limit, because it has been observed in gas-bubbles imprisoned in microscopic (liquid-filled) cavities of quartz for supposedly millions of years (Delsaulx 1877, 2).

In a lengthy 1880 paper, another Belgian Jesuit, Julien Thirion, similarly argued that Brownian movement could be easily explained by the mechanical theory of heat. According to that theory, explained Thirion, all bodies are composed of molecules in a perpetual state of motion. Although these molecular motions cannot be directly observed, various phenomena and surprising experimental facts could be easily explained by appeal to their existence (Thirion 1880, 6). For instance, the new and surprising experimental facts established in William Crookes' experiments on cathode rays could be readily explained by the tenets of the kinetic-molecular conception of gases, as proposed in the mechanical theory of heat. What made this explanation even more remarkable, Thirion claimed, was the fact that the kinetic-molecular conception of gases was originally developed to explain a totally different range of phenomena—the macroscopic behavior of gases. The simplicity with which the kinetic-molecular conception accounted for these unexpected facts, the fruitfulness of the insights it suggested, and the variety of evidence it predicted and explained gave the conviction that one was not mistaken in taking it as a guide.[32]

Thirion used Brownian movement as another example of a surprising phenomenon that could be explained by the tenets of the mechanical theory of heat. Thirion explained that the theory predicted that sufficiently small particles suspended in water would be in a state of permanent oscillation. According to the mechanical theory of heat, the surface of a solid body suspended in a liquid is continually and unequally bombarded by the movement of the unobservable molecules constituting the liquid state of matter. In large particles with sufficiently large surfaces, the inequalities of molecular collisions would compensate for one another. In these particles, therefore, despite their high irregularity, the molecular collisions would produce no visible effects. In very small particles, however, surfaces would be sufficiently small that irregularities could not be compensated for. The result would be that the total pressure exerted at any moment from the molecular collisions would no longer be zero, but would vary continuously in intensity and direction. The particle's center of gravity would be continuously displaced and so the particle would oscillate continuously. The inequalities in pressure and the resulting oscillations would be more and more apparent the smaller the suspended particles were (Thirion 1880, 43–45). For Thirion, the phenomenon of Brownian movement was a remarkable empirical verification of this prediction by the kinetic-molecular conception of liquids. What made the prediction even more remarkable was the fact that the

---

[32] "Si cette science maîtresse avait encore besoin de preuves, il nous semble qu'elle les trouverait ici solides et nombreuses. La simplicité avec laquelle elle rend compte de ce grand nombre de faits inattendus, la fécondité des aperçus qu'elle suggère, la variété des détails qu'elle prévoit et qu'elle explique, donnent à l'esprit la conviction qu'il ne s'est point fourvoyé en la prenant pour guide" (Thirion 1880, 39).

molecular conception of liquids was not developed to accommodate this kind of phenomenon. It was a happy coincidence that such a phenomenon could be detected experimentally.[33]

8.7 Brownian Movement and the Kinetic-Molecular Theory of Matter: Controlling the Evidence and the Kinetic-Molecular Hypothesis

The French physicist Louis Georges Gouy (1854–1926) is credited as the first to firmly connect Brownian movement with the molecular motions postulated by the kinetic-molecular theory of matter.[34] In this section, I show that Gouy's success stems from the fruitful combination of the experimental strategy of varying the circumstances with the theoretical and hypothetical reasoning on the causal origin of the phenomenon. More specifically, Gouy (a) used the invariance of Brownian movements to the variation of various suspected factors to reject claims identifying the cause with influences external to the suspending liquid, and (b) showed how hypotheses regarding the internal constitution of liquids—which were developed independently in the context of the kinetic theory of matter, and which were already employed successfully to explain various phenomena—were sufficient to explain the experimental facts of Brownian movement.

Gouy performed many experiments on the phenomenon during the late 1880s and was able to conclusively establish its essential features. He presented his results in a short note published in the *Journal de Physique* (Gouy 1888). He claimed that Brownian movement was characteristic of all microscopic solid particles suspended in liquids. Initially he worked with suspensions of gamboge and China ink in water. The water-drop containing the particles was covered with a slip, and the preparation was enclosed with paraffin to avoid evaporation and external influences. Using an immersion lens, Gouy observed a striking trembling motion of the suspended particles. Every particle seemed to move independently of its neighbors, and experienced a series of displacements difficult to describe because they were

---

[33] "[C]e ne sont pas des phénomènes qui se présentent à nous et qu'il faut expliquer, *ce sont des conséquences d'une théorie édifiée pour expliquer d'autres phénomènes*. Si l'expérience venait à montrer que ces conséquences ne se vérifient pas, il en faudrait conclure que la théorie est au moins inexacte, peut-être tout à fait erronée. *Heureusement* l'expérience fait tout le contraire" (Thirion 1880, 41–42, my emphasis). In addition, "Tous ces faits, observés par R. Brown, peuvent vraiment être considérés comme une vérification anticipée d'un théorème trouvé un demi-siècle plus tard" (Thirion 1880, 50).

[34] "On the contrary, it was established by the work of M. Gouy (1888), not only that the hypothesis of molecular agitation gave an admissible explanation of the Brownian movement, but that no other cause of the movement could be imagined, which especially increased the significance of the hypothesis. This work immediately evoked a considerable response, and it is only from this time that the Brownian movement took a place among the important problems of general physics" (Perrin 1910, 4–5). See also Poincaré (1905, 199).

essentially irregular. In particles with elongated form or some mark in their surface, Gouy detected an irregular rotational movement. The movements were more vivid the smaller the size of the particles, they increased with temperature, and they were more active in less viscous liquids (Gouy 1888, 561–62).

The careful observation of the phenomenon left no doubt, according to Gouy, that the movements were not the result of vital forces, external vibrations, temperature differences, or other accidental currents in the liquid. Rather, they were a normal phenomenon, occurring at a constant temperature, and attributable to the internal constitution of liquids. The independence of the movements from the nature of the particles; their irregular nature; their persistence in time even when precautions to exclude all external influences were taken—all of these results showed the cause to be the internal agitation of the liquid. Brownian movement provided a "direct and visible" proof of the molecular-kinetic hypotheses regarding the nature of heat: "Brownian movement, therefore, shows us, of course not the movement of molecules, but something very close to it, and it provides us a direct and visible proof of the correctness of current hypotheses on the nature of heat. If one adopts these views, the phenomenon, whose study is long from over, surely takes a higher order of importance for molecular physics."[35]

Gouy's (1889, 1895) next two papers on the topic present his experimental strategy and theoretical reasoning in more detail. He experimentally identified the phenomenon's essential characteristics and inquired into its causal origins. Brownian movement, he remarked, was essentially irregular and seemed to be governed only by chance. It consisted in a series of little impulses that were oriented indistinguishably in all directions and that were not subject to any law. The movement was a sort of oscillation in place, although in the long run it could produce noticeable displacements in a suspended particle's position. The rapidity and amplitude of the movement depended above all on the size of the particles, becoming greater as the particles got smaller. The movement was not influenced by the form, the state, or the chemical and physical nature of the suspended particles. It was more intense in suspending liquids with greater degrees of fluidity. Although the movement was irregular, with each particle moving independently of its neighbors, the phenomenon as a whole had an obvious regularity, in that it was always found exhibiting the same essential characteristics (Gouy 1895, 2–3). Gouy claimed that he had observed the movements under the most varied conditions using liquids and particles with different chemical and physical properties, but did not notice any difference in its essential features.[36] Regarding the question of the causal origin of Brownian

---

[35] "Le mouvement brownien nous montre donc, non pas assurément les mouvements des molécules, mais quelque chose qui y tient de fort près, et nous fournit une preuve directe et visible de l'exactitude des hypothèses actuelles sur la nature de la chaleur. Si l'on adopte ces vues, le phénomène, dont l'étude est loin d'être terminée, prend assurément une importance de premier ordre pour la physique moléculaire" (Gouy 1888, 563).

[36] "Les observations ont été faites avec des particules minérales ou organiques, solides ou liquides, en suspension dans des liquides variés, eau, solutions aqueuses, acides, alcools, éthers, carbures d'hydrogène, essences, etc. D'autres observations ont été faites sur les bulles gazeuses que renfer-

movement, Gouy was explicit that it "can only be answered by a detailed study of the phenomenon, under the most varied circumstances possible, by striving to reduce or increase at the outmost limits the external causes of agitation and examining the resulting effects."[37] That is, the question could be answered only using the varying-the-circumstances strategy.

First, Gouy claimed that it was easy to show that Brownian movement was not of a vital nature, because it had been observed in liquids where no living entity could exist: toxic substances, acids, and the strongest alkalis never stopped the movements. Indeed, temperatures high enough to destroy life increased the movements instead of stopping them (Gouy 1895, 2). Second, the phenomenon's generality, and the fact that it seemed to last indefinitely—it appeared in air bubbles suspended in liquids in cavities of quartz crystals for thousands of years—was sufficient to show that it was not attributable to any external and accidental causes. For those must act with a varying intensity depending on the circumstances (Gouy 1889, 103). To establish this last point decisively, however, Gouy conducted several rigorous experiments. To test claims about the causal origin of Brownian movement, Gouy examined how its essential characteristics changed while varying or excluding the different suspected causes. His detailed descriptions show the effort toward controlling the influence of disturbances external to the suspending liquid. The first claim to be tested was whether the Brownian movements were caused by external vibrations communicated to the suspending liquid, or undetected tremors coming from the ground. To avoid external disturbances, he installed the microscopy apparatus in a basement away from any source of agitation. To control for ground tremors or any external vibrations, he placed a basin of mercury next to the apparatus. The mercury's surface acted as a perfect mirror of extreme sensibility for detecting the slightest disturbances. While the mercury remained undisturbed, the Brownian movement continued showing its usual characteristics and intensity; the movement did not increase significantly when external disturbances were noticeable. Based on similar, often repeated, experiments Gouy concluded that external vibrations or ground tremors were not causes of the phenomenon (Gouy 1889, 103–4; 1895, 4).

The second claim to be tested was whether the Brownian movements were caused by currents in the liquid as a result of temperature differences. Gouy reduced these currents by immersing the preparation in a water trough, which ensured the attainment of a uniform temperature. He used an immersed lens for observation and saw no variations in the Brownian movement of the suspended particles during the

---

mement les inclusions liquides fréquentes dans certains quartz, et qui sont animées d'un mouvement tout à fait comparable à celui des particules solides ou liquides…. Le point le plus important est la régularité du phénomène des milliers de particules ont été examinées, et, *dans aucun cas*, on n'a vu une particule en suspension qui n'offrît pas le mouvement habituel, avec son intensité ordinaire, eu égard à la grosseur de la particule" (Gouy 1889, 103, my emphasis). See also (Gouy 1895, 2–3).

[37] "A la question ainsi posée, on ne peut répondre que par l'étude détaillée du phénomène, dans des conditions aussi variées que possible, en s'efforçant de réduire ou d'augmenter dans les limites le plus étendues les causes extérieures d'agitation, et examinant les effets produits" (Gouy 1895, 4).

entire procedure. In addition, currents in the liquid produced coordinated movements of adjacent Brownian particles, but they looked nothing like the individual vibrations constituting Brownian movement (Gouy 1889, 104; 1895, 4).

A third claim was whether the light required for the microscopy observations, affected the particles as it passed through the liquid—by heating them unequally, for example. The individual vibrations of the particles would then be the result of such temperature differences. To test this claim Gouy varied the nature and the intensity of light used to illuminate the preparation, and observed no difference in the particles' movements. Light, he concluded, played no perceptible role on Brownian movement (Gouy 1889, 104; 1895, 4–5).

Fourth, Gouy contended that other hypothetical causes, such as terrestrial magnetism and electric currents, had no influence on Brownian movements. For he observed no variation when placing the preparation in an electromagnetic field or when applying electric currents. The only agent to influence the movement was heat. At temperatures of 60° to 70 °C, the movement was a little more noticeable than at temperatures (Gouy 1889, 4; 1895, 5).

Gouy explicitly used the term "control" to indicate that his observations and experimental results could be easily verified independently and were, therefore, independent of any theoretical idea and interpretation:

> These observations which are easy to control, seem to establish as experimental facts and apart from any theoretical idea: *1st that Brownian movement occurs with any kind of particles, with an intensity that is the lesser the more the liquid is viscous and the more the particles are larger; 2nd that this phenomenon is perfectly regular, it occurs at a constant temperature and in absence of any external cause of movement.* (Gouy 1889, 104–5)[38]

Leaving the solid ground of observation and experiment, Gouy entered the second part of his argument, which relied on hypothetical and theoretical reasoning for the causal origin of Brownian movement. Theories and hypotheses, contended Gouy, have been abused and slandered, but their importance for scientific inquiry is indisputable. They may shed unexpected light on many questions. In addition, the history of the physical sciences showed that theoretical speculations have been the source of the finest discoveries and the greatest progress. The use of hypotheses was thus legitimate as long as they were used cautiously and *controlled* by empirical evidence: "Let's give them their due, the consideration deserved by eminent services, and that limited confidence that never sleeps and does not neglect any means of control."[39]

---

[38] "Ces observations qu'il est facile de contrôler, paraissent établir comme faits d'experiénces et en dehors de toute idée théorique: *1° que le mouvement brownien se produit avec des particules quelconques, avec une intensité d'autant moindre que le liquide est plus visqueux et les particules plus grosses; 2° que ce phénomène est parfaitement régulier, se produit à température constante et en absence de toute cause du mouvement extérieur*" (Gouy 1889, 104–5, emphasis in original).

[39] "Accordons leur ce qui leur est dû, la considération que méritent des services éminents, et cette confiance limitée qui ne s'endort jamais et ne néglige aucun moyen de contrôle" (Gouy 1895, 5).

Gouy argued that the cause of Brownian movement, which lasted indefinitely without an apparent cause, should not be sought in the nature of the particles or in any external factors. Rather it was to be found in the constitution of the suspending liquid itself. In fact, the hypotheses made in the context of the modern kinetic theory of matter were directly related to the phenomenon's explanation. More specifically, "the kinetic theory could make us predict this phenomenon, and it *explains* it to us in its essential features" (my emphasis).[40]

After showing how the kinetic-molecular hypotheses could explain the experimentally determined features of Brownian movement, Gouy conceded that the kinetic-molecular explanation faced a problem of underdetermination. It assumed that there were no unknown causes of which the Brownian movement could be an effect. He maintained, however, that supposing such causes was unnecessary if the kinetic-molecular hypotheses were sufficient to explain it. In addition, the hypotheses were not entirely beyond all means of control. They had already led to considerable insights about a variety of physical and chemical phenomena.[41] Among the successes of the kinetic theory Gouy listed the molecular explanations for heat and radiation. Furthermore, agreement on the numerical values for molecular dimensions, obtained by diverse theoretical methods, gave the kinetic theory's claims an aura of plausibility.[42]

To sum up, Gouy used the experimental strategy of varying the circumstances (a) to identify the essential characteristics of Brownian movement, (b) to identify macroscopic difference-makers that influenced its intensity—heat and the size of the Brownian particles—and (c) to exclude other factors as possible causes. The strategy left kinetic-molecular motions as the only plausible explanation. Although he admitted the problem of underdetermination, Gouy appealed (a) to the ability (or necessity, as Gouy saw it) of the kinetic-molecular motions to produce a phenomenon with the observable characteristics of Brownian movement and (b) to the plausibility of the kinetic-molecular conception of matter, given its ability to explain a variety of other phenomena. These arguments made unnecessary the appeal to other (unknown) causal factors and thus eased the underdetermination problem.

This summary of Gouy's reasoning helps us to make sense of his contention that "Brownian movement provides us with what the kinetic theory of matter was lacking: a direct experimental proof. No doubt, we cannot observe, and we will never be

---

[40] "La théorie cinétique pouvait nous faire prévoir ce phénomène, et elle nous l'explique dans ses traits essentiels" Gouy 1895, 7).

[41] "La théorie cinétique de la matière a conduit à des aperçus fort intéressants sur un certain nombre de phénomènes physiques et chimiques, et la part qu'elle a prise dans l'œuvre scientifique de notre époque est déjà considérable" (Gouy 1895, 6).

[42] "C'est aussi la conclusion à laquelle sont arrivés par d'autres voies les physiciens qui ont essayé de se faire une idée des dimensions moléculaires. Par des méthodes diverses, assez concordantes pour qu'on leur accorde crédit, ils sont arrivés à évaluer l'intervalle des molécules dans les liquides à la millième partie environ des dimensions des plus petits corps visibles au microscope. Il faudrait donc environ un milliard de molécules pour former le poids d'une de plus petites particules sur lesquelles nous observons le mouvement brownien" (Gouy 1895, 7).

able to observe the molecular movements; but at least we can observe something which results directly from them and necessarily indicates an internal agitation of bodies."[43]

This synthesis of experimental and theoretical modes of reasoning was perfected in Jean Perrin's (1870–1942) experimental work, which established molecular motions as the *proper and unique* cause of Brownian movement. Perrin determined by means of multiple, independent experiments that the internal motions of the liquid causing the experimentally established characteristics of Brownian movement were identical with the molecular motions postulated in the kinetic theory of matter (Coko 2020a). The multiple determination of molecular magnitudes proved to be the ultimate criterion for "controlling" the veracity of the kinetic-molecular explanation of Brownian movement.

8.8   Summary and Conclusions

In this chapter, I have argued that there was important and sophisticated experimental work done throughout the nineteenth century to investigate the characteristics and causal origin of Brownian movement. Investigators followed as rigorously as possible the methodological standards of their time to make causal claims and formulate causal explanations. They used two distinct methodological strategies.

The first was the experimental strategy of varying the circumstances. Suspected causal factors were varied to study the resulting effect on Brownian movements. The main goal of this strategy was to identify difference-making factors (i.e., factors having a causal influence on the phenomenon). All factors that could be varied without influencing the suspended particles' movement were excluded from playing a causal role in its production. On the other hand, all factors whose variation influenced the phenomenon were considered to have a causal role. The identification of a difference-making factor was sometimes followed by theoretical speculation about the concrete mechanism linking the difference-making factor with the observed movements.

This strategy was already implemented in the earliest identifications and investigations of the phenomenon—at first implicitly, and later, when the initial observations were challenged or led to conflicting results, more explicitly. The varying-the-circumstances strategy involved three notions of control: (1) control over the factor to be varied, (2) control over the rest of the factors which had to remain constant, and (3) control in the sense of comparing the situation with the varied factor to the experimental situation without it. We can distinguish two types of experimentation employing this strategy. First, there was "classic" (or direct)

---

[43] "[L]e mouvement brownien nous fournit ce qui maquait à la théorie cinétique de la matière: une preuve expérimentale directe. Sans doute, nous ne voyons pas et nous ne verrons jamais les mouvements des molécules; mais nous voyons du moins quelque chose qui en résulte directement et suppose d'une manière nécessaire une agitation interne des corps" (Gouy 1895, 7).

experimental intervention, where the comparison was between the situation before and the situation after the intervention (or variation of the investigated factor). Second, there was comparative experimentation, where the comparison was between two distinct experiments that were made to vary only with respect to the investigated factor. Although no distinctions between these two types of experimentation were made with respect to their underlying rationale and epistemic import, the second kind was used to investigate effects of longer duration, as opposed to instantaneous and immediate effects. It was also used in cases where direct intervention was not possible.

Using the varying-the-circumstances strategy did not lead to consensus regarding the essential characteristics and causal origin of Brownian movement. Disagreements revealed the importance of another notion of "control": that of the verification of experimental findings by other researchers, preferably independently from one another. Because most claims regarding the causal origin could not be verified independently, the strategy succeeded more in excluding various suspected factors as causes than in establishing a positive causal claim. Brownian movement proved to be what we would call today a *robust phenomenon*, remaining invariant to the variation of most experimental factors that the experimenters could directly vary and control. Today, with hindsight, we know why. Even when the causal influence of some factor, such as heat or particle size, made a difference for the observed movements, and received independent confirmation, investigators disagreed on the causal explanation offered. That is, they disagreed over how to describe the concrete mechanism responsible for the difference-making relationship.

The second strategy was the hypothetico-deductive strategy or *method of hypothesis*, recognized during the nineteenth century as the proper approach for validating explanatory hypotheses regarding unobservables. Rather than starting or relying exclusively on experimental work to identify difference-making factors or exclude alternative causal claims, its proponents tried to show that the tenets of the recently developed kinetic-molecular conception of matter provided a natural explanation for the essential characteristics of Brownian movement. What was remarkable about this explanation, researchers claimed, was the fact that the elements of the theory explaining Brownian movement were developed independently to explain an entirely different range of observable phenomena—the macroscopic behavior of gases and liquids. Seen in this vein, the existence of Brownian movement provided unexpected empirical evidence for the kinetic-molecular conception of matter. The ability of the kinetic-molecular theory to account for a range of unrelated phenomena and experimental evidence was therefore used to "control" its validity as well as the validity of the offered explanation.

Neither methodological strategy could, on its own, establish molecular motion as the cause of Brownian movement. It was only the combination of the two and their accompanying notions and practices of control that led, at the end of the nineteenth century, to the recognition of molecular motion as the most probable cause of the phenomenon. From then on, the main goal of experimental investigation on Brownian movement became that of evaluating and probing the validity of the kinetic-molecular explanation. This shift in goals wrought changes in the experimental strategies for establishing the validity of claims about unobservable entities and processes such as molecules and molecular motion. These changes also changed the understanding of what is meant by "rigorous" experimental research.

Acknowledgments Participants in the international workshop *Rigor: Control, Analysis, and Synthesis in Historical and Systematic Perspectives* provided helpful feedback on an earlier version of this chapter. Special thanks go to Christoph Hoffmann, Jutta Schickore, and Caterina Schürch for their detailed written comments. This research was supported by Israel Science Foundation (grant number: 1943/20).

References

Bernard, Claude. 1856. *Introduction à l'étude de médecine expérimentale*. Paris: J.B. Baillère et fils.

Boring, Edwin Garrigues. 1954. The Nature and History of Experimental Control. *American Journal of Psychology* 67: 573–589.

Botto, Giuseppe Domenico. 1840. Observations Microscopiques sur les Mouvements des Globules Végétaux Suspendus dans un Menstrue. *Memorie della Reale Academia delle Scienze di Torino* 2: 457–471.

Brewster, David. 1829. Observations relative to the Motions of the Molecules of Bodies. *Edinburgh Journal of Science* 10: 215–220.

Britannica, The Editors of Encyclopaedia. 2023. Brownian motion. In *Encyclopedia Britannica*. https://www.britannica.com/science/Brownian-motion. Accessed 23 April 2023.

Brongniart, Adolphe. 1827. Mémoire sur la Génération et le Développement de l'Embryon dans les Végétaux phanérogames. *Annales des Sciences Naturelles* 12: 14–53.

———. 1828. Nouvelles Recherches sur le Pollen et les Granules spermatiques des Végétaux. *Annales des Sciences Naturelles* 15: 381–393.

Brown, Robert. 1828. A brief account of microscopical observations made in the months of June, July and August 1827, on the particles contained in the pollen of plants; and on the general existence of active molecules in organic and inorganic bodies. *Philosophical Magazine Series 2* 4 (21): 161–173. https://doi.org/10.1080/14786442808674769.

———. 1829. Additional remarks on active molecules. *The Philosophical Magazine* 6 (33): 161–166. https://doi.org/10.1080/14786442908675115.

Brush, Stephen G. 1968. A History of Random Processes: I. From Brown to Perrin. *Archive for History of Exact Sciences* 5 (1): 1–36.

Cantoni, Giovanni. 1867. Su Alcune Condizioni Fisiche dell'Affinità e sul Moto Browniano. *Il Nuovo Cimento* 27 (1): 156–167.

Chatin. 1858. Extrait du Procès-verbal de la séance de la Société de pharmacie de Paris, du 7 Juillet 1858. *Journal de Pharmacie et de Chimie* XXXIV: 137–142.

Coko, Klodian. 2020a. Jean Perrin and the Philosophers' Stories: The Role of Multiple Determination in Determining Avogadro's Number. *HOPOS: The Journal of the International Society for the History of Philosophy of Science* 11: 143–193.

———. 2020b. The Multiple Dimensions of Multiple Determination. *Perspectives on Science* 4: 505–541.

———. Forthcoming. Hypothesis and Consilience in the Nineteenth Century. In *History and Philosophy of Modern Science: 1750–1900*, ed. Elise Crull and Eric Peterson. Bloomsbury Press.

Dancer, John Benjamin. 1868. Remarks on Molecular Activity as shown under the microscope. *Proceedings of the Manchester Literary and Philosophical Society* 7: 162–165.

Delsaulx, Joseph. 1877. Thermo-dynamic Origin of the Brownian Motions. *The Monthly Microscopical Journal* 18: 1–7.

Dujardin, Felix. 1843. *Nouveau Manuel Complet de l'Observateur au Microscope*. Paris: Manuels-Roret.

Feyerabend, Paul. [1975] 1993. *Against Method*. London: Verso.

Gouy, Louis Georges. 1888. Note sur le Mouvement Brownien. *Journal de Physique* 7: 561–564.

———. 1889. Sur le Mouvement Brownien. *Comptes Rendus* 109: 102–105.

———. 1895. Le Mouvement Brownien et les Mouvements Moléculaires. *Revue Générale des Sciences* 6: 1–7.

Griffith, John William, and Arthur Henfrey. 1856. *The Micrographic Dictionary: A Guide to the Examination and Investigation of the Structure and Nature of Microscopic Objects*. London: John Van Voorst.

Guareschi, Icilio. 1913. Nota sulla storia del movimento browniano. *Isis* 1 (1): 47–52.

Herschel, John F.W. 1830. *Preliminary Discourse on the Study of Natural Philosophy*. London: Longman, Rees, Orme, Brown, and Green.

Jevons, William S. 1870. On the So-called Molecular Movements of Microscopic Particles. *Proceedings of the Manchester Literary and Philosophical Society* 9: 78–84.

———. 1878. On the Movement of Microscopic Particles Suspended in Liquids. *The Quarterly Journal of Science, and Annals of Mining, Metallurgy, Engineering, Industrial Arts, Manufacturers, and Technology* VIII: 167–186.

———. 1879. Note on the Pedetic Action of Soap. In *Report of the Forty-Eighth Meeting of the British Association for the Advancement of Science*. London: John Murray.

Laudan, Larry. 1981. *Science and Hypothesis: Historical Essays on Scientific Methodology*. Springer.

Mabberley, David J. 1985. *Jupiter Botanicus: Robert Brown of the British Museum*. London: Lubrecht & Cramer Ltd.

Maiocchi, Roberto. 1990. The Case of Brownian Motion. *The British Journal for the History of Science* 23 (3): 257–283.

Mill, John S. [1843] 1974. *Collected Works of John Stuart Mill, Volume VII, A System of Logic, Ratiocinative and Inductive, being a Connected View of the Principles of Evidence, and the Methods of Scientific Investigation*. TorontoUniversity of Toronto Press.

Muncke, Georg W. 1829. Ueber Robert Brown's mikroskopische Beobachtungen, über den Gefrierpunkt des absoluten Alkohols, und über eine sonderbare Erscheinungen and der Coulomb'schen Drehwaage. *Annalen der Physik* 93 (9): 159–165. https://doi.org/10.1002/andp.18290930913.

Nye, Mary J. 1972. *Molecular Reality: A Perspective on the Scientific Work of Jean Perrin*. New York: American Elsevier Company.

Nägeli, Carl W. v. 1879. Ueber die Bewegungen kleinster Körperchen. *Sitzungsberichte der mathematisch-physikalischen Classe der k.b. Akademie der Wissenschaften zu München* 9: 389–453.

———. 1976. The Moral Freedom of Man and the Determinism of Nature: The Catholic Synthesis of Science and History in the Revue des questions scientifiques. *The British Journal for the History of Science* 9 (3): 274–292.

Ord, William. 1879. On Some Causes of Brownian Movements. *Journal of the Royal Microscopical Society* II: 656–662.

Perrin, Jean. 1910. *Brownian Movement and Molecular Reality*. Trans. Frederick Soddy. New York: Dover Publications.

Poincaré, Henri. 1905. *Science and Hypothesis*. London: The Walter Scott Publishing Company.

Ramsay, William. 1882. On Brownian or Pedetic Motion. *Proceedings of the Bristol Naturalists' Society* 3: 299–302.

Raspail, François. 1829a. Observations and Experiments tending to demonstrate that the Granules which are discharged in the explosion of a grain of Pollen, instead of being analogous to spermatic Animalcules are not even organized Bodies. *Edinburgh Journal of Science* 10: 96–106.

———. 1829b. Note on Mr. Brown's Microscopical Observations on the active Molecules of organic and inorganic bodies. *Edinburgh Journal of Science* 10: 106–108.

Russo, Federica, and Jon Williamson. 2007. Interpreting Causality in the Health Sciences. *International Studies in the Philosophy of Science* 21 (2): 157–170.
Schickore, Jutta. 2019. The Structure and Function of Experimental Control in the Life Sciences. *Philosophy of Science* 86 (2): 203–218.
———. 2022. Parasites, Pepsin, Pus, and Postulates: Jakob Henle's Essay on Miasma, Contagium, and Miasmatic-Contagious Diseases in Its Original Contexts. *Bulletin of the History of Medicine* 96 (4): 612–638.
Soler, Léna. 2012. Introduction: The Solidity of Scientific Achievements: Structure of the Problem, Difficulties, Philosophical Implications. In *Characterizing the Robustness of Science*, Boston Studies in the Philosophy of Science 292, ed. Léna Soler et al., 1–60. Dordrecht: Springer.
Steinle, Friedrich. 2002. Experiments in History and Philosophy of Science. *Perspectives on Science* 10 (4): 408–432.
———. 2016. *Exploratory Experiments: Ampère, Faraday, and the Origins of Electrodynamics.* Trans. Alex Levine. Pittsburgh: University of Pittsburgh Press.
Thirion, Julien. 1880. Les Mouvements Moléculaires. *Revue des Questions Scientifiques* 7: 5–55.
Whewell, William. 1847. *The Philosophy of the Inductive Sciences Founded Upon Their History.* Vol. II. 2nd ed. London: John W. Parker.
———. 1858. *Novum Organum Renovatum.* London: John W. Parker.
Wiener, Christian. 1863. Erklärung des atomistischen Wesens des tropfbar-flüssigen Körperzustandes, und Bestätigung desselben durch die sogenannten Molekularbewegungen. *Annalen der Physik* 118: 79–94.

Klodian Coko is a postdoctoral fellow in the Philosophy Department at the Ben-Gurion University of the Negev. His research focuses on the historical emergence and development of scientific methods and experimental practices.

Chapter 9
From the Determination of the Ohm to the Discovery of Argon: Lord Rayleigh's Strategies of Experimental Control

Vasiliki Christopoulou and Theodore Arabatzis

9.1 Introduction

Lord Rayleigh (1842–1919) was undoubtedly one of the most eminent nineteenth-century British physicists. During a career that spanned more than 50 years, he published an astonishing number of papers—446 in total. He studied in Trinity College at Cambridge University and graduated as Senior Wrangler of the Mathematical Tripos in 1865. He was renowned both for his outstanding mathematical skill and his facility with experiments, and theory and experiment went hand in hand in the vast majority of his work.

According to Arthur Schuster, an eminent physicist himself and author of Rayleigh's obituary in the *Obituary Notices* of Fellows of the Royal Society:

> Rayleigh's scientific activity may for convenience be divided into five periods. … The first period extends up to the time when he took up the Cavendish Professorship [in 1879, after Maxwell's death] …. The second period is dominated by his work on electrical standards. The third period … bridges over the interval between his departure from Cambridge and the experiments which led to the discovery of Argon …. The fourth, or "Argon," period was followed by one of great fertility, but no further distinction can be drawn, and it must be taken to extend to the time of death. (Schuster 1921, iii)

Although this periodization does not reflect the breadth of Rayleigh's research interests, it indicates both the significance and the duration of the two projects that are our primary focus here: determining the ohm and discovering argon. These were two of Rayleigh's foremost contributions, and they greatly reinforced his reputation as an exact experimenter.

The quest for rigor was a ubiquitous theme in Rayleigh's physics. In his experimental practice, that pursuit involved the application of control strategies, which

V. Christopoulou · T. Arabatzis (✉)
Department of History and Philosophy of Science, School of Science, National and Kapodistrian University of Athens, University Campus, Ano Ilisia, Athens, Greece
e-mail: tarabatz@phs.uoa.gr

© The Author(s) 2024
J. Schickore, W. R. Newman (eds.), *Elusive Phenomena, Unwieldy Things*,
Archimedes 71. https://doi.org/10.1007/978-3-031-52954-2_9

pervaded his work at various levels. They included the multiple determination of experimental results,[1] the variation of experimental tools and conditions, and "guided manipulation,"[2] as in the case of magnifying a disturbance or discrepancy. The control strategies also had a social character, because they involved other members of the scientific community. Moreover, experimental control had various aims, such as standardizing measurement units in determining the ohm and validating experimental results in the discovery of argon. With the ohm, Rayleigh and his team varied their apparatus design to control experimental conditions. Those control efforts lay at the heart of their methodology and aimed at dealing with errors. With argon, control permeated every step of the discovery process. This paper aims to investigate and contrast the strategies of control employed in those two cases, and to clarify their various purposes.

9.2 Rayleigh on Different Standards of Rigor

During the first half of the nineteenth century, physicists in Britain were not trained to be physicists. They often graduated as mathematicians, and the most prominent ones came from the Mathematical Tripos at Cambridge. Rayleigh, although a mathematically trained physicist, distinguished between the viewpoints of the pure mathematician and that of the physicist. In his mind, physicists proceeded without pursuing absolute rigor, and they used arguments that mathematicians would deem lacking rigor. As he wrote in the preface to *The Theory of Sound*:

> In the mathematical investigations I have usually employed such methods as present themselves naturally to a physicist. The pure mathematician will complain, and (it must be confessed) sometimes with justice, of deficient rigour. But to this question there are two sides. For however important it may be to maintain a uniformly high standard in pure mathematics, the physicist may occasionally do well to rest content with arguments which are fairly satisfactory and conclusive from his point of view. To his mind, exercised in a different order of ideas, the more severe procedure of the pure mathematician may appear not more but less demonstrative. (Rayleigh 1945, xxxiv–xxxv)

As John Howard argued in his foreword to Rayleigh's biography, which was written by his son, Rayleigh "practiced what is sometimes called the method of modest rigor" (Strutt 1968, xiii). Indeed, he used approximations, often successive ones, and expanded functions into series of terms, keeping only the lowest orders after reaching the desired accuracy. In particular, Howard noted that if the approximations were insufficient to fit the observed data, Rayleigh would use the term of

---

[1] The epistemic significance of multiple determination has been debated extensively in the history and philosophy of science (see Schickore and Coko 2013). Those debates have often focused on Jean Perrin's determination of Avogadro's number by several methods (see Coko 2020b).

[2] Schickore (2019) uses this term to refer to a targeted, precise intervention.

next-higher order. He also said that, although this iterative technique underlies processes in modern computing, "in Rayleigh's day any lack of rigor was considered distressing" (Strutt 1968, xiii).

Further examples may illustrate Rayleigh's stance toward those different standards of rigor, a stance he expressed in research papers, in reviews of other scientists' works, and in public pronouncements on science. For instance, in a research paper titled "On the manufacture and theory of diffraction-gratings," Rayleigh said that "In the present state of our knowledge with respect to the nature of light and its relations to ponderable matter, *vagueness in the fundamental hypotheses is rather an advantage than otherwise; a precise theory is almost sure to be wrong*" (Rayleigh 1874b, 218, our emphasis). Furthermore, in a review of Isaac Todhunter's *A History of the Mathematical Theories of Attraction and the Figure of the Earth from the Time of Newton to that of Laplace*, Rayleigh questioned Todhunter's tendency "to prefer rigour of treatment to originality of conception." He also suggested that "the strictest proof is not always the most instructive or even the most convincing," and added that "To deserve the name of demonstration an argument should make its subject-matter plain and not merely force an almost unwilling assent" (Rayleigh 1874a, 198).

Rayleigh did not specify exactly what rigor (or "absolute" rigor) meant for mathematicians. As we gather from these examples, he didn't always favor clearly-defined fundamental hypotheses or strict proofs, which are often thought to be indispensable features of a mathematical treatment. And when those features conflicted with physical considerations, Rayleigh preferred the latter.

Rayleigh's article on "Clerk-Maxwell's Papers" illuminates this preference for the physical. There he maintains that a physicist may sometimes depart from the dictates of "strict method":

> A characteristic of much of Maxwell's writing is his dissatisfaction with purely analytical processes, and the endeavour to find physical interpretations for his formulae. Sometimes the use of physical ideas is pushed further than strict logic can approve ... *the limitation of human faculties often imposes upon us, as a condition of advance, temporary departure from the standard of strict method*. The work of the discoverer may thus precede that of the systematizer; and the division of labour will have its advantage here as well as in other fields. (Rayleigh 1890b, 428, our emphasis)

Thus, according to Rayleigh, a physicist could escape the strict rules of mathematics when seeking a phenomenon's physical explanation. He held this view and argued for it throughout his scientific life.

In his Presidential Address at the anniversary meeting of the Royal Society in 1906, Rayleigh again explained his position on rigor for mathematicians and physicists. More than 30 years after publishing the *Theory of Sound*, he reiterated the same point:

> Much of the activity now displayed [in Mathematics] has, indeed, taken a channel somewhat remote from the special interests of a physicist, being rather philosophical in its character than scientific in the ordinary sense. [...] Closely connected is the demand for greater rigour of demonstration. Here I touch upon a rather delicate question, as to which pure mathematicians and physicists are likely to differ. *However desirable it may be in itself, the*

pursuit of rigour appears sometimes to the physicist to lead us away from the high road of progress. He is apt to be impatient of criticism, whose object seems to be rather to pick holes than to illuminate. *Is there really any standard of rigour independent of the innate faculties and habitudes of the particular mind?* May not an argument be rigorous enough to convince legitimately one thoroughly imbued with certain images clearly formed, and yet appear hazardous or even irrelevant to another exercised in a different order of ideas? (Rayleigh 1906, 89, our emphasis)

Rayleigh noted further that "what is rather surprising is that the analytical argument should so often take forms which seem to have little relation to the intuition of the physicist" (Rayleigh 1906, 89). He believed that, until reconciling the two approaches, "we must be content to allow the two methods to stand side by side, and it will be well if each party can admit that there is something of value to be learned from the point of view of the other" (Rayleigh 1906, 89).

Rayleigh then commented on experimenters and their occasional neglect of the mathematicians' view, stating

As more impartially situated than some, I may, perhaps, venture to say that in my opinion many who work entirely upon the experimental side of science underrate their obligations to the theorist and the mathematician. *Without the critical and co-ordinating labours of the latter we should probably be floundering in a bog of imperfectly formulated and often contradictory opinions.* Even as it is, some branches can hardly escape reproaches of the kind suggested. I shall not be supposed, I hope, to undervalue the labours of the experimenter. The courage and perseverance demanded by much work of this nature is beyond all praise. *And success often depends upon what seems like a natural instinct for the truth—one of the rarest of gifts.* (Rayleigh 1906, 89, our emphasis)

In any case, although he advocated for differing standards of rigor in mathematics and physics, the quest for it was omnipresent in his scientific practice. In experimentation in particular he associated rigor with control strategies, which he followed to secure experimental outcomes. This association manifests both in determining the ohm and in discovering argon.

The search for rigor also appeared as he explored spiritual phenomena, a topic he was interested in throughout his life. He even served as President of the Society for Psychical Research in the last year of his life (1919). Rayleigh believed that the problematic nature of such phenomena arose "from their sporadic character," as they could not be *"reproduced at pleasure and submitted to systematic experimental control"* (Rayleigh 1919, 648, our emphasis).[3] He maintained that, in general, "we are ill equipped for the investigation of phenomena which cannot be reproduced at pleasure under good conditions" (Rayleigh 1919, 650). For that reason, controlled experimentation was essential.

---

[3] Rayleigh's approach to spiritual phenomena is discussed in Noakes (2019). The role of control practices in investigations of spiritual phenomena is discussed in detail in Cristalli (Chap. 6, this volume). For the problems that arise from "singular experiments," see Baker (Chap. 2, this volume) and Schürch (Chap. 3, this volume).

9.3 The Determination of the Ohm

In the second half of the nineteenth century, there was persistent debate over determining and constructing electrical standards, including for resistance. Determining the ohm became an issue of great international importance for reasons both scientific and commercial. The process was intertwined with and significantly directed by the needs of electrical telegraphy (Lagerstrom 1992; Schaffer 1992, 1994, 1995; Hunt 1994; Olesko 1996; Gooday 2004; Kershaw 2007; Mitchell 2017).

In Great Britain, at its 1861 annual meeting, the British Association for the Advancement of Science (BAAS) formed a committee to determine the resistance unit and construct a corresponding standard. In 1863, noted physicist James Clerk Maxwell, engineer and electrician Fleeming Jenkin, and Balfour Stewart, a physicist and meteorologist who had been appointed director of Kew Observatory in 1859, began their experiments in King's College. They meant to determine a wire's resistance in absolute units in order "to construct the material representative of the absolute unit."[4]

In the following year, Maxwell, Jenkin, and Charles Hockin, another Cambridge Tripos graduate who assisted Maxwell and later Rayleigh in their resistance-unit experiments, repeated earlier experiments and reported their results. Their efforts resulted in defining the B.A. unit and in constructing a standard. This determination was soon questioned, however, and the matter was still unsettled when Rayleigh became Director of the Cavendish Laboratory in 1879.[5]

Before Rayleigh undertook the project, others raised objections to previous experiments and argued they were not in "reasonable agreement."[6] More specifically, during the 1860s and 1870s, eminent physicists performed resistance experiments but their results differed both from those of the Committee and among themselves. The most characteristic case involved the famous German physicist and experimentalist Friedrich Kohlrausch and the Danish physicist Ludvig Lorenz. Their results differed by 4%, with the B.A. unit falling in the middle.

Henry Augustus Rowland (1848–1901) has been described as the "father" of the American physics discipline.[7] In 1876, he was appointed the first professor of physics at Johns Hopkins University, a post he held until his death in 1901. In 1878, and amid the disagreement over the resistance standard, Rowland proceeded with his own experiments and a new method. To secure his results from unsuspected

---

[4] Report of the Committee appointed by the British Association on Standards of Electrical Resistance 1864, 116.

[5] For the unification of the lab through a research project, see Schuster (1911, 30), as well as Schaffer (1992, 1994).

[6] We owe this phrase to Kuhn, who argued that scientists do not seek "agreement" in numerical tables but "reasonable agreement". See Kuhn (1961, 161–162).

[7] For this characterization, see Kargon (1986, 132) and Wise's introduction in Sweetnam (2000).

constant errors,[8] he attempted to eliminate them in advance by means of the experimental design. As he noted:

> Such a great difference in experiments which are capable of considerable exactness, seems so strange that I decided to make a new determination *by a method different from any yet used, and which seemed capable of the greatest exactness; and to guard against all error, it was decided to determine all the important factors in at least two different ways, and to eliminate most of the corrections by the method of experiment, rather than by calculation.* (Rowland 1878, 145, our emphasis)

For Rowland, different methods lay at the heart of his approach against errors. He thought that his method was "capable of greater exactness than any other, and it certainly possessed the greatest simplicity in theory and facility in experiment" (Rowland 1878, 145). Using a new method, however, was also key for checking existing measurements and for detecting possible errors. Thus, he used "at least two different ways" to determine the experiment's important factors and for securing its result.[9] In addition, Rowland considered constant errors the ultimate threat to the experiment's success, and he sought to avoid them by designing the experiment in a suitable way. Rowland based his method on Kirchhoff's but made modifications. In Kirchhoff's approach "the magnitude of a continuous battery-current in a primary coil is compared with that of a transient current induced in a secondary coil when the primary circuit is removed." Rowland reversed the current's direction in the primary circuit in order to avoid the motion of the primary coil.[10]

Gabriel Lippmann (1845–1921), the notable French physicist whose work spanned many branches of physics,[11] also participated in determining the resistance unit. In 1882 Lippmann proposed a method based on earth induction,[12] consisting in balancing the maximum electromotive force of a continuously rotating earth inductor[13] against the fall in potential in a resistance produced by a measured current.[14] That is, a copper-wire frame revolved around a vertical axis with its circuit open. An electromotive force was thus produced by induction, which reached its peak when the plane of the frame aligned with the magnetic meridian. At that moment, the ends of the moving armature were connected to two wires and through them to a potential

---

[8] Here, we follow Rayleigh's terminology. Actually, Rayleigh used "constant error" and "systematic error" interchangeably, but the term "constant error" appeared more often in his writings.

[9] Of course, the idea of determining an experimental result in multiple ways did not originate with Rowland. For the history of this idea in the nineteenth century, see Coko (2015).

[10] For a description of Kirchhoff's method and Rowland's alteration of it, see Rayleigh (1882c, 135–37).

[11] Lippmann's name is principally associated with color photography by interference, an achievement for which he was awarded the Nobel Prize in 1908.

[12] This method in its original plan is attributed to Carey Foster, who proposed it in 1874. Lippmann maintained he had not heard of it when he designed his method for determining the Ohm. See Lippmann (1882a, 316).

[13] An earth inductor is a coil revolving in the earth's magnetic field, generating an induced current.

[14] For more on Lippmann's method, see Lippmann (1882b) and Mitchell (2012).

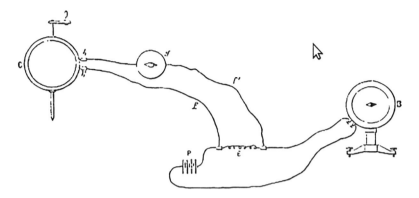

Fig. 9.1 A schematic representation of Lippmann's method (Lippmann 1882a, 315)

difference. If the electromotive force was balanced by the potential difference, no current occurred and the resistance could be estimated by deviations in a tangent galvanometer. The following schematic representation of the experiment was given by Lippmann himself (see Fig. 9.1).

Lippmann never gave the results produced by his method. Nevertheless, he believed that its strength was its directness,[15] meaning that the experimental design avoided errors and therefore avoided corrections. The result was direct control over the method. As he wrote: "Note that this method is most direct: it does not require any calculation of reduction or correction. … As a result, *the control of the method is also direct*."[16] In Lippmann's case, as in Rowland's, we see a distinction between ways of eliminating errors: by calculating corrections to measurements, and by designing experimental controls. From that distinction Lippmann advocated his own experimental method for determining the ohm.

It was characteristic of the determination process that different scientists used different methods. Éleuthère Mascart (1837–1908), the renowned French physicist also involved in the project, coauthored a famous book with Jules Joubert (1834–1910), titled *Leçons Sur L'Électricité Et Le Magnétisme*. There they listed the methods and results they had produced until 1885 (Mascart and Joubert 1897, 619–20). At least seven were based on physical processes—six on induction phenomena, and one on the mechanical equivalent of heat and calorimetry. The number of scientists was equally impressive: more than twenty individuals or teams.[17]

---

[15] Mitchell (2012) argues that Lippmann followed the tradition of Regnault in his preference for the direct method.

[16] "On remarquera que cette méthode est des plus directes: elle n'exige aucun calcul de réduction ou de correction. … Il en résulte que *le contrôle de la méthode est également direct*" (Lippmann 1882b, 1349, our emphasis).

[17] A complete enumeration of the methods for determining the resistance unit is not an easy task, and there are different accounts. Harvey L. Curtis (1875–1956), a member of the National Bureau of Standards who was widely recognized for his work on absolute electrical measurements, classified the methods up until the 1930s. See Curtis (1942, 41, 51–52).

The primary purpose in using these different methods was not to choose the best but to reinforce the trustworthiness of the results. Rayleigh did review those methods and attempt to compare them, but this effort was of secondary importance.

In this international debate, multiple determination as a control strategy was a common theme. Gustav Heinrich Wiedemann, the German physicist known for editing *Annalen der Physik und Chemie*, had himself an active role in determining the ohm. In 1882, he described the requirement for multiple determination:

> Hence at any rate it is indicated that the final determination of the ohm must not rest alone on experiments made only according to one method and carried out at one place. Further, the results of each separate method (as I have already mentioned) offer security against possible constant errors only if they are obtained from entirely independent series of experiments, made with apparatus varied in all possible ways. Since investigations are already in progress in different places, with excellent apparatus and according to different methods, we may shortly expect to be in a position to compare together the data which they yield, and so to attain as reliable a final result as possible. (Wiedemann 1882, 275)

The methods and observers should be multiple, and the apparatus should vary in all possible ways. These checks would guard against constant errors. As Wiedemann stated, "the apparatus itself must be frequently altered in various ways. *Only so can we obtain results independent of each other, which can be used for mutual control*" (Wiedemann 1882, 265, our emphasis). Thus, multiple determination of experimental results functioned as a control strategy.

Rayleigh was elected Cavendish Professor of Physics in 1879. Partly because the original apparatus was at the Cavendish laboratory, he tried to unify his laboratory in a common cause by taking up the redetermination of the unit of electrical resistance. He decided "to repeat the measurement by the method of the Committee, which has been employed by no subsequent experimenter" (Rayleigh and Schuster 1881, 1), making alterations he considered necessary. In performing their experiments, he and his team followed this method in two phases, where the composition of the team taking measurements and the apparatus they employed were different. Changing the apparatus aimed at better controlling the experiment conditions.

In the first phase they made experiments with the original apparatus, altered in certain respects to secure uniformity and more accurate measurements. Here Rayleigh's team consisted of Mrs. Sidgwick, Horace Darwin, and Arthur Schuster. Mrs. Sidgwick was Rayleigh's sister-in law, a graduate of Newnham College and an activist for women's rights in education. She assisted Rayleigh in some of his research on electrical standards. Horace Darwin (1851–1928), son of Charles, was an engineer who designed and built instruments and was a co-founder of the Cambridge Scientific Instrument Company. Others took part too, although not in the measuring process; Professor James Stuart (1843–1913) was one, the "first true professor"[18] in engineering in Cambridge. He secured the insulation of the apparatus.[19]

---

[18] This is how Stuart is described on the official site of the Cambridge Engineering Department. See http://www-g.eng.cam.ac.uk/125/1875-1900/stuart.html#:~:text=The%20first%20true%20professor%20of,and%20for%20the%20working%20classes

[19] Some critics of the Committee's apparatus had pointed out that its insulation was defective.

According to the preliminary conclusions in the experiments' first phase, it was necessary to enlarge the apparatus to improve the results' accuracy. Thus, the experiments of 1881 were repeated with a new apparatus—with linear dimensions in a ratio of about 3:2. This time, the team recording the measurements also changed as they obtained them. Its members included Rayleigh, Shuster, Mrs. Sidgwick, Lady Rayleigh, Arnulph Mallock, the experimental assistant, and J. J. Thomson, who had just received his B.A. and become Fellow of Trinity College.

Figure 9.2 provides a schematic representation of the method Rayleigh and his team used to determine the ohm. It also shows the enlarged apparatus from the second phase of their experiments. The method was to cause a coil to revolve around a vertical axis and then to observe a magnet's deflection from the magnetic meridian as it hung suspended from the center. The amount of deflection was independent of the earth's magnetic field and varied inversely as the resistance of the circuit. Throughout their work Rayleigh and his team used control strategies, which extended from their initial plan and experimental design to the measuring process and validation of the experimental results. The strategies' principal aim was to standardize the measurement unit.

Control came in different forms in each experimental phase. Controlling the experimental conditions was one form. Rayleigh and his team tried to avoid disturbances (e.g., they performed experiments during the night) and tried to eliminate the effects of things such as short circuits, ground tremors, and observer eye fatigue. Even the experimenters were targets of control.

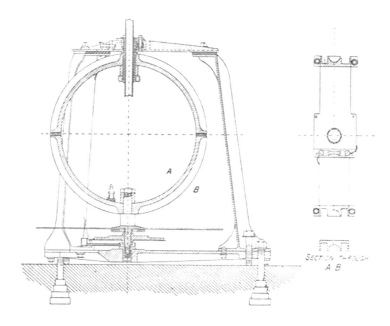

Fig. 9.2 A representation of the apparatus for determining the ohm (Rayleigh 1882a, 39)

Moreover, the direction of the earth's magnetic action varied constantly, and so it was necessary to correct for that variation during the experiment. In this case, the source of interference was itself variable and could not be eliminated; for that reason, it had to be controlled via a measuring process. Rayleigh and his team used a second magnetometer to make direct comparisons between the two devices, whereas the Committee had compared their magnetometer with photographic records of the earth's magnetism obtained by the Kew Observatory at the time of their experiment.[20] Rayleigh and his team were therefore attempting to calibrate their instrument to avoid errors from potential variations in the magnetic field.[21] As Rayleigh explicitly stated:

> It is perhaps worth remarking that owing to the absence of any *controlling instrument* equivalent to our auxiliary magnetometer, the Committee of the British Association had no opportunity of discovering the presence of air currents, as any changes in the zero position would naturally have been ascribed by them to a causal change in the direction of the earth's magnetic force. (Rayleigh and Schuster 1881, 30, our emphasis)

Rayleigh also tried to better control the apparatus' prime mover. The Committee had used a Huygens' gearing,[22] driven by hand in conjunction with a governor. Rayleigh thought that an engine acting by a jet of water upon revolving cups would be an improvement.[23] To achieve a constant head of water,[24] with Darwin's help he connected the engine to a cistern at the top of the building. Although he intended to use a governor of his own invention, he found it unnecessary in the end, as the observer "could easily control the speed" by having the water power a little in excess and using the stroboscopic method (Rayleigh and Schuster 1881, 8–9).

One other general principle Rayleigh followed was to "magnify the disturbances," in order to more closely view any possible causes. Maxwell had advocated this approach as well. In 1876, in a paper titled "General considerations concerning Scientific Apparatus," Maxwell explained that the disturbing agents in an experiment may become the subject of other experiments. In his words:

> We may afterwards change the field of our investigation and include within it phenomena which in our former investigation we regarded as disturbances. The experiments must now be designed so as to bring into prominence the phenomena which we formerly tried to get rid of. (Maxwell 2010, 505)

---

[20] See Jenkin and Thomson (1873, 104–105) and Rayleigh and Schuster (1881, 9).

[21] Calibration as a means to secure stability in Rayleigh's experimental practice is also discussed in Schaffer (1994).

[22] A few years later, Rayleigh explained what a "Huygens' gearing" was, and how it could be used in research in electromagnetism. See Rayleigh (1890a).

[23] He wrote in the report "This, it appeared to me, might advantageously be replaced by a water-motor." See Rayleigh and Schuster (1881, 5).

[24] The head of water is a measure for the power of a pump, that is, the highest height at which a pump can convey the water against the action of gravity.

Rayleigh knew Maxwell's work and was probably aware of this guiding principle. It is clear, in any case, that the point is at the core of Rayleigh's experimental practice, as we shall see below in analyzing the discovery of argon.

It is also evident that Rayleigh's team used multiple determination of self-induction, a principal factor for their result's accuracy. Maxwell had done the same.[25] However, Rayleigh thought that, in the Committee's experiments, the value of the coefficient for self-induction had been underestimated. He and his team determined it by different means, including calculating it directly from the dimensions of the coil, basing it on measurements with an electric balance, and deriving it from the principal observations themselves.[26]

Furthermore, at the end of the first part of the experiments' first reports,[27] Rayleigh suggested that most existing determinations introduced time by a swing of the galvanometer needle. Although he did not question the reliability of those determinations, he pointed out that "it is, to say the least, satisfactory to have them confirmed by a method in which the element of time enters in a wholly different manner" (Rayleigh and Schuster 1881, 20). In the second report, Rayleigh included a brief comparison of their own result with values obtained previously by Kohlrausch[28] and Rowland and Joule,[29] and he commented on their (dis)agreement and their expected accuracy.[30]

Thus, Rayleigh's control strategies included using multiple experimental methods and comparing their results. For him, this aspect of control was crucial, as it secured an experiment's outcome. In 1882 he devoted an article to the subject, entitling it "Comparison of Methods for the Determination of Resistances in Absolute Measure." There he reviewed the six available methods for determining the resistance unit, pointing out their relative merits and demerits. Those methods were based on different experimental apparatuses and used different formulas for the value of the electrical resistance according to which procedure was followed. Rayleigh focused on methods involving an induced electromotive force, not considering Joule's calorimetric method. The others included three we have already mentioned (Kirchhoff's, Lippmann's, and that of the BAAS Committee), along with three others: two by Weber (employing transient currents and damping, respectively)[31] and Lorenz's method. Rayleigh was convinced that "it is only by the coincidence of results obtained by various methods that the question can be satisfactory settled" (Rayleigh 1882c, 139).

---

[25] Note, though, that they did not include that determination in their report. It was to be found in Maxwell's paper on the "Electromagnetic Field." See Rayleigh and Schuster (1881, 11).

[26] Maxwell had also determined it by different means.

[27] The first report of 1881 was written in two parts, one by Rayleigh and the other by Schuster.

[28] Kohlrausch had followed "Weber's *Method by Damping*." See Rayleigh (1882c, 145).

[29] Here, Rayleigh referred to the experiments on the mechanical equivalent of heat involving measurements of absolute resistance.

[30] See Rayleigh (1882a, 47–51).

[31] Weber had actually proposed four methods. See Wiedemann (1882).

It is worth noting here that Rayleigh also cared about evaluating each method's accuracy. At the end of the article, he suggested that Lorenz's method offered the best chance of success. In that method, "A circular disk of metal, maintained in rotation about an axis passing through its centre at a uniform and known rate, is placed in the magnetic field due to a battery-current which circulates through a coaxal coil of many turns" (Rayleigh 1882c, 145–46). Rayleigh believed that, with this way of performing the experiment, the errors of the principal quantities to be measured did not affect the final result as much as they did with other methods. He reached his conclusion regarding the propagation of errors by applying differential calculus. Before pursuing that method (Rayleigh and Sidgwick 1883), however, he thought that "the value now three times obtained in the Cavendish Laboratory by distinct methods should be approximately verified (or disproved) by other physicists" (Rayleigh 1882c, 150).[32]

Collective knowledge and experience were indispensable elements of the control process for validating Rayleigh's experimental results. Indeed, standardization demanded consensus among the members of the scientific community—and not among them only, but also among the "practical men," the practitioners working in electrical telegraphy.[33] As several scholars have argued, consensus in determining the ohm was a complex matter, involving national rivalries and personal agendas. Agreement was not established solely on scientific grounds or on the accuracy of the determinations as such.[34] At any rate, multiple determination was a guiding principle for Rayleigh, and stemmed from his beliefs about sound experimental methodology.

Schuster mentioned that Rayleigh "never felt satisfied until he had confirmed his results by different methods, and had mastered the subject from all possible points of view" (Schuster 1921, xxvi). Rayleigh thought that all experimenters should follow this principle in physics, and in 1882 he discussed it in his Address to the Mathematical and Physical Science section of the British Association meeting. In his words:

> The history of science teaches only too plainly the lesson that no single method is absolutely to be relied upon, that sources of error lurk where they are least expected, and that they may escape the notice of the most experienced and conscientious worker. *It is only by the concurrence of evidence of various kinds and from various sources that practical certainty may at last be attained, and complete confidence justified.* Perhaps I may be allowed to illustrate my meaning by reference to a subject which has engaged a good deal of my attention for the last two years—the absolute measurement of electrical resistance. (Rayleigh 1882b, 119–20, our emphasis)

It is noteworthy that Rayleigh drew his example from the project of determining the ohm. He did not search for a concurrence of evidence solely with his own experiments; he also appealed to the scientific community, and this appeal served as a basis for controlling his own results. At the very end of his 1882 Address, he stated that:

---

[32] Rayleigh here referred also to the experiments made by Glazebrook following Rowland's method. Cf. Schaffer (1994, 282).

[33] See Hunt (1994).

[34] See Schaffer (1994) on 'fiat' agreement, and Gooday (2004).

If there is any truth in the views that I have been endeavouring to impress, our meetings in this section are amply justified. If *the progress of science demands the comparison of evidence drawn from different sources, and fully appreciated only by minds of different order*, what may we not gain from the opportunities here given for public discussion, and, perhaps, more valuable still, private interchange of opinion? Let us endeavour, one and all, to turn them to the best account. (Rayleigh 1882b, 124, our emphasis)

Rayleigh's expression "minds of different order" referred to different kinds of physicists. In particular he distinguished between two kinds: the experimenters and the mathematicians. He claimed that each values different sorts of evidence and argumentation. The experimenters, according to Rayleigh, "disregard arguments which they stigmatise as theoretical," while the mathematicians "overrate the solidity of the theoretical structures and forget the narrowness of the experimental foundation upon which many of them rest" (Rayleigh 1882b, 122). For Rayleigh, however, each approach mattered: using different experimental methods and multiple observers, finding agreement among experimental results, and appealing to different sorts of arguments (theoretical and experimental) all had their place in securing an outcome's validity.

Rayleigh's involvement in determining electrical standards was not limited only to experiments with the Committee's method, or to reviews of other methods. In 1882 he and Mrs. Sidgwick also began experiments by Lorenz's method, reporting their results the following year. They worked on related topics as well, such as the electro-chemical equivalent of silver and the absolute electromotive force of Clark cells. Regarding silver's equivalent, in 1897 Rayleigh recollected that, when they undertook the task, the previous results' uncertainty was at least 1%. He also restated his conviction about the necessity of using different methods and different observers for securing experimental results:

Security is only to be obtained by the coincidence of numbers derived by different methods and by different individuals. It was, therefore, a great satisfaction to find our number (*Phil. Trans.* 1884) (0.011179) confirmed by that of Kohlrausch (0.11183), resulting from experiments made at about the same time. (Rayleigh 1897a, 332)

As we have already mentioned, this was a guiding principle in his experimental practice. And it is also a principle at work in the discovery of argon.

9.4 The Discovery of Argon

Determining the ohm was the project that gave Rayleigh a reputation as an exact experimenter. He is perhaps best remembered, however, for the discovery of argon, a new element and hitherto unknown constituent of the atmosphere.[35] He won the 1904 Nobel Prize in Physics for "his investigation on the densities of the most important gases, and for his discovery of Argon, one of the results of those

---

[35] The discovery of argon has been the topic of many studies in the history and philosophy of science. For full references to this literature, see Arabatzis and Gavroglu (2016).

investigations."[36] The same year, William Ramsay won the Prize in Chemistry for his "discovery of the inert gaseous elements in air, and his determination of their place in the periodic system,"[37] with argon being the first. Rayleigh and Ramsay, working at first independently and then in concert, took on the task of isolating the gas and studying its properties.

As Arabatzis and Gavroglu have argued,[38] the discovery of argon was not an event but an extended process. As such it comprised not only detecting but also identifying and assimilating argon into the conceptual framework of nineteenth-century chemistry. Throughout the process Rayleigh used various control strategies: from detecting discrepancies between the densities of "atmospheric" and "chemical" nitrogen, to isolating and identifying a new constituent of the atmosphere, and subsequently to exploring its properties. Here as elsewhere, the main aim of experimental control was to validate the experimental results.

The starting point for the discovery process was Prout's law, which says that the atomic weights of the elements were whole multiples of the atomic weight of hydrogen. As early as 1882, Rayleigh had expressed his willingness to redetermine the densities of the "principal gases"[39] to test that law. He started by determining the relative densities of hydrogen and oxygen and then proceeded to the density of nitrogen. Given that he originally aimed to test Prout's hypothesis, Rayleigh determined the ratio of atomic weights of oxygen to hydrogen via the densities of those gases. But he also tried an independent and novel determination, one based on the composition of water.[40] Rayleigh was not the only one who used more than one method to determine atomic weights. For instance, the American chemist Theodore William Richards (1868–1928) used five to determine copper's atomic weight.[41]

In experimenting with the density of nitrogen, Rayleigh used two principal methods to prepare the gas.[42] In the first, atmospheric air was "freed from CO_2 by potash" and then the oxygen was removed by "copper heated in hard glass over a large Bunsen" burner. It was then passed over "red-hot copper in a furnace" before being treated with "sulphuric acid, potash and phosphoric anhydride" (Rayleigh 1892, 512). Regnault had followed this method in experimenting with the densities of the principal gases.

---

[36] Award ceremony speech. Available at https://www.nobelprize.org/prizes/chemistry/1904/ceremony-speech/

[37] Award ceremony speech. Available at https://www.nobelprize.org/prizes/chemistry/1904/ceremony-speech/

[38] Arabatzis and Gavroglu (2016).

[39] That is, hydrogen, oxygen, and nitrogen. See Rayleigh (1882b, 1904).

[40] See Rayleigh (1888) and Clarke ([1882] 1897).

[41] See Ihde (1969).

[42] Different methods of preparation were not only applied to nitrogen. As mentioned above, in his earlier experiments on the relative densities of hydrogen and oxygen Rayleigh also used different methods for the preparation. See Rayleigh (1887). As he pointed out in another paper concerning the densities of carbon oxide, carbonic anhydride, and nitrous oxide, "agreement … is some guarantee against the presence of impurity." See Rayleigh (1897b, 348).

The main difference between the first and second method is the use of ammonia. In the second the oxygen was combined with the hydrogen of ammonia, through which the air passed before the furnace with the red-hot copper. Rayleigh used the method on Ramsay's suggestion. In his reports, Rayleigh referred to nitrogen of different origins with different names. He called the gas obtained by the first method "atmospheric nitrogen," whereas that prepared with ammonia was "chemical nitrogen."

Although the results of the second method[43] were in close agreement, Rayleigh still used the other. As he observed in his Nobel Lecture, multiple methods were always desirable:

> Turning my attention to *nitrogen,* I made a series of determinations using a method of preparation devised originally by Harcourt and recommended to me by Ramsay […] Having obtained a series of concordant observations on gas thus prepared I was at first disposed to consider the work on nitrogen as finished. Afterwards, however, I reflected that the method which I had used was not that of Regnault *and that in any case it was desirable to multiply methods,* so that I fell back upon the more orthodox procedure according to which, ammonia being dispensed with, air passes directly over red hot copper. (Rayleigh 1904, 212–13, our emphasis)

To his surprise, he found a discrepancy of 1/1000 in nitrogen's density as given by those two methods. He could not attribute the discrepancy to experimental error because the measurements for each method did not present deviations greater than 1/10000.[44] With this order-of-magnitude difference, Rayleigh claimed that experimental error could not explain the discrepancy. Thus, his claim stems from his confidence that the experimental conditions were stable and well-controlled. As it turned out, the discrepancy between "atmospheric" and "chemical" nitrogen was the initial step that later led to discovering a new constituent of air: the inert gas argon.

Faced with the discrepancy, Rayleigh published a letter in *Nature*[45] inviting criticism from chemists and asking for their help. At the time he regarded the situation "only with disgust and impatience," although his call for help may seem striking in itself.[46] At any rate, his rush to publish the letter may be due to a lack of confidence in his chemical knowledge.[47] As with determining the ohm, however, the call reveals a communal aspect to his control processes. Rayleigh expected that chemists would make suggestions, and then he could examine them. It was not only an invitation for

---

[43] The numbering of the methods here follows the one that Rayleigh gave in his reports. He considered the first method more established, something apparent in the following quotation and the fact that he called it "the more orthodox procedure."

[44] In earlier work on the densities of hydrogen and oxygen, Rayleigh had also aimed at an accuracy of 1/10000. Thus, the magnitude of experimental error was established both from previous research and from experiments on nitrogen. On this point see also Spanos (2010, 362).

[45] Rayleigh (1892). See also Rayleigh (1895).

[46] Note, though, that this was actually rather common in the eighteenth and early nineteenth century. See Schickore (2023) and Schürch (Chap. 3, this volume).

[47] As Rayleigh's son suggested in his account of the discovery of argon. See Strutt (1968, 189).

public discourse—he appreciated private communication as well. Thus, he placed control in the hands of the community and did so early in his research. He hoped others would help him explain the unequal measurements. He only obtained, however, "useful suggestions, but none going to the root of the matter" (Rayleigh 1895, 189).

His next step was to magnify the discrepancy. In the preparation of "chemical" nitrogen by ammonia, only one-seventh of the final quantity was "derived from the ammonia," with the rest from atmospheric air (Rayleigh 1895, 189).[48] Thus, the most obvious way to achieve such a magnification was to get all the nitrogen from ammonia. Here is how Rayleigh explained that process:

> One's instinct at first is to try to get rid of a discrepancy, but I believe that experience shows such an endeavour to be a mistake. What one ought to do is to magnify a small discrepancy with a view to finding out the explanation; and, as it appeared in the present case that the root of the discrepancy lay in the fact that part of the nitrogen prepared by ammonia method was nitrogen out of ammonia, although the greater part remained of common origin in both cases, the application of the *principle* suggested a trial of the weight of nitrogen obtained wholly from ammonia. (Rayleigh 1895, 189, our emphasis)

In his Nobel Lecture he repeated the same point: "It is *a good rule in experimental work to seek to magnify a discrepancy* when it first presents itself, rather than to follow the natural instinct of trying to get quit of it" (Rayleigh 1904, 213, our emphasis). Whether a rule or principle, "magnifying the discrepancies" was indispensable to Rayleigh's experimental practice. This form of control amounted to "guided manipulation," which aimed at finding an appropriate explanation for the discrepancy. In this case the discrepancy was magnified about five times, firmly establishing the initial experimental outcome and indicating the need for further research.[49]

The next stage in the discovery process was to explain the discrepancy. Was the "atmospheric" nitrogen heavier than the "chemical" because of impurities? If so, in which nitrogen were the impurities to be found? Were there lighter impurities in the "chemical" nitrogen, or heavier impurities in the "atmospheric" nitrogen? Was there some other form of nitrogen, like N_3 or nitrogen in a partially "dissociated state"?

Rayleigh altered the preparation of nitrogen to confirm his initial result and clarify the discrepancy's cause. In the next 2 years he produced "atmospheric nitrogen" by replacing hot copper with hot iron or ferrous hydrate, and "chemical nitrogen" by using nitric oxide, nitrous oxide, and ammonium nitrite, along with substituting hot copper with hot iron. The result did not change.

Regarding the possibility of lighter impurities, the possibility of hydrogen as their source struck Rayleigh as the most worth investigating. If that was the case, however, and hydrogen was present, the copper oxide should consume it. Rayleigh approached the matter experimentally. To exclude the possibility of a lighter hydrogen-based impurity in the "chemical nitrogen," a certain amount was

---

[48] Note, however, that in his Nobel Lecture some years later, Rayleigh claimed that a larger part of the nitrogen (one-fifth) was obtained from ammonia.

[49] Here, we follow Rayleigh's account in his Nobel Lecture.

introduced into the heavier "atmospheric nitrogen." It made no difference to the result and the hypothesis was rejected.

At first at least, Rayleigh leaned toward the possibility that nitrogen was being produced in a "dissociated state." But he changed his mind because of skeptical suggestions from his "chemical friends." There was chemical evidence that if nitrogen was dissociated, it was likely the atoms would not continue to exist for long. Rayleigh also checked the hypothesis of dissociated nitrogen by subjecting both gases to the action of silent electric discharge. Their weights remained unchanged, indicating the hypothesis was probably wrong. Finally, he made another experiment to secure the conclusion. He stored a sample of "chemical nitrogen" for 8 months to check its density. He found no sign of increase.[50]

As is evident, every step in the detection process was cross-checked, either with different experimental methods or with a combination of theory and experiment. The methodology of multiple preparations motivated Rayleigh and Ramsay to the conclusion that "chemical" nitrogen was a uniform substance. The properties of the samples produced different showed it had to be one and the same substance. On this point they stated: "That chemical nitrogen is a uniform substance is proved by the identity of properties of samples prepared by several different processes and from several different compounds" (Rayleigh and Ramsay 1895, 180). Rayleigh and Ramsay also maintained it was difficult to see how a gas of chemical origin could be a mixture. If that was the case, there should have been two kinds of nitric acid (when that acid was used in the preparation). They argued further that the claim that nitrogen is a mixture could not be reconciled with the work of Belgian chemist Jean Stas and others on the atomic weight of nitrogen.[51] Thus, control via multiple preparations went hand in hand with control via consistent agreement with the works of other chemists.[52]

In addition, the question of whether "atmospheric" nitrogen was a mixture of nitrogen and another substance was also investigated in detail, along with its isolation. They first tried to isolate it using two methods and then used atmolysis to ascertain its nature.

On the one hand, Rayleigh approached the question as Cavendish had done in 1785, more than a century earlier. He turned his attention to Cavendish after Dewar made a suggestion in 1894.[53] Nitrogen was removed with the aid of oxygen, subjecting the mixture to an electric spark (see Fig. 9.3). The process always left residue, which could be isolated.

---

[50] See Rayleigh (1894, 104–8).

[51] See Rayleigh and Ramsay (1895, 135–36).

[52] See Rayleigh and Ramsay (1895, 180).

[53] On this point there was a difference of opinion. According to Ramsay's recollection, he was the one who had first drawn Rayleigh's attention to Cavendish. However, Rayleigh himself, together with his son, claimed that Dewar was the first to say it. Rayleigh's son also supplied testimonies from other scientists, such as Dewar and Boy, confirming that claim. See Rayleigh (1895, 191) and Strutt (1968, 194–95).

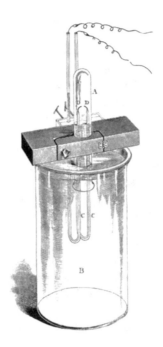

Fig. 9.3 Isolation of argon with the aid of oxygen, subjecting the mixture to an electric spark (Rayleigh and Ramsay 1895, 142)

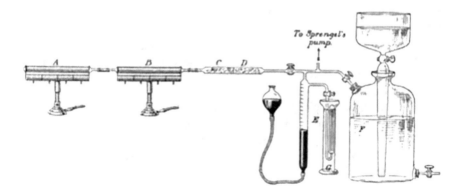

Fig. 9.4 Isolation of argon by means of red-hot magnesium (Rayleigh and Ramsay 1895, 144)

On the other hand, Ramsay also followed another method: absorbing the nitrogen by means of magnesium at full heat (see Fig. 9.4).

Rayleigh and Ramsay gave seven reasons to justify their conclusion that atmospheric nitrogen was a mixture of nitrogen and argon. One was based on the double isolation just mentioned, along with their belief that "It is in the highest degree improbable that two processes [Cavendish's and Ramsay's], so different from each other, should each manufacture the same product" (Rayleigh and Ramsay 1895, 180).[54] This philosophical commitment was key to their method.

Rayleigh and Ramsay examined every alternative hypothesis that they or others thought of, and eliminated all but one: they concluded that the origin of the discrepancy must be a new constituent of the atmosphere. Considering alternative hypotheses was another way to control their explanation's validity. They tested the alternatives with auxiliary experiments and/or theoretical considerations. The latter method was a way to control theory, since it rested on reasons to exclude possible explanations and not on any material manipulation.

Control practices were also present in exploring argon's properties. To determine its density, for example, Rayleigh and Ramsay used different methods and directed a number of experiments toward that end. The gas(es) obtained from Cavendish's method, along with those from Ramsay's, were examined to determine their densities.

A first estimation of the gas's density from Cavendish's method used the initial measurements of the densities of nitrogen from different origins. They were able to calculate its density as long as they assumed the densities differed because of argon. Because it was difficult to directly determine the density of argon owing to the small quantities collected, they filled a large globe with an oxygen–argon mix of known proportions and determined its density. In every measurement, experimental objects and conditions were standardized and then correction applied for certain constant errors. The amount of the residual nitrogen was estimated through spectrum analysis.[55]

Rayleigh and Ramsay also determined argon's density using Ramsay's magnesium method. Using another gas as reference, they had three auxiliary experiments to test the accuracy with which the density of the unknown gas could be determined. They chose the density of dried air as their reference value and compared the mean of their measurements with that obtained by "several [other] observers" (Rayleigh and Ramsay 1895, 149). The control process again rested on multiple determination and knowledge established by other scientists. Rayleigh and then Ramsay proceeded to directly determine argon's density, and concluded it was "at least 19 times as heavy as hydrogen" (Rayleigh and Ramsay 1895, 150).

Spectroscopy was another means used to identify the atmosphere's new constituent. After isolating it on a larger scale by the magnesium method, two other scientists—William Crookes and Arthur Schuster, working independently—subjected the gas to spectrum analysis. They meant to identify it and determine whether it was a mixture or not.[56] To achieve the best results they used electrodes of different materials. Both sources of argon gave identical spectra.

---

[54] A similar "argument from coincidence" has been more recently employed by Ian Hacking to support "truth in microscopy." See Hacking (1983, 200–2).

[55] See Rayleigh and Ramsay (1895, 166; footnote added in April 1895).

[56] At the time, spectroscopy was a controversial technique among chemists. See Arabatzis and Gavroglu (2016).

Crucial to the new element was its ratio of specific heats, as it was directly related to its number of atoms. To determine the ratio, Rayleigh and Ramsay performed experiments on the velocity of sound in argon. They used a familiar apparatus, but in a way that "differed somewhat from the ordinary pattern." To test "the accuracy of this instrument," "fresh experiments were made with air, carbon dioxide and hydrogen," and their results were compared to those of other observers. By this control process they "established the trustworthiness of the method," which then led to a ratio of specific heats that was "practically" identical with "the theoretical ratio for a monatomic gas" (Rayleigh and Ramsay 1895, 174–76).

Rayleigh and Ramsay therefore employed several control strategies in discovering argon, and these strategies were integral to the discovery. They also played different roles in different research stages. A primary form of experimental control was the multiple methods. This was not just because Rayleigh and Ramsay participated in that project independently. Rather, as we explained, multiple determinations were essential throughout the discovery process—from Rayleigh working by himself to detect the initial discrepancy, to Rayleigh and Ramsay working together to identify argon and explore its properties.

9.5   Concluding Remarks

Other historians and philosophers of science have discussed each episode we have treated here. The reason to bring them together in this paper is to explore the control strategies used by Rayleigh across his research. We have thereby revealed a pervasive pattern in how he conducted experiments. Both stories began as a project of redetermination: of the B.A. unit of resistance, and of the densities of the principal gases. But the stories developed differently. In determining the ohm, the project never changed direction and terminated with a measured value; with argon, the research agenda shifted radically. After the first experiments revealed the initial discrepancy, the aim was no longer to determine the density of nitrogen. It was to establish the discrepancy beyond doubt and identify its cause. This explanatory quest ended with identifying a new constituent of the atmosphere, after all other explanations had been rejected. Thereafter their research focused on determining the new gas's properties.

Experimental control itself had different aims in those two cases. With the ohm, control served to standardize a unit, while with argon control strategies were used to validate initial experimental results, identify argon's sources, and investigate its properties. Nevertheless, in both cases the control strategies shared some common features, such as varying the experimental conditions and obtaining agreement among results produced in different ways. Those features stemmed from Rayleigh's general methodological approach and from his attitude about experimental practice. As we saw, Rayleigh advocated the use of multiple determination whenever feasible. This focus was evident at various levels in determining the ohm, and at nearly every step in discovering argon.

Crucial for Rayleigh also was a check or comparison via multiple determination, along with the evaluation of each method. Although agreement between items of evidence was essential to secure experimental results in both cases, it was especially in the case of argon that the agreement became the ground on which both the validity of the result and the identification of the new element rested.

Furthermore, multiple determination had several epistemic aims. First, it was a means to secure experimental results against undetected systematic errors. If results from independent methods and/or observers agreed within certain limits of accuracy, researchers could assume that no important sources of error had been left out of consideration. Rayleigh used this argument in both of the cases that we have examined here. Moreover, the significance of multiple determination as a security factor against error was widely recognized among the other scientists involved in those cases.

The epistemic aims continued, though. Agreement of experimental results was also the ground on which a fact could be distinguished from an artefact. On that basis, for instance, Rayleigh and Ramsay argued that argon was a new element in the atmosphere and was not a "manufactured" product of the experimental process. Here, an existence claim was based on the epistemic strategy of multiple determination.

Another key feature of Rayleigh's experimental methodology also evident in both cases was magnifying a discrepancy or disturbance. The gist of this procedure was to control the experimental conditions via a "guided manipulation," so as to magnify the initial discrepancy and distinguish it from experimental artefacts or background noise. In this way it facilitated the search for an explanation.[57]

Finally, in both cases, Rayleigh favored involving different experimenters. In his view, the communal aspect of control was essential. Previous knowledge of the scientific community, along with accumulated experimental results, were points of reference for his checks or comparisons.[58]

All in all, various control practices were key features of Rayleigh's experimental research. Among them we may underscore the following: controlling the experimental conditions, varying experimental parameters so as to find out the underlying causes and determine their contribution to the final result, and multiple determination at every step. These practices were accompanied by an open attitude toward the scientific community, which could offer supplementary control for any results. At any rate, one thing is certain: control strategies and multiple determinations were not idle philosophical constructs, but rather indispensable elements of Rayleigh's experimental practice.[59] As he insisted many times, a multiply determined outcome was more secure than one derived from a single method. His systematic use of

---

[57] This rule of "magnifying the discrepancy" or the disturbance was applied by Rayleigh both to distinguish a fact from an artefact, and to identify an experimental error and its contribution to the final result. However, artefacts and systematic experimental errors are not the same thing.

[58] For the communal aspect of control, see also Schürch (Chap. 3, this volume).

[59] Cf. Coko (2020a, 508), who points out that multiple determination "is not a philosopher's invention, but a strategy employed by scientific practitioners themselves."

multiple determination indicates that researchers, well before the early twentieth century, recognized the epistemic force of agreement among independently produced experimental results. Jean Perrin's use of multiple determination for demonstrating the existence of atoms, which has become a canonical case in the history and philosophy of science literature, had a worthy precedent in Rayleigh.

Acknowledgments We are grateful to Bill Newman and Jutta Schickore for the invitation to participate in the conference and the workshop that gave rise to this volume. We would like to thank the conference and workshop participants for their helpful comments and questions. We are particularly indebted to Evan Arnet, Jutta Schickore, and Friedrich Steinle for their perceptive and constructive suggestions on improving our paper. Research for this paper was supported by the Hellenic Foundation for Research and Innovation (H.F.R.I.) under the "First Call for H.F.R.I. Research Projects to support Faculty members and Researchers and the procurement of high-cost research equipment grant" (Project Number: 875).

References

Arabatzis, Theodore, and Kostas Gavroglu. 2016. From Discrepancy to Discovery: How Argon Became an Element. In *The Philosophy of Historical Case Studies*, ed. T. Sauer and R. Scholl, 203–222. Cham: Springer.

Clarke, Frank Wigglesworth. [1882] 1897. *The Constants of Nature, Part 5: A Recalculation of the Atomic Weights*, 2nd edition. Washington: Smithsonian Institution.

Coko, Klodian. 2015. *The Structure and Epistemic Import of Empirical Multiple Determination in Scientific Practice*. Unpublished PhD dissertation, Indiana University.

———. 2020a. The Multiple Dimensions of Multiple Determination. *Perspectives on Science* 28 (4): 505–541.

———. 2020b. Jean Perrin and the Philosophers' Stories: The Role of Multiple Determination in Determining Avogadro's Number. *HOPOS: The Journal of the International Society for the History of Philosophy of Science* 10 (1): 143–193.

Curtis, Harvey L. 1942. A Review of the Methods for the Absolute Determination of the Ohm. *Journal of the Washington Academy of Sciences* 32 (2): 40–57.

Gooday, Graeme. 2004. *The Morals of Measurement: Accuracy, Irony, and Trust in Late Victorian Electrical Practice*. Cambridge: Cambridge University Press.

Hacking, Ian. 1983. *Representing and Intervening: Introductory Topics in the Philosophy of Natural Science*. Cambridge: Cambridge University Press.

Hunt, Bruce J. 1994. The Ohm is Where the Art is: British Telegraph Engineers and the Development of Electrical Standards. *Osiris* 9: 48–63.

Ihde, Aaron J. 1969. Theodore William Richards and the Atomic Weight Problem: He Applied Physical Chemical Principles to Critical Chemical Problems. *Science* 164 (3880): 647–651.

Jenkin, Fleeming, and William Thomson, ed. 1873. *Reports of the Committee on Electrical Standards Appointed by the British Association for the Advancement of Science, revised by Sir W. Thomson [and others]; with A Report to the Royal Society on Units of Electrical Resistance, and the Cantor Lectures, by Prof. Jenkin; ed. by F. Jenkin*. London: E. & FN Spon.

Kargon, Robert H. 1986. Henry Rowland and the Physics Discipline in America. *Vistas in Astronomy* 29: 131–136.

Kershaw, Michael. 2007. The International Electrical Units: A Failure in Standardisation? *Studies in History and Philosophy of Science Part A* 38 (1): 108–131.

Kuhn, Thomas S. 1961. The Function of Measurement in Modern Physical Science. *Isis* 52 (2): 161–193.

Lagerstrom, Larry R. 1992. *Constructing Uniformity: The Standardization of International Electromagnetic Measures, 1860–1912*. Unpublished PhD dissertation, University of Berkeley.

Lippmann, G. 1882a. Sur les méthodes à employer pour la détermination de l'Ohm. *Journal de Physique Théorique et Appliquée* 1 (1): 313–317.

———. 1882b. Méthode électrodynamique pour la détermination de l'Ohm. Mesure expérimentale dela constante d'une bobine longue. *Comptes Rendus* 95: 1348–1350.

Mascart, Éleuthère, and Jules Joubert. 1897. *Leçons sur l'électricité et le magnétisme*. Tome I, II. Paris: Masson.

Maxwell, James Clerk. 2010. General considerations concerning Scientific Apparatus. In *The Scientific Papers of James Clerk Maxwell*, ed. W.D. Niven, vol. 2, 505–522. Cambridge: Cambridge University Press.

Mitchell, Daniel Jon. 2012. Measurement in French Experimental Physics from Regnault to Lippmann. Rhetoric and Theoretical Practice. *Annals of Science* 69 (4): 453–482.

———. 2017. Making Sense of Absolute Measurement: James Clerk Maxwell, William Thomson, Fleeming Jenkin, and the Invention of the Dimensional Formula. *Studies in History and Philosophy of Science Part B: Studies in History and Philosophy of Modern Physics* 58: 63–79.

Noakes, Richard. 2019. *Physics and Psychics: The Occult and the Sciences in Modern Britain*. Cambridge: Cambridge University Press.

Olesko, Kathryn M. 1996. Precision, Tolerance, and Consensus: Local Cultures in German and British Resistance Standards. In *Scientific Credibility and Technical Standards in 19th and early 20th Century Germany and Britain*, ed. J.Z. Buchwald, 117–156. Dordrecht: Springer.

Rayleigh, Lord, 1874a. A History of the Mathematical Theories of Attraction and the Figure of the Earth from the Time of Newton to that of Laplace. By I. Todhunter M.A. F.R.S. Two Volumes. *The Academy* V: 176–77. Reprinted in Rayleigh, *Scientific Papers* Vol. I, 196–98. Cambridge: Cambridge University Press, 1899.

———, 1874b. XII. On the Manufacture and Theory of Diffraction-gratings. *The London, Edinburgh, and Dublin Philosophical Magazine and Journal of Science* 47 (310): 81–93. Reprinted in Rayleigh, *Scientific Papers*, Vol. I, 199–221. Cambridge: Cambridge University Press, 1899.

———, 1882a. XIII. Experiments to Determine the Value of the British Association Unit of Resistance in Absolute Measure. *Philosophical Transactions of the Royal Society of London* 173: 661–97. Reprinted in Rayleigh, *Scientific Papers*, Vol. II, 38–77. Cambridge: Cambridge University Press, 1900.

———, 1882b. *Address to the Mathematical and Physical Science Section of the British Association*. Spottiswoode. Reprinted in Rayleigh, *Scientific Papers*, Vol. II, 118–24. Cambridge: Cambridge University Press, 1900,

———, 1882c. XXXVIII. Comparison of Methods for the Determination of Resistances in Absolute Measure. *The London, Edinburgh, and Dublin Philosophical Magazine and Journal of Science* 14 (89): 329–46. Reprinted in Rayleigh, *Scientific Papers*, Vol. II, 134–50. Cambridge: Cambridge University Press, 1900.

———, 1887. On the Relative Densities of Hydrogen and Oxygen. Preliminary Notice. *Proceedings of the Royal Society of London* 43: 356–63. Reprinted in Rayleigh, *Scientific Papers*, Vol. III, 37–43. Cambridge: Cambridge University Press, 1902.

———, 1888. On the Composition of Water. *Proceedings of the Royal Society of London* 45: 424–30. Reprinted in Rayleigh, *Scientific Papers*, Vol. III, 233–37. Cambridge: Cambridge University Press, 1902.

———, 1890a. III. On Huygens's Gearing in Illustration of the Induction of Electric Currents. *The London, Edinburgh, and Dublin Philosophical Magazine and Journal of Science* 30 (182): 30–32. Reprinted in Rayleigh, *Scientific Papers*, Vol. III, 376–78. Cambridge: Cambridge University Press, 1902.

———, 1890b. Clerk-Maxwell's Papers. *Nature* 43: 26–27. Reprinted in Rayleigh, *Scientific Papers*, Vol. III, 426–28. Cambridge: Cambridge University Press, 1902.

———, 1892. Density of Nitrogen. *Nature* 46: 512–13. Reprinted in Rayleigh, *Scientific Papers*, Vol. IV, 1–2. Cambridge: Cambridge University Press, 1903.

———, 1894. I. On an Anomaly Encountered in Determinations of the Density of Nitrogen Gas. *Proceedings of the Royal Society of London* 55: 340–44. Reprinted in Rayleigh, *Scientific Papers*, Vol. IV, 104–108. Cambridge: Cambridge University Press, 1903.

———, 1895. Argon. *Science* 1 (26): 701–12. Reprinted in Rayleigh, *Scientific Papers*, Vol. IV, 188–202. Cambridge: Cambridge University Press, 1903.

———, 1897a. The Electro-chemical Equivalent of Silver. *Nature* 56: 292. Reprinted in Rayleigh, *Scientific Papers*, Vol. IV, 332. Cambridge: Cambridge University Press, 1903.

———, 1897b. On the Densities of Carbonic Oxide, Carbonic Anhydride, and Nitrous Oxide. *Proceedings of the Royal Society of London* 62: 204–9. Reprinted in Rayleigh, *Scientific Papers*, Vol. IV, 347–52. Cambridge: Cambridge University Press, 1903.

———, 1904. Extracts from Nobel Lecture. In Rayleigh, *Scientific Papers*, Vol. V, 212–15. Cambridge: Cambridge University Press, 1912.

———, 1906. Address of the President, Lord Rayleigh, O. M., D. C. L., at the Anniversary Meeting on November 30. *Proceedings of the Royal Society of London B* 79: 83–94.

———, 1919. Presidential Address. *Proceedings of the Society for Psychical Research* XXX: 275–90. Reprinted in Rayleigh, *Scientific Papers*, Vol. VI, 642–53. Cambridge: Cambridge University Press, 1920.

———, 1945. *The Theory of Sound*. New York: Dover Publications. Original edition, 1877.

Rayleigh, Lord, and William Ramsay. 1895. Argon, a New Constituent of the Atmosphere. *Philosophical Transactions of the Royal Society of London A* 186: 187–241. Reprinted in Rayleigh, *Scientific Papers*, Vol. IV, 130–87. Cambridge: Cambridge University Press, 1903.

Rayleigh, Lord, and Arthur Schuster. 1881. On the Determination of the Ohm in Absolute Measure. *Proceedings of the Royal Society of London* 32: 104–41. Reprinted in Rayleigh, *Scientific Papers*, Vol. II, 1–37. Cambridge: Cambridge University Press, 1900.

Rayleigh, Lord, and H. Sidgwick. 1883. Experiments, by the Method of Lorentz, for the Further Determination of the Absolute Value of the British Association Unit of Resistance, with an Appendix on the Determination of the Pitch of a Standard Tuning-fork. *Philosophical Transactions of the Royal Society of London* 174: 295–322. Reprinted in Rayleigh, *Scientific Papers*, Vol. II, 155–83. Cambridge: Cambridge University Press, 1900.

Report of the Committee Appointed by the British Association on Standards of Electrical Resistance. 1864. *Report of the Thirty-Third Meeting of the British Association for the Advancement of Science* 33: 111–76.

Rowland, Henry A. 1878. ART. XLII. Research on the Absolute Unit of Electrical Resistance. *American Journal of Science and Arts (1820–1879)* 15 (88): 281. Reprinted in *The Physical Papers of Henry Augustus Rowland*, 144–78. Baltimore: The Johns Hopkins Press, 1902.

Schaffer, Simon. 1992. Late Victorian Metrology and its Instrumentation: A Manufactory of Ohms. In *Proceedings SPIE 10309, Invisible Connections: Instruments, Institutions, and Science*. https://doi.org/10.1117/12.2283709.

———, 1994. Rayleigh and the Establishment of Electrical Standards. *European Journal of Physics* 15 (6): 277–285.

———, 1995. Accurate Measurement is an English Science. In *The Values of Precision*, ed. M.N. Wise, 135–172. Princeton, NJ: Princeton University Press.

Schickore, Jutta. 2019. The Structure and Function of Experimental Control in the Life Sciences. *Philosophy of Science* 86 (2): 203–218.

———, 2023. Peculiar Blue Spots: Evidence and Causes around 1800. In *Evidence: The Use and Misuse of Data*, ed. The American Philosophical Society, 31–55. Philadelphia, PA: American Philosophical Society.

Schickore, Jutta, and Klodian Coko. 2013. Using Multiple Means of Determination. *International Studies in the Philosophy of Science* 27 (3): 295–313.

Schuster, Arthur (Sir). 1911. *The Progress of Physics During 33 Years (1875–1908)*. Cambridge: Cambridge University Press.

——— (Sir). 1921. John William Strutt, Baron Rayleigh, 1842–1919. *Proceedings of the Royal Society of London Series A* 98, 695: i–xxxvii+xxxviii–lvii.

Spanos, Aris. 2010. The Discovery of Argon: A Case for Learning from Data? *Philosophy of Science* 77 (3): 359–380.

Strutt, Robert John. 1968. *Life of John William Strutt, Third Baron Rayleigh*. Madison, WI: The University of Wisconsin Press. Original edition, 1924.

Sweetnam, George Kean. 2000. *The Command of Light: Rowland's School of Physics and the Spectrum*. Philadelphia, PA: American Philosophical Society.

Wiedemann, G. 1882. On the Methods Employed for Determining the Ohm. *The London, Edinburgh, and Dublin Philosophical Magazine and Journal of Science* 14 (88): 258–276.

Vasiliki Christopoulou is a Ph.D. Candidate in the Department of History and Philosophy of Science at the National and Kapodistrian University of Athens. Her research focuses on the history of modern physical science, particularly on experimental practice in nineteenth-century physics.

Theodore Arabatzis is Professor of History and Philosophy of Science at the National and Kapodistrian University of Athens. His research interests include the history of modern physical sciences and historically informed philosophy of science. More detailed information about his work can be found at http://scholar.uoa.gr/tarabatz/home.

Chapter 10
Controlling Away the Phenomenon: Maze Research and the Nature of Learning

Evan Arnet

10.1 Introduction

Let me begin by clarifying two different senses, or perhaps scales, of control at play in this study. The first is familiar—experimental control, in which researchers use control measures of one kind or another within the confines of an experiment or short series of experiments. The second sense understands control as a historically extended process. From this perspective, control happens alongside enduring research programs. This second sense involves more than just introducing a control arm. An experimental context or system is successively scrutinized, thereby stabilizing or stripping away all the loose interfering parts of the world until all that remains is the object of interest. The state in which the experimental setup occurs is such that it can answer certain questions with authority, at least to the satisfaction of some inquirers, but this is always contestable—by new information (or the recovery of old), shifting norms of best practices, or a revised understanding of the components.

A historical perspective on control is unusual for two reasons. The first is that, with the partial exceptions of Jutta Schickore and Hans-Jörg Rheinberger, this is simply not how philosophers of science talk about control (Schickore 2019; Rheinberger 1997). That is, when we talk about control at all.[1] The second reason is that this sense of control pushes back against a classical Hackingesque account of the phenomena, and emphasizes not the creation of laboratory phenomena but rather the elimination of interfering nature (Hacking 1983, Ch. 13). What is important to

---

[1] Schickore reviews the surprisingly scant literature prior to this volume (Schickore 2019). See also Sullivan 2022; Guttinger 2019.

E. Arnet (✉)
Indiana University Bloomington, Bloomington, IN, USA
e-mail: earnet@indiana.edu

© The Author(s) 2024
J. Schickore, W. R. Newman (eds.), *Elusive Phenomena, Unwieldy Things*,
Archimedes 71. https://doi.org/10.1007/978-3-031-52954-2_10

scientists working in these ways is precisely that the phenomenon is not "created," but rather that it has a fidelity to nature that can be achieved only in the structured context of the experiment. To his credit, Hacking recognizes this, for instance when he states regarding the Hall effect that, "nowhere outside the laboratory is there such a pure arrangement" (Hacking 1983, 226). The development of control during a research program as seen here involves two simultaneous stories: one articulates the control practices in the experimental context and why they are effective, and the other articulates the central object of inquiry itself. Control as a historical process is also deeply social, with different experimenters testing different background assumptions and loose ends and then combining them into a kind of virtual understanding of the experimental context.

The "phenomenon" I shall discuss here is animal learning, in particular that studied with maze research. No one doubted that animals could learn. The challenge was to isolate the process of animal learning in its pure form, to understand "learning" as such. The central tension in learning experiments is that what counts as the phenomenon, its "pure" form, and what counts as the interference or impurity, are contestable categories. And, so the methodological argument goes, in trying to control every way an animal might "cheat" at maze learning, scientists all but eliminated the phenomenon of animal learning entirely. Later investigators raised this criticism explicitly because of, not in spite of, the rigorous laboratory approaches of early-twentieth-century comparative psychologists.

10.2   Phenomena and Control as a Historical Process

Jutta Schickore recently brought forth "control" as an object of historical and philosophical interest after a period of comparative neglect (Schickore 2019). Drawing from the life sciences, she argues for a distinction between "probes" and "checks, Over the course of developing an experimental research program, scientists create a confounder repertoire of factors that may interfere with the relationship of interest, namely, that between the independent and dependent variables.[2] Schickore notes that identifying the confounders and actually controlling for them are separate tasks. Checks constitute the more iconic function of control as a comparison or contrast against which the variable of interest can stand out. Probes are the necessary ancillary investigations to find and manage possible sources of interference in an experiment. More generally, probing helps one to focus on the phenomenon of interest by identifying and clearing away confounders, and the check elucidates causation in single decisive instances.

---

[2] A confounder repertoire in my understanding is best viewed as something virtual—an abstraction or reconstruction from a tradition of experimental work—as opposed to an actual complete list of confounders that exists in written form or in the mind of any single experimenter. One of Schickore's normative recommendations is for scientists to make the confounder repertoire more explicit in submitted research.

Jacqueline Sullivan extends Schickore's account into animal learning research. Investigating the development of a touchscreen operant for rats beginning in the 1990s, Sullivan attends specifically to the "dynamics" of control (Sullivan 2021, 2022). As they design the new apparatus, researchers study how varying different components affects the behavioral apparatus as a first stage of "probing" control, in Schickore's terminology. They also build in established stock controls, such as limiting extraneous auditory stimuli, for confounds well known in the field. Once the new apparatus is launched it becomes a community project, with multiple researchers critiquing, investigating, and controlling for possible sources of experimental error. The probing function of control in Sullivan's analysis is an ongoing process— it is intended not for the specification of a single experimental setup, but rather for the improvement and specification of a scientific instrument, namely, rodent touchscreen operant chambers.[3] In the shadow of its development exists a parallel lineage of control probes identifying and addressing possible confounds. The system itself is an object of investigation as scientists experimentally explore whether rodents have preferences for parts of the touchscreen, certain training regimes, or rewards.[4]

Control as understood by Schickore and Sullivan has both technical and investigatory components. As a technical practice, control involves the concrete implementation of tools, practices, and procedures to mitigate the effect of variables other than the one of interest. Alongside emerges a reflection on control practices and on methodology generally that helps to guide experiments (Schickore 2017). The technical side of control can be understood as the physical mastery of a space or system. Part of the rodent touchscreen operant chamber's appeal in the first place is that it enables a standardized setup in which a similarly standardized battery of tests can be administered relatively free from the physical involvement of the experimenter (Bussey et al. 2008; Horner et al. 2013; Dumont et al. 2021). As an investigatory practice, scientists probe the system under investigation to identify what confounds they must worry about. The notion of a "system" must be understood broadly, for, as Sullivan points out, the process is both social and collaborative, with different members of the community exploring different aspects of the rodent touch screen operant chamber and combining their findings into an overall assessment (Sullivan 2022).

This account motivates an understanding of control as a historically extended process. While a "check" or a comparison may exist in the context of a single experiment, the development of the confounder repertoire and the physical mastery of the space or system take time. Moreover, control is often an iterative process in which scientists identify and then correct for new sources of error, thus forming lineages

---

[3] For related research not on control but on the broader notion of an experimental system, see Rheinberger 1997. Like Sullivan on the operant chamber, Rheinberger emphasizes that experimental systems are historical, constituting not simply a snapshot setup but a lineage of experiments and relatively faithfully reproduced material assemblages.

[4] Similar points have been made regarding instruments being objects of reflective investigation, especially in microscopy (Schickore 2001, 2007; Rasmussen 1993, 1999; Baird 2004; Dörries 1994).

of investigation with progressive control practices. Consider, for example, the controversial Donohue–Levitt hypothesis, which says that a reduction in crime follows the legalization of abortion, on the assumption that reproductive autonomy allows parents more control over the (possibly criminogenic) situations in which their children grow up (Donohue III and Levitt 2001). The hypothesis has seen critiques pointing to possible confounds, such as changes in cocaine usage, and in response has marshaled additional data and implemented more careful statistical controls. In turn, others have responded with rejoinders (Donohue and Levitt 2004, 2020; Joyce 2004; Shoesmith 2017; Moody and Marvell 2010). To be clear, not everything done in pursuit of the Donohue–Levitt hypothesis, like looking for evidence of the effect in different countries, is best thought of as control practices. Nonetheless, a historical lineage of investigation emerges even in non-experimental cases like this, and the probing of confounds becomes an important part of these lineages.

One way to envision control as a historical process is as an expanding circle, consisting of one relatively narrowly understood central object of investigation, like a hypothesis, an instrument, or a topic such as "touchscreen learning." From it branch lines of inquiry that serve to identify and control for confounds and sources of irregularity. These branches are auxiliary investigations or modifications to the experimental setup that serve to eliminate alternative explanations or partial explanations of an observation. The research work is simultaneously creative labor, in devising possible confounds and control strategies to address them, and labor in a more straightforward sense, in doing the work to tie all these loose ends. The advantage provided by a community of investigators, both in terms of diversity of ideas and the labor of investigation, is clear (Sullivan 2022; Longino 1990, 2022).

What do control practices aim at ultimately? In the cases of experimental control where a comparison is made, or control as "check" in Schickore's terminology, the intent is often to crisply illustrate a single causal variable. From a strictly logical perspective, if it were truly certain that two experimental setups differed in only a single variable, then regardless of how messy and cluttered the experimental setups might be, they would provide clear evidence of causation. Dealing with the confounds, however, helps manage the tangle of causes surrounding an observation or an object of interest. For example, to support the Donohue–Levitt hypothesis, confounds for the observed correlation between the legalization of abortion and a reduction in crime must be cleared away. In Sullivan's example, the other factors that make a difference in the touchscreen learning experimental setup are identified, measured, and eliminated or corrected for to provide a clearer sense of the touchscreen operant chamber's utility. As another example, in the case we are about to discuss, the aim was to isolate animal learning, or at least a crucial aspect of animal learning.

The overall picture emerging here is one of elaborate simplification or isolation of the object of inquiry. In a now-classic work, Ian Hacking set out an account of "phenomena" as "something public, regular, possibly law-like, but perhaps exceptional" (Hacking 1983, 222; 1988, 1991). Notably, phenomena are not waiting out there in nature to be discovered or observed, but are rare and generally occur only in the contrived setup of the laboratory. As Hacking puts it, "The truths of science have long ceased to correspond to the world, whatever that might mean; they answer to the phenomena created in the laboratory" (Hacking 1991, 239). Hacking in turn draws from the French philosopher and historian of science Gaston Bachelard, who underscored the constructive and technological orientation of science particularly via his concept of *phenomenotechnique*, albeit in a way quite different from late-twentieth-century constructivist accounts (Bachelard 2006; Rheinberger 2005; Castelão-Lawless 1995). Inspired by contemporary work in physics, Bachelard noted that scientists are trying to realize their theoretical reality. Their guiding theory postulates certain entities and mathematical regularities of behavior, and the challenge is how, through technical mastery, to create a situation in which that theory of reality could be instantiated, observed, tested, and manipulated (Bachelard 2006). For neither Bachelard nor Hacking are phenomena "made up" in a pejorative sense; rather, they are the carefully constituted objects of scientific inquiry. Nancy Cartwright, in a similarly classic work, speaks of nomological machines. These are "fixed (enough) arrangements of components, or factors, with stable (enough) capacities that in the right sort of stable (enough) environment will, with repeated operation, give rise to the kind of regular behavior that we describe in our scientific laws" (Cartwright 1997, 66; 1999). These philosophers, by emphasizing that science is in an important way the study of complicated systems and phenomena built by scientists themselves, allowed this constructive process to become part of a larger understanding of science. They thus pushed back against a naïve understanding of brute scientific observation.

Their accounts all occurred in the context of contemporary debates and my intent here is not to relitigate those issues.[5] Moreover, these philosophers are simply correct in observing that experiments are constructed systems, and emphasizing this point has been enormously fruitful over the past several decades. Focusing on control, however, reminds us why scientists build such elaborate setups, for at root it is the world that is heterogenous and complicated and that must therefore be disciplined by contrivance to be rendered clear and predictable. Controls aim not to constitute phenomena, but to expose them, to get them alone. The hope, if not necessarily the reality, is to pull an unbroken thread from the warp and woof of the world into the confines of the laboratory.[6] This tendency appears in the early to mid-twentieth-century maze research, which sought through careful instrument and experiment design to isolate animal learning as such.

---

[5] For discussion and contextualization of the philosophy of the experiment, see Simons and Vagelli 2021.

[6] This is meant as a description of the actors, but rhetoric aside, even from an analytic perspective this need not be a realist assertion. An inquirer could place certain boundaries around a "thread" of the world (individuation) such that something like this same thread stably exists among the buzzing causal confusion of the world as well as among other experimental investigations. Yet it could still be the case both that this individuation is fundamentally arbitrary and that the thread as such is not characterized in any accurate sense beyond some of its behavior. For an elaboration of some of these ideas, see Arabatzis 2006.

However, the conceptual understanding of the object of inquiry and the control practices needed to study it are always interrelated. Along with shifting accounts of animal learning came new perspectives on early-twentieth-century animal learning experimentation, including the concern that they had controlled away the phenomenon entirely. This point illustrates a complicating factor in a historical process understanding of control, as the clarity provided by the expanding circle of research—extending out from a central hypothesis or phenomenon—can be ruined by conceptual revision of just what that central object of inquiry is. Control practices can then be recast as interference.[7]

10.3   Isolating Animal Learning in the Maze

Maze research has been a dominant approach in investigating the behavioral and cognitive features of organisms, especially in the early twentieth century. It occurred in lineages of research with maze designs being developed, copied, adapted, and modified in response to new critiques or to support new research programs.[8] Seminal figures of early American psychology, including John Watson, Clark Hull, and Edward Tolman, were all maze researchers. Nonetheless, their fascination was not with the maze as such, but rather with the maze as an instrument to unlock the secrets of animal learning. As Hoffmann discusses in this volume, organisms represent a particular challenge for control practices, for the animal enters the experimental setup with its own disposition and agency. The maze was a central way to structure the animal's behavior.

Edward Tolman concluded his 1937 American Psychology Association presidential address this way: "Let me close, now, with a final confession of faith. I believe that everything important in psychology (except perhaps such matters as the building up of a super-ego, that is, everything save such matters as involve society and words) can be investigated in essence through the continued experimental and theoretical analysis of the determiners of rat behavior at a choice-point in a maze. Herein I believe I agree with Hull and also with Professor Thorndike" (Tolman 1938, 34). Notably, Tolman was no radical, and was a chief representative of the more cognitive wing of early-twentieth-century behaviorism (Tolman 1932). My interest here is the program of control centered on maze research, which took its object of interest to be animal learning and which sought to corner the pure phenomenon in a maze alongside the experimental animal.

We begin across the Atlantic in late nineteenth-century Britain. Conwy Lloyd Morgan was a psychologist and philosopher whose 1894 text, *Introduction to*

---

[7] For an illustrative example along these lines dealing not with control *per se* but rather with discovery, see Arabatzis and Gavroglu 2016.

[8] For an exploration of how artifacts can be analyzed from an evolutionary perspective, despite the fact that they do not reproduce the way organisms do, see Wimsatt and Griesemer 2007; Baird 2004. For a sampling of the incredible diversity of maze research, see Bimonte-Nelson 2015.

Comparative Psychology, arguably inaugurated Anglo-American comparative psychology as an experimental discipline (Boakes 1984; Wilson 2002; Arnet 2019a; Dewsbury 1984). Following Herbert Spencer, George Romanes, and others, Morgan had a general theory of learning, not necessarily in the sense that animal behavior was not impacted by the environment, but in the sense that there were a small number of core underlying learning faculties, corresponding to instinct, intelligence, and reason. These he took to be hierarchically arranged in accordance with a progressive theory of mental evolution (Clatterbuck 2016; Arnet 2019a). From this perspective, it is sensible to hypothesize that all non-human animals can learn in fundamentally the same way. Morgan is best known for Morgan's canon, an interpretive rule of comparative psychology encouraging investigators to default to psychological processes lower in the "psychological scale" for the inference of mind from behavior.[9] His canon was but the first of several conservative moves regarding the animal mind within early comparative psychology, and other researchers, such as physiological psychologist Jacques Loeb, were even more deflationary. Morgan was also known for his advocacy of trial-and-error learning approaches, which provided a powerful explanatory approach for animal behavior that had previously been explained via abstract reasoning (Morgan 1896). Morgan's drive for rigor, conservativism, and experimentalism helped set the stage for early American comparative psychology (Galef Jr. 1988).

As we approach the dawn of the twentieth century, three factors converge in animal psychology. The first is a generalized understanding of learning (Seligman 1970). The second is a new laboratory experimentalism (Capshew 1992). The third is a deflationary tendency toward the animal mind (Dewsbury 2000; Arnet 2019a). To be clear, though, comparative psychology was not monolithic. This convergence is exemplified in the figure of Edward Thorndike, whose 1898 dissertation was a landmark in experimental comparative psychology (Galef Jr. 1988, 1998; Jonçich 1968; Thorndike 1898). Thorndike used "puzzle boxes," which were cages that required the animal to engage in some behavior, like pulling a wire loop, in order to get out. He designed them such that they would not trigger on wild or instinctive behavior but, ostensibly, on some purer form of learning. Thorndike clarifies that his design goal was to "get the association process…free from the helping hand of instinct" (Thorndike 1898, 9). For this reason, "Especial care was taken not to have the widest openings between bars at all near the lever, or wire-loop, or what not, which governed the bolt on the door. For the animal instinctively attacks the large openings first" (Thorndike 1898, 9). Thorndike then graphed the learning over time. If the animal were reasoning, Thorndike assumed, then at some point it would have an "ah-ha" moment that would appear on the graphs as a precipitous drop in escape time. Seeing no such drop, Thorndike concluded that there is nothing going on in the animal beyond associative learning. Reflecting on his methodology, Thorndike wrote, "The general argument of the monograph is used in all sort of scientific work

---

[9] The literature on Morgan's canon is extensive. See Costall 1993; Allen-Hermanson 2005; Thomas 2001; Sober 1998; Fitzpatrick 2008; Radick 2000; Arnet 2019a.

and is simple enough. It says: 'If dogs and cats have such and such mental functions, they will do so and so in certain situation and will not do so and so; while, on the other hand the absence of the function in question will lead to the presence of certain things and the absence of certain other things'" (Thorndike 1899, 414–15). Thorndike's view is a generalized account of learning and its epistemic implications on full display—asking how a general capacity will be made manifest in a specific situation, without assuming that the situation will in any way change how learning works.

Thorndike posited just three general types of learning: trial-and-error, imitation, and learning by ideas (Thorndike 1901, 2). In research on monkeys he argued that while humans alone learn by ideas, learning by ideas is itself an elaboration and refinement of associative learning (Thorndike 1901). Ultimately, Thorndike was able to explain almost all learning in terms of the development of associations between stimuli and responses, with animals differing both in how quickly they were able to develop those associations and in how many associations they developed. Thorndike seems to have had a formative role in at least some maze research, with his puzzle boxes sharing striking similarity to the minimalist mazes Yerkes later used in his own work (Yerkes 1901; 1902; Yerkes and Huggins 1903).

The more traditional origin for maze research is at Clark University in Massachusetts (Miles 1930; Traetta 2020). There two graduate students, Linus Kline and Willard Small, were engaged in rat learning experiments based on the model of the English Hampton Court Hedge Maze—an idea partially inspired by their adviser, Edmund Sanford, who had a strong evolutionary orientation (Goodwin 1987). In Kline's "Suggestions toward a laboratory course in comparative psychology," where he outlined his structure for a laboratory class, Kline wrote, "A careful study of the instincts, dominant traits and habits of an animal as expressed in its free life—in brief its natural history—should precede as far as possible any experimental study. Procedure in the latter case, *i.e.* by the experimental method, must of necessity be largely controlled by the knowledge gained through the former, *i.e.* by the natural method" (Kline 1899, 399, emphasis in original). Far from Thorndike's attempt to control for instinctive behavior, a maze was chosen precisely because of its similarity to the warrens used by rats. Crucially, then, the maze as envisioned by Small and Kline tested rat behavior specifically, with a fidelity to their natural environments. They also tested different species of rats and investigated the differences between them, in an approach that by and large did not continue out of Small's early experiments.[10]

While others adopted their experimental setup, they did not necessarily take up their theoretical commitments—most notably, the up-and-coming behaviorists John Watson and Harvey Carr. One central difference is that Watson wanted to study rats on their first encounter with the maze, at the beginnings of an association process, whereas Small allowed the rats to freely explore the maze prior to investigation.

---

[10] But see Florence Richard, who tested the differences between white (standard laboratory rat) and black rats (likely the Fancy rat) (Richardson 1909).

The implicit critique, one stated years earlier by Morgan, is that if one is interested in the process of learning, then simply looking at an already formed behavior is inadequate (Morgan 1894). In the famous "kerplunk" experiments of Watson and Carr, rats learned to run a Hampton Court Maze and then were confronted with a shortened version. The unfortunate subjects ran into the wall of the maze with an audible "kerplunk" (Carr and Watson 1908; Watson 1907). The hypothesis was that the rat was associating a series of kinesthetic and motor movements with each other. The associated movements unroll automatically as the rat races through the maze, leading to collision when the environment changes. Sensory cues were, if involved at all, decidedly secondary. In parallel, Watson and Carr proposed a general theory of learning, where animals like rats learn primarily by the random physical exploration of space, after which they chain their movements together. This view contradicts Small's cognitively-oriented suggestions about memory and mental processes. It also explained individual variation in rat behavior, namely as expected variation in a random process. Watson and Carr's conclusion was further supported by blinding some of the rats, which began a long and somewhat unsettling trajectory of sensory deprivation experiments. Small himself had been impressed by the efficient learning of a blind rat, although his rat was naturally blind. As usual, though, history is complicated. Far from being a direct follower of Thorndike, the early Watson alludes to the same naturalistic rationale for the selection of the maze (Watson 1907, 3).[11]

Here I want to step back and attend to the two different control regimes that are beginning to emerge but that have not yet been fully articulated. The temptation is to retreat to familiar discourse in comparative psychology that wrestles with how natural or artificial experiments are. Critiques along these lines were made of Thorndike's work by influential psychologists such as Wesley Mills and Conwy Lloyd Morgan (Mills 1899; Morgan 1898). This temptation would be, I believe, a mistake. Small, Watson and Carr, and even Thorndike, are fixated on something that (they believe) is out there in world. What they disagree on is how that "thing" is conceptualized, and correspondingly what the control practices required to study it are. For Small, although he prioritizes motor senses in the rat and particularly de-emphasizes vision, sensation is partly constitutive of rat learning (Small 1899). To know how a rat learns is to at least in part understand how it deploys its senses. Small was thus still very much concerned with experimental control and wrote in 1901 that "the aim in these experiments, as indicated above, was to make observations upon the free expression of the animal mental processes under as definitely controlled conditions as possible; and, at the same time, to minimize the inhibitive influence of restraint, confinement, and unfamiliar or unnatural circumstances" (Small 1901, 206).

Carr and Watson, motivated by successful maze learning in blind and anosmic rats, wanted to strip away the senses (sometimes literally) to get at the distilled form of learning underneath. Despite the popularity of their modified Hampton Court

---

[11] More generally, Watson's preference for a Pavlovian as opposed to a Thorndikean account of learning is discussed by Gewirtz (2001).

Maze, the Clark tradition of maze work manifested in Small's research on the rat never quite made it out of Clark. It was Watson and Carr who become influential, and their research program regarding the senses a rat needs to complete the maze took off. Importantly, their claim was not simply that a rat *can* complete a maze by merely chaining together proprioceptive cues—that is, cues related to the position of its body—but that this is in fact how rats *do* learn mazes, even with their other senses intact.[12] The proposal that the vision and olfaction of rats are essentially peripheral to their learning was treated as radical, even by generally sympathetic contemporaries (Vincent 1915d), but the work of Watson and Carr had an important impact on the core experimental logic and research questions of early maze investigations.

There are two overarching features to much of this early work. The first is that rat learning is conceptualized in terms of what specific problems the rat can complete in the context of the maze (e.g., whether it can complete a maze with certain design features or remember whether food is on the left or right in a series of trials; see Hunter and Nagge 1931; Carr 1917). A maze of one design or another is almost always the context for testing animal capacities. Whether or not the maze still resembled rodent warrens, which resemblance was a key property of the Hampton Court design, is no longer a matter of concern.[13] The second overarching feature is that the cues available to the rat are carefully restricted.

Even for those who hypothesized that rats were using more than proprioceptive cues, the research program was often one of disaggregating the role of different senses in rat learning. Central to this project was preventing the animal from using environmental cues outside the maze—and thus began an elaborate tradition of experimental control involving myriad modifications to animals, the maze, and surrounding environments. In her research, Stella Vincent removed whiskers to evaluate their impact on maze performance. She also painted the correct path white and erroneous paths black to test the involvement of vision, among other controls (Vincent 1915d; 1912).[14] Notably, Vincent explicitly saw the introduction of additional visual information into the maze design as a control for the hypothesis that rats are using kinesthetic information alone. Warner and Warden sought to standardize the maze design itself (Warden 1929; Warner and Warden 1927). Florence Richardson, who studied with Watson, performed similar sensory deprivation experiments, but also manipulated the complexity of the task by using "problem boxes" apart from the maze (Richardson 1909). Walter S. Hunter developed a temporal maze in which spatial clues were eliminated (Hunter 1920).

If we take a step back, we see first the fixation on animal learning, and more precisely Watson and Carr's kinesthetic hypothesis as the initial object of interest. We then see the expanding circle of experimentation, which manipulates

---

[12] This assumption was challenged from the outset; see, for example, Washburn 1908.

[13] Not everyone was so enamored. B. F. Skinner partially eschewed mazes in his later research, and influential critiques came from maze researchers such as Walter S. Hunter (Hunter 1926; Skinner 1938).

[14] Also see Vincent 1915a, 1915b, 1915c.

surrounding variables such as the maze design and available sensory cues. Alongside these more dramatic practices, a collection of stock control measures, such as starting experiments as the same time, keeping the experimental area free of wild rats and their odor trails, placing rewards in the same place, and maintaining stable food reward amounts and varieties between rats, develops as well.[15] This is the collection of a thousand and one little things that maintain the integrity of an experimental system.[16]

All of these control practices in combination serve to purify the *phenomenon* of animal learning.[17] That is, they free it from intervening variables. This is especially true for Watson and Carr, who were trying to establish that the association of random movements into fluid behavior was a vast part of what animal learning is. Control practices are relational, relying on a specific understanding of the target of investigation. In the tradition established by Watson and Carr, kinesthetic association as the nucleus of animal learning remains constant even if rats do occasionally incorporate other senses—and therefore their experimental program of stripping away the senses is a compelling one. If learning is not additive in this way, and instead learning is holistically different when more senses are involved, then the relationship between sensory deprivation (the control practice) and animal learning (the target of control) is also different. Just what is being exhibited by sensory-deprived rats (e.g., *one way* an animal can learn a maze, versus how an animal *always* learns a maze) is a site of conflict and fertile ground for introducing additional probes and checks.

10.4 Reconceptualizing Animal Learning

Nonetheless, for all the experimental rigor and complexity on display, this early-twentieth-century emphasis on proprioception and the concatenation of random movements in animal learning faded away. There is a larger story here, but I shall skip to the latter half of the century and focus on two researchers, David Olton and William Timberlake.[18] Both were trained at the University of Michigan, a leading location for neo-Hullian learning theory (Arnet 2019b; Shapiro n.d.).[19] Both focused on rat behavior. And both explicitly wrestled with the legacy of early maze research.

---

[15] Sullivan refers to this collection of standard control practices as "canonical" (Sullivan 2022, 1207).

[16] These matters also relate to whether an experimental system can effectively be maintained and reproduced between different uses, different researchers, and different labs. In this sense it is a detailed description that facilitates sameness of setup, but in the context of a single trial or experiment, sameness comes from inventorying and clearing away other possible intervening variables.

[17] For a similar point, see Steinle on the epistemic goal of exploratory experimentation (1997).

[18] For more general historical discussion of twentieth-century comparative psychology, see Burkhardt 2005; Braat et al. 2020; Dewsbury 1984; Watrin 2017; Watrin and Darwich 2012.

[19] Although Olton also focused on neurophysiology.

Their analysis recasts early experimental work as stripping away not simply intervening variables, but also crucial aspects of animal learning itself.

Olton is best known for his research on spatial memory and "place learning," here understood as the use of discriminative stimuli associated with a particular location (e.g., the way you could know that you were in a certain room by a familiar picture on the wall; see Shapiro n.d.). In a 1976 paper, Olton and Robert Samuelson introduced the radial arm maze (Olton and Samuelson 1976). The eight identical spokes of the maze radiating from a central point were intended to force the rat to use spatial cues from the surrounding environment to orient itself in the maze.[20] In their initial study, a food reward was placed on the end of each spoke, a rat placed in the center, and then the movement through the maze monitored. Rats were tested in how quickly they could get to each food reward without revisiting a spoke (which would not have new food). Orienting themselves with respect to the older maze research tradition, Olton and Samuelson wrote, "In spite of the ubiquitous nature of place learning, most experiments have treated place learning as a factor to be controlled and have chosen to assess rats' cognitive abilities by making place learning impossible" (Olton and Samuelson 1976, 97). The most dramatic example is the research of Walter S. Hunter mentioned above, who had rats memorize series of left and right alternations all done in a single box, thereby trying to eliminate spatial clues entirely. Olton and Samuelson instead made place learning the object of investigation in their study. But Olton was interested in more than just a change in focus. He contended that even Watson and Carr's original aspiration to understand animal learning was undermined by controlling away spatial cues as interference, when those cues were actually essential to understand how rats learn and navigate. In his historical work, he called out the maze explicitly:

> Structural characteristics of the apparatus [maze] suppress some kinds of behaviors and enhance others. Thus the maze itself influenced the types of behaviors and the types of theories that were developed from these observations. On the other hand, mazes reflected the theoretical biases of their users. Experimenters had a predilection to address certain types of issues, and the mazes were constructed with these issues in mind. (Olton 1979, 583)

The early maze tradition, especially as a reaction to Watson and Carr, had primarily been one of narrowing and isolating the capacities of the animal in order to see what functions remain when the animal is stripped of its senses and environmental cues. Researchers assumed that this practice did not distort the phenomenon; that is, they believed that the effect of additional sensory cues is at most additive to the underlying skeleton of learning. (In the extreme case of Watson and Carr's early research, the additional senses hardly did anything at all.) In addition, in terms of understanding animal capacities, the Watson and Carr-led maze research tradition sought to understand what an animal can still accomplish given certain restrictions, but it did not seek to know what the animal can do with full recourse to their faculties. As Olton put it later, "On the one hand, rats consistently demonstrate a preference for

---

[20] For example, if the maze were placed in a room in which there was a sink, a cabinet, and a poster visible on the walls from the maze, then these would be spatial cues.

solving discrimination tasks on the basis of spatial cues; on the other hand, experiments just as consistently prevent rats from exploiting this preference" (Olton 1978, 341). Olton and Samuelson argued that it is because of a history of certain kinds of control practices that scientists allegedly had a stunted vision of rat learning and cognition. In their 1976 paper, Olton and Samuelson conclude "the introduction of a spatial location paradigm may change [increase] our estimate of rat's cognitive capacities"(114). They also tied their work to ecological considerations, such as foraging behavior, based on the kinds of foraging strategies and food finding capacities rats exhibited more generally (Olton 1978). What is invariant between expected natural behavior and the laboratory is no longer something as abstract as the general structure of learning, but a specific foraging strategy (win-shift) that can be triggered in the lab.

Nonetheless, Olton's core interest remained spatial memory. The reconceptualization of animal learning as such is made far more explicit by another researcher, William Timberlake.

Timberlake began as a learning theorist but quickly took to more ethology-inflected work and integrated it into his behavior systems approach (Arnet 2019b). He was part of a larger movement looking to bring evolution and ecology to American laboratory psychology, and to studies of learning in particular. This movement also included researchers such as Sara Shettleworth, Robert Bolles, Martin Seligman, and Michael Domjan. I shall focus on two of Timberlake's criticisms for maze research.

The first was a recovery of the initial ecological focus of the maze, testing questions such as whether a rat would be motivated to run a maze even with no reward (Timberlake 1983b). In a reinforcement approach to animal learning, it was assumed that the incentive to learn an artificial system such as a maze was controlled for by not providing the animal food so that it hungered, and then allowing food alone to serve as an interventionist variable. Timberlake challenged this assumption by providing evidence that rats have intrinsic motivation to engage in edge-following behavior (i.e., to run mazes); this challenges the very idea that animal motivation was being controlled as assumed (Timberlake 2002).[21] Great experimentalists such as B. F. Skinner had, in Timberlake's estimation, partly put themselves on the map by getting the animal to cooperate in the artificial circumstances of the laboratory. This achievement occurs against the backdrop of the tongue-in-cheek Harvard Law of Animal Behavior: "Under carefully controlled experimental conditions, the animal will behave as it damn well pleases." Experimentally cooperative animal behavior, however, was not the product of luck; according to Timberlake, it was generated by the experimentalist's carefully fitting the setting and apparatuses to the physical and behavioral features of the organism.[22] There are reasons that pigeons were expected to peck keylights, that auditory stimuli were placed in ranges that did not

---

[21] Although early-twentieth-century researchers had explored this concept as part of broader studies of rat motivation, albeit without Timberlake's ecological tie-in. Timberlake himself references this literature (Timberlake 1983b, 170–71). See also Tolman 1930.

[22] For a peek behind the curtain, see Skinner 1956; Hoffman, this volume.

startle animals, and that passages were approximately rat-sized. From lever shape to maze design, instruments are tuned to organisms. While demonstrative of experimental skill, Timberlake contended that the epistemological effect of tuning was to smooth the ecological traits into the experimental backdrop where they can no longer be seen. He also argued that tuning facilitated the extremely general and abstract accounts of learning that characterized early-twentieth-century learning theory. By design, experiments and instruments were intended to suppress behavioral instincts and idiosyncrasies in order to produce something standardizable and easy to study. They sought a fit between organism and experiment and, once achieved, Timberlake thought that it was too easy to forget the species-specific design principles that made it possible. His own research interpreted laboratory practices as selectively eliciting aspects of the animals' evolved patterns of motivation and behavior (Timberlake 2002).

Second, along with other ecologically influenced psychologists, Timberlake wanted to reconceptualize animal learning and critiqued general process accounts in which learning is understood as abstract and domain-general. Timberlake in particular adopted a behavior systems approach, a development of the framework originally suggested by the famed ethologist Niko Tinbergen (Timberlake 1993, 1983a; Bowers 2017, 2018). On this view, organismal behavior was understood as a structured and hierarchical system of motivations and associated behaviors that had formed in the environment and evolutionary history of the organism. From the ecological perspective, learning is no longer domain-general, as traditionally characterized in behaviorist and learning-theoretic approaches. Instead, the sensitivity and richness of animal learning is, by evolution and development, rooted in the actual environment of the animal. These characteristics then make it into the lab. When undergoing reinforcement learning, animals readily "misbehave" by exhibiting species-typical behavior even when these behaviors are not reinforced (Breland and Breland 1961), and more ethology-inflected psychologists argue that it is easiest for animals to learn along the contours of their existing patterns of behavior. Timberlake writes, "Researchers…are studying niche-related behavior in specific species, whether they planned to or not" (Timberlake 2002, 372). Consequently, any attempt to fully abstract away from this in order to achieve a pure form of animal learning would contradict what constitutes learning, at least from the perspective of Timberlake and other like-minded psychologists.

Neither Olton nor Timberlake are, to be sure, opposed to control practices generally. Olton and Samuelson used Old Spice deodorant as an olfactory control in their early research, on the assumption that if rats were following odor trials—an alternative hypothesis to spatial learning—their performance would be worse if the apparatus were doused in Old Spice. Instead, Olton, Samuelson, and others are recasting the relationship between control practices and the phenomenon of interest. Olton contends that researchers cannot see the importance of spatial cues in learning because they have for years simply controlled it away as part of their experimental setup. He reintroduces spatial cues to experimental paradigms in order to illuminate a (hopefully) more complete animal learning. Similarly for Timberlake: mazes, as instruments composed of walls and edges, had been hiding the natural tendency of rats to follow trails and edges as opposed to using other patterns of search and loco-motion. Stock control practices of food deprivation, to ensure that rats are hungry for food rewards, therefore masked the rats' intrinsic motivation (Timberlake 1983b, 2002; Hoffman et al. 1999; Timberlake et al. 1999).

Along with reconceptualizing learning to emphasize the evolved behaviors and tendencies the animal brought into the experimental situation, Timberlake sought to reopen long-closed features of experimental design. For example, building off the work of Pavlov, in mid-twentieth-century animal psychology, the predictive stimu-lus was generally seen to be a neutral stimulus with the animal's reaction to the stimulus being shaped purely by the experimental context. (A predictive stimulus simply indicates the coming of another stimulus, e.g., a light turning on to signal the arrival of food.) In experimental work, Timberlake found that when a live rat was used as the predictive stimulus, and then food was presented, it elicited social feed-ing behaviors from the subject rat (Timberlake and Grant 1975). His point was not that traditional stimuli such as key-lights are artificial, but rather that they are not neutral and may intersect with the dispositions of the research animal. Put differ-ently, they may be confounds.

Unlike some ethology-inflected researchers, Timberlake was enthusiastic about the structure and control provided by laboratory investigations. He was, after all, trained in the American laboratory tradition of comparative psychology. Where he and researchers such as Watson and Carr would disagree is over the phenomena they see at the heart of the experimental system. On Timberlake's account, there is no abstract general process structure of learning to be found; the learning theorist is instead trying to investigate a sophisticated hierarchically organized structure of motivation and behavior, the behavior system, which the animal brings into the lab. Animal learning is not simply association or reinforcement but modification of the animal behavior system. Control practices for Timberlake must be understood in relation to the specific animal under study, and universality, where it exists at all, is a function of shared evolutionary history and biological needs, rather than the nature of learning.

Olton's research on spatial learning becomes enormously influential in the field; Timberlake's specific behavior systems approach somewhat less so. But perhaps the biggest beneficiary of Olton's and Timberlake's fight with the past has been the rat. Olton and Timberlake were part of a larger shift in late twentieth-century psychol-ogy that emphasized the sophistication and nuance of animals. This shift appreci-ated animals' cognitive and behavioral capacities and saw their dispositions in light of their evolutionary history.[23] Early researchers were not interested in all that an animal could do. The late nineteenth century was, after all, overflowing with accounts of the incredible observed behavior of animals that had nonetheless been successfully explained by simple approaches such as trial-and-error learning (Romanes 1882; Morgan 1894). And so scientists such as Watson and Carr sought the underlying architecture of learning, controlling away ecological variability and

---

[23] The essential text for the revival of interest in the animal mind is Donald Griffin's *The Question of Animal Awareness* (Griffin 1976).

the functional capacities of the experimental animal—sometimes through paint and protocols, sometimes through surgery. Experimental design and associated control practices emerge as a few factors among many that lead to conservativism about the animal mind in early American comparative psychology.

10.5 Conclusion

Scholars such as Ian Hacking and Hans-Jörg Rheinberger have emphasized the built nature of the experiment. Phenomena are not simply stumbled across, but have to be carefully created in the confines of the laboratory. Hacking pairs this with his famous characterization of the "self-vindication" of experimental work, in which a form of coherence is achieved through different aspects of the experimental setup being fit and calibrated to each other (Hacking 1992). This is a powerful analysis. We can see such fit implicitly at play in maze research, especially in the critiques of Olton and Timberlake on the theoretical limitations imposed by maze design.

Taking a perspective of control as an intended purificatory process, in which scientists attempt to stabilize intervening variables to expose the contours and nature of a phenomenon or an intervention, can foreground other aspects of experimental work. First, it highlights that the fit or coherence is more than the smooth operation of the experimental system. It is a critical coherence based on scientists scrutinizing the system in the hope of detecting confounds that are hard to detect precisely because the experimental system operates smoothly whether they are present or not. Second, and true to the specifically historical perspective adopted by this paper, control helps to make sense of how scientists relate experiments to each other. For instance, Stella Vincent filled a gap in Watson and Carr's research by seeing the impact made by a purely visual variable. Over the short term, the cumulative effect of control practices is relatively linear and progresses by explaining how, in the contexts of experimental traditions, an expanding circle of control helps to clear away intruding causes and thus expose phenomena. Longer-term, however, there may be more drastic shifts in the understanding of the phenomenon of interest, and consequently in the relationship between control practices and phenomena. Olton's asterisk-shaped maze and experimental setups sought to recover previously controlled-away spatial cues. Timberlake showed that stimuli previously regarded as neutral can actually be confounds. Perhaps surprisingly, early conceptualizations of learning stemming from maze research were critiqued specifically because of their tightly implemented control practices and experimental rigor—they had controlled away the phenomenon.

Acknowledgments This article benefited from the immense assistance and insights of the other contributors to this volume and the Mellon-Sawyer Seminar more broadly. Particular thanks go to Kärin Nickelsen and Caterina Schürch for their incisive feedback on an earlier draft, and to Jutta Schickore, Claudia Cristalli, Bill Newman, and Jared Neumann for being fellow travelers in the investigation of scientific methodology. It is only with their help that I have (hopefully) shed light on the smallest corner of experimental control.

References

Allen-Hermanson, Sean. 2005. Morgan's Canon Revisited. *Philosophy of Science* 72 (4): 608–631.

Arabatzis, Theodore. 2006. *Representing Electrons: A Biographical Approach to Theoretical Entities*. Chicago: University of Chicago Press.

Arabatzis, Theodore, and Kostas Gavroglu. 2016. From Discrepancy to Discovery: How Argon Became an Element. In *The Philosophy of Historical Case Studies*, ed. Tilman Sauer and Raphael Scholl, 203–222. Boston Studies in the Philosophy and History of Science. Cham: Springer International Publishing. https://doi.org/10.1007/978-3-319-30229-4_10.

Arnet, Evan. 2019a. Conwy Lloyd Morgan, Methodology, and the Origins of Comparative Psychology. *Journal of the History of Biology* 52 (3): 433–461. https://doi.org/10.1007/s10739-019-09577-2.

———. 2019b. William Timberlake: An Ethologist's Psychologist. *Behavioural Processes* 166 (September): 103895. https://doi.org/10.1016/j.beproc.2019.103895.

Bachelard, Gaston. 2006. Noumenon and Microphysics. *The Philosophical Forum* 37 (1): 75–84. https://doi.org/10.1111/j.1467-9191.2006.00230.x.

Baird, Davis. 2004. *Thing Knowledge: A Philosophy of Scientific Instruments*. Berkeley: University of California Press.

Bimonte-Nelson, Heather A. 2015. *The Maze Book: Theories, Practice, and Protocols for Testing Rodent Cognition*. New York, NY: Humana.

Boakes, Robert A. 1984. *From Darwin to Behaviourism: Psychology and the Minds of Animals*. Cambridge: Cambridge University Press. http://agris.fao.org/agris-search/search.do?recordID=US201300673256.

Bowers, Robert Ian. 2017. Behavior Systems. In *Encyclopedia of Animal Cognition and Behavior*, ed. Jennifer Vonk and Todd Shackelford, 1–8. Cham: Springer International Publishing. https://doi.org/10.1007/978-3-319-47829-6_1232-1.

———. 2018. A Common Heritage of Behaviour Systems. *Behaviour* 155 (5): 415–442.

Braat, Michiel, Jan Engelen, Ties van Gemert, and Sander Verhaegh. 2020. The Rise and Fall of Behaviorism: The Narrative and the Numbers. Preprint. January 29, 2020. http://philsci-archive.pitt.edu/16865/.

Breland, Keller, and Marian Breland. 1961. The Misbehavior of Organisms. *American Psychologist* 16 (11): 681–684. https://doi.org/10.1037/h0040090.

Burkhardt, Richard W. 2005. *Patterns of Behavior: Konrad Lorenz, Niko Tinbergen, and the Founding of Ethology*. Chicago: University of Chicago Press.

Bussey, Timothy J., Tina L. Padain, Elizabeth A. Skillings, Boyer D. Winters, A. Jennifer Morton, and Lisa M. Saksida. 2008. The Touchscreen Cognitive Testing Method for Rodents: How to Get the Best out of Your Rat. *Learning & Memory* 15 (7): 516–523. https://doi.org/10.1101/lm.987808.

Capshew, James H. 1992. Psychologists on Site: A Reconnaissance of the Historiography of the Laboratory. *American Psychologist* 47 (2): 132–142. https://doi.org/10.1037/0003-066X.47.2.132.

Carr, Harvey. 1917. A Preliminary Study. *The Journal of Animal Behavior* 7: 365.

Carr, Harvey, and John B. Watson. 1908. Orientation in the White Rat. *Journal of Comparative Neurology and Psychology* 18 (1): 27–44.

Cartwright, Nancy. 1997. Where Do Laws of Nature Come From?*. *Dialectica* 51 (1): 65–78. https://doi.org/10.1111/j.1746-8361.1997.tb00021.x.

———. 1999. *The Dappled World: A Study of the Boundaries of Science*. Cambridge: Cambridge University Press.

Castelão-Lawless, Teresa. 1995. Phenomenotechnique in Historical Perspective: Its Origins and Implications for Philosophy of Science. *Philosophy of Science* 62 (1): 44–59. https://doi.org/10.1086/289838.

Clatterbuck, Hayley. 2016. Darwin, Hume, Morgan, and the Verae Causae of Psychology. *Studies in History and Philosophy of Science Part C: Studies in History and Philosophy of Biological and Biomedical Sciences* 60 (December): 1–14. https://doi.org/10.1016/j.shpsc.2016.09.002.

Costall, Alan. 1993. How Lloyd Morgan's Canon Backfired. *Journal of the History of the Behavioral Sciences* 29 (2): 113–122. https://doi.org/10.1002/1520-6696(199304)29:2<113:: AID-JHBS2300290203>3.0.CO;2-G.

Dewsbury, Donald A. 1984. *Comparative Psychology in the Twentieth Century*. Stroudsburg: Hutchinson Ross Publishing Company.

———. 2000. Issues in Comparative Psychology at the Dawn of the 20th Century. *American Psychologist* 55 (7): 750–753. https://doi.org/10.1037/0003-066X.55.7.750.

Donohue, John J., III, and Steven D. Levitt. 2001. The Impact of Legalized Abortion on Crime*. *The Quarterly Journal of Economics* 116 (2): 379–420. https://doi.org/10.1162/00335530151144050.

Donohue, John J., and Steven D. Levitt. 2004. Further Evidence That Legalized Abortion Lowered Crime: A Reply to Joyce. *The Journal of Human Resources* 39 (1): 29–49. https://doi.org/10.2307/3559004.

Donohue, John J., and Steven Levitt. 2020. The Impact of Legalized Abortion on Crime over the Last Two Decades. *American Law and Economics Review* 22 (2): 241–302. https://doi.org/10.1093/aler/ahaa008.

Dörries, Matthias. 1994. Balances, Spectroscopes, and the Reflexive Nature of Experiment. *Studies in History and Philosophy of Science Part A* 25 (1): 1–36. https://doi.org/10.1016/0039-3681(94)90018-3.

Dumont, Julie R., Ryan Salewski, and Flavio Beraldo. 2021. Critical Mass: The Rise of a Touchscreen Technology Community for Rodent Cognitive Testing. *Genes, Brain and Behavior* 20 (1): e12650. https://doi.org/10.1111/gbb.12650.

Fitzpatrick, Simon. 2008. Doing Away with Morgan's Canon. *Mind & Language* 23 (2): 224–246. https://doi.org/10.1111/j.1468-0017.2007.00338.x.

Gewirtz, Jacob L. 2001. J. B. Watson's Approach to Learning: Why Pavlov? Why Not Thorndike? *Behavioral Development Bulletin* 10: 23–25. https://doi.org/10.1037/h0100478.

Goodwin, C.J. 1987. In Hall's Shadow: Edmund Clark Sanford (1859–1924). *Journal of the History of the Behavioral Sciences* 23 (2): 153–168. https://doi.org/10.1002/1520-6696(19870 4)23:2<153::aid-jhbs2300230205>3.0.co;2-4.

Griffin, Donald Redfield. 1976. *The Question of Animal Awareness: Evolutionary Continuity of Mental Experience*. New York, NY: Rockefeller University Press.

Guttinger, Stephan. 2019. A New Account of Replication in the Experimental Life Sciences. *Philosophy of Science* 86 (3): 453–471. https://doi.org/10.1086/703555.

Hacking, Ian. 1983. *Representing and Intervening: Introductory Topics in the Philosophy of Natural Science*. Cambridge; New York: Cambridge University Press.

———. 1988. The Participant Irrealist at Large in the Laboratory. *The British Journal for the Philosophy of Science* 39 (3): 277–294.

———. 1991. Artificial Phenomena. *The British Journal for the History of Science* 24 (2): 235–241. https://doi.org/10.1017/S0007087400027096.

———. 1992. *The Self-Vindication of the Laboratory Sciences,[w:] A. Pickering (red.), Science as Practice and Culture*, 29–64. London: The University of Chicago Press, Chicago.

Hoffman, Cynthia M., William Timberlake, Joseph Leffel, and Rory Gont. 1999. How Is Radial Arm Maze Behavior in Rats Related to Locomotor Search Tactics? *Animal Learning & Behavior* 27 (4): 426–444. https://doi.org/10.3758/BF03209979.

Horner, Alexa E., Christopher J. Heath, Martha Hvoslef-Eide, Brianne A. Kent, Chi Hun Kim, Simon R.O. Nilsson, Johan Alsiö, et al. 2013. The Touchscreen Operant Platform for Testing Learning and Memory in Rats and Mice. *Nature Protocols* 8 (10): 1961–1984. https://doi.org/10.1038/nprot.2013.122.

Hunter, Walter S. 1920. The Temporal Maze and Kinaesthetic Sensory Processes in the White Rat. *Psychobiology* 2: 1–17. https://doi.org/10.1037/h0073855.

Hunter, Walter S. 1926. "A Reply to Professor Carr on 'The Reliability of the Maze Experiment'."

Hunter, Walter S., and Joseph W. Nagge. 1931. The White Rat and the Double Alternation Temporal Maze. *The Pedagogical Seminary and Journal of Genetic Psychology* 39 (3): 303–319. https://doi.org/10.1080/08856559.1931.10532407.

Jonçich, Geraldine. 1968. *The Sane Positivist: A Biography of Edward L. Thorndike*. Middletown: Wesleyan University Press.

Joyce, Ted. 2004. Did Legalized Abortion Lower Crime? *Journal of Human Resources* XXXIX (1): 1–28. https://doi.org/10.3368/jhr.XXXIX.1.1.

Galef Jr. Bennett G. 1988. "Evolution and Learning before Thorndike: A Forgotten Epoch in the History of Behavioral Research." In Evolution and Learning, edited by R. C. Bolles and M. D. Beecher, 39–58. Hillsdale: Lawrence Erlbaum Associates, Inc.

Galef, Bennett G., Jr. 1998. Edward Thorndike: Revolutionary Psychologist, Ambiguous Biologist. *American Psychologist* 53 (10): 1128–1134. https://doi.org/10.1037/0003-066X.53.10.1128.

Kline, Linus W. 1899. Suggestions toward a Laboratory Course in Comparative Psychology. *The American Journal of Psychology* 10 (3): 399–430. https://doi.org/10.2307/1412142.

Longino, Helen E. 1990. *Science as Social Knowledge: Values and Objectivity in Scientific Inquiry*. Princeton: Princeton University Press.

———. 2022. What's Social about Social Epistemology? *The Journal of Philosophy* 119 (4): 169–195. https://doi.org/10.5840/jphil2022119413.

Miles, Walter R. 1930. On the History of Research with Rats and Mazes: A Collection of Notes. *The Journal of General Psychology* 3 (2): 324–337.

Mills, Wesley. 1899. The Nature of Animal Intelligence and the Methods of Investigating It. *Psychological Review* 6 (3): 262–274. https://doi.org/10.1037/h0074808.

Moody, Carlisle E., and Thomas B. Marvell. 2010. On the Choice of Control Variables in the Crime Equation. *Oxford Bulletin of Economics and Statistics* 72 (5): 696–715.

Morgan, Conwy Lloyd. 1894. *An Introduction to Comparative Psychology*, Contemporary Science Series. London: Walter Scott Publishing Co.

———. 1896. *Habit and Instinct*. London: E. Arnold.

———. 1898. Review of Thorndike's Animal Intelligence. *Nature* 58 (1498): 249–250.

Olton, David S. 1978. Characteristics of Spatial Memory. In *Cognitive Processes in Animal Behavior*, ed. Stewart H. Hulse, Harry Fowler, and Werner K. Honig, 341–373. Routledge.

———. 1979. Mazes, Maps, and Memory. *American Psychologist* 34 (7): 583–596. https://doi.org/10.1037/0003-066X.34.7.583.

Olton, David S., and Robert J. Samuelson. 1976. Remembrance of Places Passed: Spatial Memory in Rats. *Journal of Experimental Psychology: Animal Behavior Processes* 2 (2): 97–116. https://doi.org/10.1037/0097-7403.2.2.97.

Radick, Gregory. 2000. Morgan's Canon, Garner's Phonograph, and the Evolutionary Origins of Language and Reason. *The British Journal for the History of Science* 33 (1): 3–23. https://doi.org/10.1017/S0007087499003842.

Rasmussen, Nicolas. 1993. Facts, Artifacts, and Mesosomes: Practicing Epistemology with the Electron Microscope. *Studies in History and Philosophy of Science Part A* 24 (2): 227–265.

———. 1999. *Picture Control: The Electron Microscope and the Transformation of Biology in America, 1940–1960*. Stanford: Stanford University Press.

Rheinberger, Hans-Jörg. 1997. *Toward a History of Epistemic Things: Synthesizing Proteins in the Test Tube (Writing Science)*. Stanford: Stanford University Press.

———. 2005. Gaston Bachelard and the Notion of 'Phenomenotechnique. *Perspectives on Science* 13 (3): 313–328. https://doi.org/10.1162/106361405774288026.

Richardson, Florence. 1909. A Study of the Sensory Control in the Rat. *The Psychological Review: Monograph Supplements* 12 (1): i–124. https://doi.org/10.1037/h0093009.

Romanes, George J. 1882. *Animal Intelligence. The International Scientific Series*. London: Kegan Paul, Trench & Co.

Schickore, Jutta. 2001. Ever-Present Impediments: Exploring Instruments and Methods of Microscopy. *Perspectives on Science* 9 (2): 126–146. https://doi.org/10.1162/106361401317447255.

———. 2007. *The Microscope and the Eye: A History of Reflections, 1740–1870*. Chicago: University of Chicago Press.

———. 2017. *About Method: Experimenters, Snake Venom, and the History of Writing Scientifically*. Chicago: University of Chicago Press.

———. 2019. The Structure and Function of Experimental Control in the Life Sciences. *Philosophy of Science* 86 (2): 203–218.

Seligman, Martin E. 1970. On the Generality of the Laws of Learning. *Psychological Review* 77 (5): 406–418. https://doi.org/10.1037/h0029790.

Shapiro, Matthew L. n.d. "Olton, David (1943–1994) l." Encyclopedia.Com. Accessed 14 Sept 2022. https://www.encyclopedia.com/psychology/encyclopedias-almanacs-transcripts-and-maps/olton-david-1943-1994

Shoesmith, Gary L. 2017. Crime, Teenage Abortion, and Unwantedness. *Crime & Delinquency* 63 (11): 1458–1490. https://doi.org/10.1177/0011128715615882.

Simons, Massimiliano, and Matteo Vagelli. 2021. Were Experiments Ever Neglected? Ian Hacking and the History of Philosophy of Experiment. *Philosophical Inquiries* 9 (1): 167–188. https://doi.org/10.4454/philinq.v9i1.339.

Skinner, B.F. 1938. *The Behavior of Organisms: An Experimental Analysis*. Oxford: Appleton-Century.

———. 1956. A Case History in Scientific Method. *American Psychologist* 11: 221–233. https://doi.org/10.1037/h0047662.

Small, Willard S. 1899. Notes on the Psychic Development of the Young White Rat. *The American Journal of Psychology* 11 (1): 80–100.

———. 1901. "Experimental Study of the Mental Processes of the Rat. II." The American Journal of Psychology 12 (2): 206–239. https://doi.org/10.2307/1412534.

Sober, Elliott. 1998. Morgan's Canon. In *The Evolution of Mind*, ed. D.D. Cummins and C. Allen, 224–242. New York, NY: Oxford University Press.

Steinle, Friedrich. 1997. Entering New Fields: Exploratory Uses of Experimentation. *Philosophy of Science* 64 (S4): S65–S74.

Sullivan, Jacqueline A. 2021. Understanding Stability in Cognitive Neuroscience through Hacking's Lens. *Philosophical Inquiries* 9 (1): 189–208. https://doi.org/10.4454/philinq.v9i1.346.

———. 2022. Novel Tool Development and the Dynamics of Control: The Rodent Touchscreen Operant Chamber as a Case Study. *Philosophy of Science* 89: 1203–1212. https://doi.org/10.1017/psa.2022.63.

Thomas, Roger K. 2001. *Llloyd Morgan's Canon: A History of Its Misrepresentation*. History and Theory in Psychology.

Thorndike, Edward L. 1898. Animal Intelligence: An Experimental Study of the Associative Processes in Animals. *The Psychological Review: Monograph Supplements* 2 (4): i–109. https://doi.org/10.1037/h0092987.

———. 1899. A Reply to ' The Nature of Animal Intelligence and the Methods of Investigating It.'. *Psychological Review* 6 (4): 412–420. https://doi.org/10.1037/h0073289.

———. 1901. The Mental Life of the Monkeys. *The Psychological Review: Monograph Supplements* 3 (5): i–57. https://doi.org/10.1037/h0092994.

Timberlake, William. 1983a. The Functional Organization of Appetitive Behavior: Behavior Systems and Learning. *Advances in Analysis of Behavior* 3: 177–221.

———. 1983b. Appetitive Structure and Straight Alley Running. In *Animal Cognition and Behavior*, ed. Roger L. Mellgren, 165–222. Advances in Psychology, vol. 13. Amsterdam: North-Holland. https://doi.org/10.1016/S0166-4115(08)61797-5.

———. 1993. Behavior Systems and Reinforcement: An Integrative Approach. *Journal of the Experimental Analysis of Behavior* 60 (1): 105–128. https://doi.org/10.1901/jeab.1993.60-105.

———. 2002. Niche-Related Learning in Laboratory Paradigms: The Case of Maze Behavior in Norway Rats. *Behavioural Brain Research* 134 (1): 355–374. https://doi.org/10.1016/S0166-4328(02)00048-7.

Timberlake, William, and Douglas L. Grant. 1975. Auto-Shaping in Rats to the Presentation of Another Rat Predicting Food. *Science* 190 (4215): 690–692.

Timberlake, William, Joseph Leffel, and Cynthia M. Hoffman. 1999. "Stimulus Control and Function of Arm and Wall Travel by Rats on a Radial Arm Floor Maze." *Animal Learning & Behavior* 27 (4): 445–460. https://doi.org/10.3758/BF03209980.

Tolman, E.C. 1930. Maze Performance a Function of Motivation and of Reward as Well as of Knowledge of the Maze Paths. *The Journal of General Psychology* 4 (1–4): 338–342.

———. 1932. *Purposive Behavior in Animals and Men*. Berkeley: University of California Press.

———. 1938. The Determiners of Behavior at a Choice Point. *Psychological Review* 45 (1): 1–41. https://doi.org/10.1037/h0062733.

Traetta, Luigi. 2020. At the Beginning of Learning Studies There Was the Maze. *Open Journal of Medical Psychology* 9 (04): 168.

Vincent, Stella B. 1912. *The Functions of the Vibrissae in the Behavior of the White Rat*. Vol. 1. Chicago: University of Chicago.

———. 1915a. The White Rat and the Maze Problem: The Introduction of a Tactual Control. *Journal of Animal Behavior* 5: 175–184. https://doi.org/10.1037/h0075779.

———. 1915b. The White Rat and the Maze Problem: The Introduction of an Olfactory Control. *Journal of Animal Behavior* 5: 140–157. https://doi.org/10.1037/h0070053.

———. 1915c. The White Rat and the Maze Problem: The Number and Distribution of Errors- A Comparative Study. *Journal of Animal Behavior* 5: 367–374. https://doi.org/10.1037/h0075635.

———. 1915d. The White Rat and the Maze Problem: The Introduction of a Visual Control. *Journal of Animal Behavior* 5 (1): 1–24. https://doi.org/10.1037/h0072410.

Warden, C.J. 1929. A Standard Unit Animal Maze for General Laboratory Use. *The Pedagogical Seminary and Journal of Genetic Psychology* 36 (1): 174–176.

Warner, Lucien Hynes, and Carl John Warden. 1927. "The Development of a Standardized Animal Maze." In Archives of Psychology, vol. 15, no. 92, edited by R. S. Woodworth, 1–59. New York, NY: Archives of Psychology.

Washburn, Margaret Floy. 1908. Review of *Review of Orientation in the White Rat*, by Harvey Carr and John B. Watson. *The Journal of Philosophy, Psychology and Scientific Methods* 5 (10): 275–277. https://doi.org/10.2307/2011775.

Watrin, João Paulo. 2017. The 'New History of Psychology' and the Uses and Abuses of Dichotomies. *Theory & Psychology* 27 (1): 69–86. https://doi.org/10.1177/0959354316685450.

Watrin, João Paulo, and Rosângela Darwich. 2012. On Behaviorism in the Cognitive Revolution: Myth and Reactions. *Review of General Psychology* 16 (3): 269–282. https://doi.org/10.1037/a0026766.

Watson, John B. 1907. Kinæsthetic and Organic Sensations: Their Role in the Reactions of the White Rat to the Maze. *The Psychological Review: Monograph Supplements* 8: i–101. https://doi.org/10.1037/h0093040.

Wilson, David A.H. 2002. Experimental Animal Behaviour Studies: The Loss of Initiative in Britain 100 Years Ago. *History of Science* 40 (3): 291–320. https://doi.org/10.1177/007327530204000302.

Wimsatt, William C., and James R. Griesemer. 2007. Reproducing Entrenchments to Scaffold Culture: The Central Role of Development in Cultural Evolution. In *Integrating Evolution and Development: From Theory to Practice*, ed. Roger Sansom and Robert N. Brandon, 227–323. Cambridge, MA: MIT Press.

Yerkes, Robert M. 1901. The Formation of Habits in the Turtle. Popular Science Monthly, March.

———. 1902. "The Formation of Habits in the Turtle" *Popular Science Monthly,* March. "Habit Formation in the Green Crab, Carcinus Granulatus." Biological Bulletin III: 241–244. https://doi.org/10.2307/1535878.

Yerkes, Robert M., and Gurry E. Huggins. 1903. Habit Formation in the Crawfish Cambarus Affinis. *The Psychological Review: Monograph Supplements* 4 (1): 565–577.

Evan Arnet is a philosopher and historian researching the long-term epistemology of science and was a Mellon-Sawyer Fellow at the Department of History and Philosophy of Science and Medicine, Indiana University (Bloomington). He is particularly interested in the role of reflection and critique in the development and refinement of scientific methodology, especially in the animal behavior sciences. He is currently a research development specialist at Indiana University.

Chapter 11
Controlling Animals: Carl von Heß, Karl von Frisch, and the Study of Color Vision in Fish

Christoph Hoffmann

> But when I came to the aquarium in the evening at 9 o'clock, to my surprise I found the whole basin empty of fish; only 3–4 Blennius lay on the sandy bottom, the other Blennius I found lying in niches and caves by searching the walls with the arc lamp. Even with strong illumination they remained almost motionless. The Julies, however, had all hidden in the sand lying about 10 cm deep in the basin and could be driven out neither by strong illumination of the sand nor by poking with long sticks (Hess 1909, 15)[1]

11.1 Introduction

Experiments with animals are peculiar. Georges Canguilhem emphasized four circumstances: the "specificity of living forms," the "diversity of individuals," the "totality of the organism," and the "irreversibility of vital phenomena," such as aging and learning (Canguilhem 2008, 3–22). Because of these peculiarities, findings from animal experiments cannot easily be transferred from one species to another and it is also difficult to compare one set of findings to another or to repeat them. Researchers have developed many measures in response to these challenges, including, for example, the standardization of experimental animals all the way from their genetic makeup to their rearing and housing conditions.

---

[1] "Als ich aber Abends um 9 Uhr ins Aquarium kam, fand ich zu meiner Ueberraschung das ganze Bassin leer von Fischen; nur 3–4 Blennius lagen auf dem Sandboden, die anderen Blennius fand ich beim Absuchen der Wände mit der Bogenlampe in Nischen und Höhlen der Felsen liegend. Auch bei starker Belichtung blieben sie hier fast regungslos liegen. Die Julis aber hatten sich sämmtlich in dem etwa 10 cm hohen Sande des Bassins verkrochen und waren weder durch starke Belichtung des Sandes noch durch Aufstöbern mit langen Stangen heraus zu treiben."

C. Hoffmann (✉)
Department of Cultural and Science Studies, University of Lucerne, Lucerne, Switzerland
e-mail: christoph.hoffmann@unilu.ch

© The Author(s) 2024
J. Schickore, W. R. Newman (eds.), *Elusive Phenomena, Unwieldy Things*,
Archimedes 71. https://doi.org/10.1007/978-3-031-52954-2_11

We could extend Canguilhem's list, however. In experimenting with living animals, there are always at least two organisms involved: the researcher who performs activities and has intentions, and the animal with its own activities and circumstances, which may not be transparent to the researcher.[2] In this respect, Ian Hacking's oft-quoted remark that "experimentation has a life of its own" (Hacking 1983, 150) takes on a second meaning. It reminds us that, in contrast to inanimate matter, living beings are not always compliant. In fact, sometimes the challenge is simply to create situations in which the animals behave as desired. While this point is still in line with Canguilhem's account (although he depicts a surprisingly passive picture of the animal in the experiment) a second point seems beyond the reach of his argument: human researchers cannot share the experience of the animal directly. There remains a gap between the animal's observable responses and the researcher's conclusions. From a philosophical point of view, therefore, inferences about animal perceptions and sensations, like inferences about the perceptions and sensations of other humans, are underdetermined.

In the following, I shall discuss how the particularities of research with living beings expand our understanding of experimentation as a controlled procedure and determine the possibility of controlling the insights gained through experiments with animals. For this purpose, I refer to studies on color vision in fish, which the physiologist and eye specialist Carl von Heß and the zoologist Karl von Frisch realized between 1909 and 1914 (for more on this debate, see Munz 2016, ch. 2; Dhein 2021, 743–749; Dhein 2022, 32–36). Their experiments led to divergent results. Heß concluded that fish lack any color sense, whereas Frisch concluded the opposite. The resulting debate between Heß and Frisch, which more famously later included color vision in bees, led to continuous mutual criticism of methodological weaknesses and unsubstantiated conclusions, with the effect that the modes and functions of control were more explicitly expressed than in less controversial studies.

I develop my argument in three steps. I first trace the modes and functions of control in the experiments of Heß and Frisch. I then discuss the special circumstances of their work with animals. I focus on the fundamental fact, rarely discussed, that the experimental animal must be made to follow the experimenter's intentions. Finally, I am interested in how Heß and Frisch dealt with the problem that conclusions about whether fish perceive colors are systematically uncertain because researchers have access to fish sensory perceptions only indirectly, as mediated by behavioral responses.

---

[2] See a basic discussion of this situation in Köchy (2018). Köchy focuses on behavioral research with its particular critical entanglement of observer and observed. His considerations are only partly applicable to the situation in sensory research.

11.2   Varieties of Control

The central point in discussions between Heß and Frisch concerned whether or not fish discriminate light of a certain wavelength from light of another wavelength by its color. Heß defended the position that fish perceive shades of gray and not color. As a former assistant of physiologist Ewald Hering, he was familiar with the study of color blindness in humans. He therefore knew that in perception, each spectral color corresponded to a shade of gray of specific brightness (on a scale from black to white), by which even color-blind persons can distinguish it.[3] With this in mind, Heß built his investigations on two basic experiments (see Hess 1909, 3–9 and 11–14). In the first, he projected a spectrum onto the test tank's side wall. He then observed in which parts of the spectrum the fish preferred to stay. For the second experiment, he illuminated one tank wall with colored and white light, or with two lights of different colors, such that each covered exactly one half of the wall. Heß then measured how much the strength of one light source had to change so that the fish were either evenly distributed in the tank or moved from one half of the tank to the other. The result of this experiment was a "brightness equation." Later we shall see how the resulting findings let Heß to the conclusion that fish are fully color blind. These experiments usually work only with juvenile fish, which are photo-trophic positive (i.e. attracted by light). To study adult fish, Heß used experiments where fish received colored food for a certain time (see Hess 1911, 421–427). He then tested whether the fish could distinguish paper dummies of the same color from an equally bright gray background or from various gray dummies.

   Frisch, in contrast, believed that fish perceive color. Initially he carried out exper-iments relying on the fact that some fish species can adapt their appearance to the environment (see Frisch 1912b, 188–197; Frisch 1913a, 113–117; Frisch 1914a, 53–62). He worked mainly with the minnow (German *Elritze* or *Pfrille*), a species that can change its brightness very quickly, although its adaptation to color happens more slowly. Frisch placed two fish on a gray and a yellow ground that he assumed the fish perceived as equal bright, as determined by their similar outer appearance. Then he observed whether the "gray fish" and the "yellow fish" would change color over a given period of time. If the coloration of the "yellow fish" increased in con-trast to the "gray fish," this change should support color perception. In a second line of research, Frisch expanded on Heß's experiments with colored food (see Frisch 1914a, 43–53). In the beginning, like Heß, he used dyed food during training and then paper dummies in the actual experiments. Because the training situation dif-fered markedly from the experiments, Frisch started presenting the food on colored glass tubes, which were placed in the test tank together with other gray or colored tubes. The actual experiments used the same glass tubes but without the food.

---

[3] Brightness here refers to the perceived brightness of an illuminated surface or light source. Of course, the perceived brightness depends on the intensity of the illumination or light source, but if this is kept constant, there are specific differences in the perceived brightness between rays of dif-ferent wavelengths.

Heß's and Frisch's experiments became the subject of control in many ways. They rarely used the term "control," however (German: *Kontrolle, kontrollieren*), either alone or in combination with other expressions. Instead, they usually used related expressions ranging from *verifying, checking, testing,* and *copying* experiments, to *repeating* experiments or *trying* them once again. This terminology notwithstanding, there are many hints in their reports about procedures that, from an analytical point of view, are to be understood as control activities.[4]

Typically, one encounters the term "control" with respect to experimental instruments. Presenting a new arrangement for measuring brightness equations, Heß explained how to "control the correct setting" (Hess 1914, 255). Less typically, the instruments also include the fish themselves. In all experiments the organisms have a double status: with respect to the question of whether they perceive colors, the fish are the research object; with respect to the behavioral responses used to answer that question, the fish are the means of research. In Hans-Jörg Rheinberger's terminology, in these experiments every fish was both an epistemic thing and a technical object (Rheinberger 1997, 24–37). It is precisely this circumstance that Heß had in mind when criticizing Frisch for omitting an "indispensable control experiment" from the adaptation studies (Hess 1912b, 634). To assess the results' reliability, Frisch should first have investigated how sensitive his "measuring instrument" was to differences in brightness—Heß meant the minnow and, more precisely, the minnow's capacity to adapt to the ground (ibid.). Because Heß showed that the fish reacted to two grounds of different brightness with a similar adaptation, another possibility was equally conceivable: that the yellow and gray ground in Frisch's tanks, despite Frisch's assertion otherwise, were not of similar brightness (ibid., 636). For Heß, consequently, the increasing coloration of the "yellow fish," and the lack of increase for the "gray fish," had to do with a different stimulus basis (for Frisch's response, see Frisch 1913a, 110–118).

Sometimes the term "control" appeared when some circumstance had to be excluded as an experimental variable. Frisch's experiments with fish kept in a monochromatic green and red environment (experiments realized at the Zoological Station in Naples in 1911), were deemed worthless by Heß because an "essential control experiment" had been omitted: Rather than keeping a third group of fish in a transparent test tank, Frisch should have ensured that the light illuminating that tank had the same brightness as the light in the test tanks when surrounded by green and red solutions (Hess 1912b, 632). Frisch himself spoke of "control aquaria" and "control animals" when he referred to the fish in the transparent, "white" test tanks (Frisch 1912b, 209; Frisch 1913b, 153). In addition, Frisch worked with three more groups of fish that had been blinded before starting the experiments and then placed in green, red, and transparent tanks. Although Frisch did not use the term on this occasion, the groups of blinded fish had the character of control groups.

---

[4] In translations from German-language sources, the English term "control" is reserved for the corresponding terms *Kontrolle, kontrollieren.*

In this manner he attempted to discover whether the change in coloration was regulated by skin pigment or the sense of sight.

The term "control" seems particularly pertinent in connection with the repetition of experiments. For example, Heß emphasized that certain of his spectrum experiments were "often controlled measurements" (Hess 1909, 5). Repetition experiments normally help to check research results. When Heß (1913, 439) called the results of all minnow adaptation experiments "completely wrong," Frisch performed the experiments again—with positive results (Frisch 1914a, 55). The repetition of a competitor's experiments, in contrast to the repetition of one's own, casts doubt on their validity. Heß based his negative judgment of Frisch's adaptation experiments on the fact that his replication failed (Hess 1913, 404–414). Replication did not mean that an experiment was reproduced exactly, in every circumstance. That was common only in repeating one's own experiments. In repeating their opponent's experiment, on the other hand, Heß and Frisch almost always modified them. Heß and Frisch gave no reasons for the modifications but we can deduce from their comments that they already considered the experimental design to be flawed, to the extent that an unchanged repetition to them must have seemed pointless.

When Frisch studied the adaptation of minnows to their environment, he initially worked with two fish in two separate aquaria placed on different gray or yellow papers in daylight (Frisch 1912b, 188). Heß, in turn, aiming to "check Frisch's claims," worked with groups of minnows and first placed the aquariums in "dim daylight" on surfaces of different colorless brightness, illuminated from below. Later he placed three groups in three different aquariums in daylight, one on red, one on white, and one on black paper (Hess 1912b, 634–637, quote 634).[5] In his response, Frisch noted these differences and implied that they explained the varying results (Frisch 1913a, 111–113). The feeding experiments offer another example. Frisch (1914a, 44, fn. 1) discounted the refutation of his experiments by pointing out that Heß had used a "diverging experimental design." But Frisch himself also modified his opponent's feeding experiments. Instead of the red food of Heß he used yellow food, taking into consideration previous results suggesting that the red end of the visible spectrum might be shortened for fish (Frisch 1912b, 220). Taking these results together, one can conclude two things. First, Frisch saw little sense in identical repetitions of a competitor's experiment if it seemed to promise only limited insights. Second, as his criticism of Heß's modifications show, he insisted that only an exact repetition could question a statement based on experimentation.

Although Heß and Frisch do not use the term "control" in this context, experiments can also be understood as control experiments, which serve to deepen new ideas, to stabilize and explore originally random observations, or to check theoretical objections by competitors. An example of such an experiment goes back to Frisch's argument that the coloration of many fish during the spawning season

---

[5] Frisch also used red, green, blue, and violet papers in some experiments, and once also compared three minnows placed on yellow, red and gray paper (Frisch 1912b, 192–193). However, in this series of adaptation experiments, he always worked with one fish per aquarium, always placed the transparent aquaria on colored papers, and always worked with daylight.

supports color perception (Frisch 1913a, 121–126). Heß (1913, 401) immediately countered Frisch's point with experiments on how objects change their color at different depths due to absorption. He concluded that "the so-called 'wedding dress' of freshwater fish cannot possibly be conceived as a decoration calculated for the eye in the cases treated here."[6] I shall return to this point in Sect. 11.4.

Finally, Heß and Frisch allowed the scientific community to check their experimental procedures and conclusions. Although they did not use the term in this context, the measures they took aimed to control experimental results by intersubjective agreement. Both asked colleagues to judge "blindly," without prior knowledge, about the color of the ground on which a particular fish was placed in adaptation experiments (Hess 1913, 409–410; Frisch 1914a, 55). Both of their reports included meticulous descriptions of the rules by which they evaluated and compared the coloration of fish (Hess 1912b, 637; Frisch 1913a, 115). And both occasionally included excerpts from the original experimental records in their publications (Frisch 1913a, 116–117; Frisch 1914a, 49–52 and 53–57; Hess 1909, 29–30; Hess 1913, 410–411). In addition, intending to broaden his audience, Heß designed a simple setup that "can be easily applied even by laymen without knowledge of color theory and can often produce surprisingly beautiful results without special optical devices, even without a darkened room" (Hess 1914, 254). Frisch, for his part, preferred public demonstrations. In June 1914, he presented his experiments with bees and fish at the annual congress of the German Zoological Society, allowing the scientific community—at least in principle—to verify his results with their own eyes (Frisch 1914b). The bee experiments went well but most of the fish experiments had to be canceled: all the specimens had died during the congress (Doflein 1914, 710). Frisch remarked in his autobiography: "Unfortunately, the tap water of Freiburg did not agree with the trained fish I had brought with me from Munich, making them sick and useless for my demonstration" (Frisch 1957/1967, 57).

11.3   Cooperation Through Control

Kelle Dhein emphasizes that Frisch gained much more control over the experimental situation in his studies of color vision in fish, and later in bees, than had other researchers previously (Dhein 2022, 34–36). As Munz (2016, 68) points out, Frisch benefited from Heß's criticisms, just as Heß himself tried to eliminate errors in his own experiments. Nevertheless, as we just read above, experiments with living beings are seldom fully plannable. Plants and animals subvert researchers' intentions and precautions—they can interact with experimental circumstances and require special attention. The "high level of control" Frisch "exerts in his experiments" (Dhein 2022, 34) was always fragile in practice. Heß and Frisch's

---

[6] By "wedding dress" Frisch meant the eye-catching colorations of mostly male specimens during the spawning season.

publications are full of revealing remarks: they both mention often that many specimens died during the experiments, and Heß said there was only a short time during which the "material" used was sufficiently "fresh" to produce reliable results (Hess 1909, 2; 1914, 247). Another problem was that fish of the same species varied among themselves. This point came to light in a surprising way, as researchers working with specimens from different regions discovered that whether or not minnows adapted to the environment color depended on their place of origin — which partly explained the different results of Heß and Frisch (Haempel and Kolmer 1914). Frisch also reported that the fish training was important: minnows with several rounds of trials reacted more rapidly (Frisch 1912b, 193–194). Finally, minnows (but also other fish) had a feature unfavorable to adaptation experiments: they also changed color and brightness when disturbed by external influences, such as noise or the transfer from one aquarium to another (Frisch 1912b, 186–187; see also Hess 1913, 405).

Such problems have long been discussed in the history and philosophy of the life sciences in connection with animal experimentation. It is easy to overlook another central challenge in animal experiments, however, and one that has been addressed only rarely. Even when it is acknowledged, it is often only in passing. Obtaining results requires more than doing things like experimenting with the right species variation, accounting for individual differences, standardizing experimental conditions, using statistics, and so on. To obtain results, first and foremost, the animals must cooperate. This is not the case in all experiments, but in those where an animal must solve a task or exhibit a certain behavior, they must actively participate. The experiment cannot be accomplished otherwise. Cooperation in these cases is not another experimental variable, but rather a condition necessary to generate experimental knowledge.

Instructive in this respect is a look into Ivan Pavlov's chapter on the "Allgemeine Technik der physiologischen Versuche und Vivisektionen" [General technique of physiological experiments and vivisections], published in 1911 in Robert Tigerstedt's *Handbuch der physiologischen Methodik*. Pavlov described tools such as grippers, head restraints, and muzzle locks, all of which helped to immobilize animals and ensure smooth access to them. He differentiated between unruly animals—"It is only through necessity that one is induced to experiment on cats—impatient, screaming, nasty animals"—and compliant ones like the rabbit—"a gentle, passive animal that rarely cries out or protests" (Pavlov 1911, 7). And he praised his favorite animal, the dog, which is "irreplaceable" in "chronic experiments," that is, experiments conducted with the same animal for a long time (ibid.). In his words: "He [the dog] is, as it were, a participant in the experiments that are performed on him, and contributes immensely to the success of the experiments by his understanding and willingness" (ibid.).

Pavlov's account is certainly euphemistic, but at the same time it shows that the status of the animal in the experiment changes fundamentally when the classical vivisectionist approach, which always results in the death of the experimental animal (in Pavlov's terminology, the "acute" experiment), is supplemented and largely replaced by an experimental practice that requires, at least for a time, that the animal survive while remaining in as normal a state as possible. As Daniel Todes notes, a single experiment on pancreas secretion usually lasted five to ten hours, during which the animal was to remain as calm as possible. All excitement was to be avoided because physiological factors had to be isolated from mental factors (Todes 2002, 135). For "Druzhok," the first dog operated on successfully for the experiments, we learn that he "adapted rapidly to this requirement, lying peacefully on the table and 'taking no particular interest in anything' during the experiment. Better yet, he greatly facilitated the research by frequently sleeping for five to seven hours at a stretch" (ibid., 136). Once again, we need not believe Pavlov completely, but this kind of experiment was not feasible without the animal "assisting" the experimenter. In fact, this "assistance" begins with the animal surviving an experiment as long as intended. Not dying represents the animal's minimal active "service" to the experiment, as it were.

Cooperation becomes even more important when the animal subjects' behavior is unconstrained. In his *Intelligenzprüfungen* (Intelligence tests) on primates, which relied on results from 1914 to 1918 at the Anthropoid Station on Tenerife, Wolfgang Köhler considered "a foolish but eager animal" to be "more usable" than a "slacker," or a lazy animal that is not stupid but "never seriously responds to an examination task" (Köhler 1921/1932, 95). If researchers reinforce participation with food rewards (as is even more common today), they must expect that "a completely satiated monkey does not make any considerable efforts for its own sake" (ibid., 97). This point is trivial, but underlines once again that the experimenter in (non-vivisectional) animal experiments depends on the cooperation of the so-called "material." Nicolas Langlitz reports on the research group around the Japanese primatologist Tetsuro Matsuzawa, noting that their laboratory experiments undertaken since the 1970s "crucially depended on the chimpanzees' natural dispositions and—to use Matsuzawa's expression—their 'free will' to participate in experiments (which does not necessarily mean that their participation conformed to the researchers' study protocols)" (Langlitz 2017, 107). In other words, the laboratory was not and is not a place where animals are always "passive guests," as Robert Kohler (2002, 192) asserted.[7] Rather, in all experiments requiring a minimum of cooperation, animals must be encouraged to do what is desired.

Heß and Frisch confronted the fact that their experimental subjects *did have* a life of their own. In saying this I do not assume any human intentionality or consciousness. Rather, the fish belonged to an ecosystem in which they pursued their way of life. Recall the passage from Heß's first experimental report, quoted at the beginning of the article. When he wanted to extend his experiments from juvenile to adult fish, Heß had to work with a different experimental setup. He planned to project a spectrum into one of the public aquaria of the Naples Zoological Station, where he had conducted his first investigations, in order to observe where in the spectrum the fish recognized descending particles of food. While spectrum experiments were

---

[7] In marked contrast, Kohler (1993, 282) notes of *Drosophila*'s arrival in the laboratory that they were "active players in the relationship with experimental biologists, capable of unexpectedly changing the conventions of experimental practice."

promising only in sufficient darkness, the fish in the aquarium were inactive at night. Instead of foraging, they hid and rested.

Animating a living creature to give evidence for the questions of a research program is not a trivial matter. To achieve this goal, Heß and Frisch followed two strategies. In the first, they designed the experimental situation so that the fish automatically, in keeping with their natural behavior, reacted in ways that helped to answer the question. We see this strategy in Heß's experiments with phototrophic positive juvenile fish in the spectrum, where the animals gather wherever they find it brightest. His determination of brightness equations offers another example: the fish's tendency to gather in the brighter part of the test tank allowed him to detect minuscule increments of brightness differences perceived by the fish. Frisch too relied on this strategy when exploiting the capacity of minnows to adapt their coloration to the environment. In the second strategy, the fish were externally motivated to participate in investigating the research question. Heß and Frisch used this technique in their experiments on the association of food with color. The trick was to introduce a suitable stimulus, colored food, which the fish fully internalized by habituation or training.

The first strategy worked by creating experimental situations in which known reactions or behavior patterns can be observed in isolation, as often as required. In contrast, the second worked by adapting known behavior patterns—in this case, the consumption of food—to produce observations useful to the research question. In both, the cooperation of the fish was triggered by the experimenter's intervention. Somewhat paradoxically, however, in the first case this happened rather passively: cooperation was achieved by controlling the environment, without manipulating behavior directly. In the second case, the experiments worked by actively training the fish so that they would integrate a new stimulus into their usual behavior. Here, cooperation was achieved by controlling the behavior.

Active behavioral and passive environmental control of living beings in the aquarium, which is itself a limited and controlled space, represent two different styles of experimentation, amounting to two different ways of eliciting cooperation. For the experimental setup, passive control need not but often does involve much greater technical effort. Heß's experiments with spectra and the production of brightness equations demanded finely adjustable electric light sources, prisms, color filters, mirrors, spectrometers, and much other equipment. But implementing active control is no less complex. Rather than measuring tools and elaborate installations, what was needed in this case—apart from glass tubes, food, dyes, and standardized gray and colored papers—were two things: animals sufficiently capable of learning,[8] and researchers sufficiently capable of training them. Frisch may have been better at this, but not because he was educated as a zoologist whereas Heß was a physiologist. Frisch had simply cared for animals since childhood (see Frisch 1957/67, 19–23).

---

[8] My thanks to Kärin Nickelsen for bringing this to my attention.

In general, Heß considered that experiments based on training were not suitable for studying color vision in animals. In a handbook chapter on approaches to studying the sensation of light and color, he acknowledged that the training-based method can be helpful in mammals and birds, but "[u]ncritical application of the method to lower animals has often led astray" (Hess 1921/1937, 308). Going further, he argued that in his experiments—this time referring to the study of color vision in honeybees—the animals, "free of any compulsion," follow "exclusively their innate inclination to brightness," whereas animals in training-based experiments had first to learn something new, implying that this was a less natural situation (Hess 1922a, 94). Frisch tacitly agreed with Heß insofar as he distinguished experiments based on training, which he had developed to perfection, from experiments "under natural circumstances" (Frisch 1922/1932, 134). Unlike Heß, however, Frisch understood "natural circumstances" to be the observation of animals in the wild, rather than an arrangement in which animals reacted to light sources in a restricted space.

11.4 Constitutive Ignorance

Animal cooperation does not guarantee the successful execution of an experiment. At the very least, control over the behavior is always temporary. For example, Frisch reported that fish trained on colors recognized the "deception" after several experiments where no food was given. After that they no longer attended to the dummies (Frisch 1914a, 46). In studying color vision in fish, however, Heß and Frisch faced the additional problem of never being completely sure of experimental results. Frisch's answer to Heß's objection, that absorption prevents perceiving the coloration of many fish during the spawning season at a depth of only a few meters, is revealing. Frisch replied that Heß transferred "observations made on the human eye [namely observations made by Heß's eyes] to the fish eye without further ado, which is not permissible" (ibid., 64). Between what Heß and Frisch saw and what the "fish eye" perceived, there remained a gap: while one can control such things as the experimental environment, the fish's behavior, and many factors influencing the fish's responses, one cannot control the quality of the fish's perceptions, sensations, or experiences.

This problem leads from control practices in experiments to the control of conclusions based on experimental findings. From a philosophical point of view, Heß's and Frisch's results appear underdetermined—there is no immediate relationship between the fish's observable behavior and its presumed causes. As with all such experiments on non-human animals, researchers could make only plausible guesses to explain behavioral responses, or *why* an animal reacts in a certain way. Heß tackled this point directly in his first comprehensive review of the sense of light and color in animals. He started with a fundamental consideration:

> The question as to whether it is possible to identify a color sense existing in animals is still often answered with no. One justifies such an opinion with the fact that the animals are not able to give us information about their optical perceptions, as the human being is able to do

by designating different colors. One forgets here that the designation can give us only imperfect information about the visual qualities of a human being: If someone calls an object red which appears red to us, we still do not know whether his visual qualities correspond to ours. (Hess 1912a, 1).[9]

From this starting point, Heß concluded that a reliable basis for studying the color sense, in humans or animals, could be attained only by measuring which lights appear differently or the same to an eye. However, although animals could show behaviorally that light appeared to them the same or different, unlike humans they could not indicate *what* appeared to them to be the same or different about the lights. Was it the brightness, or the color? Or something completely different? To solve this problem, Heß resorted to studying the human eye. From Hering's research on human color blindness, he knew that total absence of color vision is accompanied by a characteristic shift in the perception of the brightest part of the spectrum from yellow to yellow-green, and a shortening of the spectrum at the red end (Hess 1909, 35). If an animal eye showed the same characteristics, Heß also considered it totally color blind. From his brightness equations he believed this to be true for fish, concluding his first communication with the statement that "the relative [perceived] brightnesses […] correspond almost or completely with those in which the totally color-blind human being" perceives colored surfaces (Hess 1909, 35). After studies on more animal species, especially bees, crustaceans, and worms, Heß found that, on the whole, the eyes of fish and invertebrates resembled those of a color-blind human eye in terms of their "visual qualities" (Hess 1912a, 151).

The way Heß inferred the conditions of animal eyes from the human eye indicated that, for him, all eyes—fish, bee, or human—were alike in their basic properties.[10] For Frisch, on the other hand, Heß's experiments proved only that "the brightness values of the colors for fish and invertebrates are the same as for the totally color-blind human being" (Frisch 1923, 471). Thus, a typical "characteristic of the total color blindness of man is found," but the truly essential characteristic is not the brightness distribution of colors in the spectrum. It is that colors be "distinguished only according to their brightness, not according to their quality" (Frisch 1919, 123).

---

[9] "Die Frage, ob es möglich sei, über einen bei Tieren etwa vorhandenen Farbensinn Aufschluß zu bekommen, wird noch vielfach mit Nein beantwortet. Man begründet eine solche Stellungnahme damit, daß die Tiere nicht imstande seien, uns über die Art ihrer optischen Wahrnehmungen Auskunft zu geben, wie es der Mensch durch die Bezeichnung der verschiedenen Farben zu tun vermöge. Man vergißt hier, daß die Bezeichnung uns über die Sehqualitäten eines Menschen nur unvollkommen Auskunft geben kann: Wenn jemand einen für uns roten Gegenstand gleichfalls rot nennt, so wissen wir noch nicht, ob seine Sehqualitäten mit den unsrigen übereinstimmen."

[10] Kelle Dhein (2021, 747) emphasizes that both Heß and Frisch followed a reductionist style of reasoning, differing only in the extent to which a "holistic perspective on animal capacities" was integrated into their experiments. I think that Heß and Frisch also differ in the extent to which they reduce the sensory abilities of animals to those of human beings. For Heß, the animal senses were ultimately more primitive versions of the human counterparts, whereas for Frisch they were senses with properties of their own.

In turn, Heß criticized (here referring to bees, but similarly valid for the experiments with fish) that Frisch's experiments, if they had produced any tenable results at all, in his eyes "at best provide information about the fact that one of two or more surfaces appears *different* to the bees than the others, but not about *how* the two surfaces are different for them; but this alone is of interest for the color sense question. "(Hess 1922a, 94) And, in fact, Frisch could not be sure about this. In tests, for example, fish trained on yellow also visited red dummies, and fish trained on red visited yellow dummies (Frisch 1914a, 45). This result could mean several things: that yellow and red appeared to the fish as the same color; that they appeared to the fish as the same grayish brightness; or that they appeared to the fish as different, but the difference was not large enough to play a role in behavior. Nevertheless, it was obvious to Frisch that color was decisive for the fish's behavior in all his experiments.

In his very first report on the color sense in fish at the Congress of German-Speaking Zoologists in 1911, he combined his results with a more teleological rationale.[11] After he had introduced Heß's experiments, Frisch remarked:

> Many fish possess the ability to change color to a very marked degree. When you see how some of them take on the most splendid colors as a result of a nervous influence on the pigment cells at spawning time, it is hard to believe that they are color blind. (Frisch 1911, 222).[12]

Frisch admitted that "it would not be right to attribute much significance" to such "intuitive arguments [Gefühlsargumente]" (ibid.). Nevertheless, in his publications on color vision in fish and later in bees, he frequently pointed out that the so-called "wedding dress" of fish or the relationship between flower colors and the pollination of flowers by bees make the assumption of a color sense seem compelling. Frisch concluded his lecture "Über Färbung und Farbensinn der Tiere [On Coloration and the Sense of Color in Animals]" of mid-1912 with the following words:

> And I believe that each of us would only acquaint himself with a certain discomfort with the prospect that all the splendor of flowers is a coincidence; that *by chance* the flowers that are pollinated by the wind are so inconspicuously colored, *by chance* the flowers that are set up for pollination by insects are so conspicuous, and so often conspicuous by their *color*; that everything here might as well be gray in gray. (Frisch 1912a, 38).[13]

Frisch's considerations show that he ultimately made a philosophical argument. One could observe bees visiting flowers. Similarly, one could observe the mutual

---

[11] In his conception of nature, Frisch joined a basic evolutionary stance to notions of purposefulness. Frisch also did not completely reject Lamarckian thinking.

[12] "Viele Fische besitzen in sehr ausgesprochenem Maße die Fähigkeit des Farbwechsels. Wenn man sieht, wie manche von ihnen zur Laichzeit als Folge einer nervösen Beeinflussung der Pigmentzellen die prächtigsten Farben annehmen, fällt es einem schwer zu glauben, sie seien farbenblind."

[13] "Und ich glaube, jeder von uns würde sich nur mit einem gewissen Unbehagen mit der Ansicht vertraut machen, dass die ganze Blütenpracht ein Zufall sei; dass *zufällig* die Blüten, die vom Wind bestäubt werden, so unscheinbar gefärbt sind, *zufällig* die Blüten, die auf Bestäubung durch Insekten eingerichtet sind, so auffallend, und zwar so oft durch ihre *Farbe* auffallend sind; dass hier alles ebensogut grau in grau sein könnte" (emphasis in original).

advantage of this activity: the bees gathered nectar, and the plants with striking flower colors depended on cross-pollination. That flower-visiting insects such as bees therefore possessed a sense of color, however, followed only if Frisch assumed a purposeful organization of nature—an organization in which it was difficult to imagine chance. Referring to Frisch's argument, Heß spoke of the "suggestive effect" inherent "in long-established trains of thought" (Hess 1922b, 1239).

Frisch himself was not very convinced of the epistemological significance of such more philosophical considerations. Remember that he spoke of "intuitive arguments." Nevertheless, they must have been indispensable to him, as their regular repetition indicates.[14] Just like Heß, he closed the gap between what could be observed in a controlled manner—manifold behavioral reactions—and what one wanted to find out—how do animals perceive light of a certain wavelength?—with a supplementary assumption. However, neither Heß nor Frisch would have spoken of supplementary assumptions. For them, a fish's ability to perceive or not perceive colors was a matter of plausible conclusions.

A fully satisfactory answer to the question of whether fish perceive colors and, if so, how those colors must be imagined, would require several things. For one, we would need to control the environment and the behavior, but we would also need immediate access to the perceptions of other living beings. So far this has not been possible. This may explain why the debate between Heß and Frisch was so persistent and never yielded a resolution.[15] The debate ended only with Heß's death in June 1923. From today's point of view, Frisch's results have, by and large, been confirmed. But as ethologist Niko Tinbergen noted in the 1940s and many others after him, whether fish and bees react as if they are colorblind or as if they perceive colors ultimately depends on the situation (Dhein 2021, 746).

11.5 Summary

Experiments with inanimate matter and those with living beings may differ less than suggested by Canguilhem. Among the modes and functions of control that Heß and Frisch employed, none was specifically adapted to organisms. Like many other experimenters discussed in this volume, they tried to identify, separate, and stabilize experimental conditions, and they looked for disruptive factors. Even Frisch's work with control groups, common in biological and medical research today, is not limited to experiments with living beings. Today, researchers usually speak of "groups" when they simply mean "sets of trials." The specificity of experimenting with living

---

[14] In many articles on the subject, Frisch mentioned the protective or reproductive role of coloration in fish and the relationship between flower colors and pollination by bees. The philosophical underpinnings of these remarks become particularly clear in his popular writings; see Frisch (1918, 1954/1966, 73–81).

[15] By resolution I mean a closure based on agreement between both sides. See McMullin (1987, 77–78).

beings does not necessarily arise from the variability of their properties and behaviors. A materials scientist can report similar experiences when studying the properties of new composites, for example. Rather, the most distinctive property of these experiments seems to be that living beings are not readily at hand for the researcher. One can stock them in large quantities, but the "material" (as scientists often call research animals) must still cooperate in many cases, and especially in the more interesting ones.[16]

Evan Arnet (this volume) provides an example of how animals can be motivated on their own to participate in an experiment—in this case, rats completing learning experiments in a maze. In that context, the hope of controlling motivation by food deprivation proved to be doubtful, if not an illusion. Going further, one could say that maze-based learning experiments use a particularly well-adapted form of environmental control, in which the animals show with good grace the behavior necessary for the research goal. At the same time, the example of the rat shows that it is not the scientist alone who controls and ensures the cooperation of the animal. Rats may like to poke around, but other tasks are not part of their behavior. For example, rats and mice are the experimental animals for alcohol research in the United States. But because they tend to prefer water, researchers must induce in them an alcohol dependence (see Ankeny et al. 2014, 493–494). An important criterion for the validity of the results obtained in this context is whether the animals are more or less forced to consume alcohol.

I wanted to underline this peculiarity of animal experiments when I emphasized that animals in experimentation lead "a life of their own." However, we should not assume that animals participate in experiments in the same way as the experimenter. Susan Leigh Star and James Griesemer (1989, 401) once asked the question: "How does one persuade a reluctant and clever animal to participate in science?" Star and Griesemer's famous essay on the boundary object is about how actors living in different social worlds can be stimulated to cooperate for a common cause—in this case, a zoological collection. That animals apparently also participate in this enterprise, however, and are listed alongside the scientific director and other contributors, obscures the fact that the animals are not "persuaded" but overwhelmed. In fact, the human actors in Star and Griesemer's story considered the animals mainly as "recalcitrant" (ibid., 402). Generally speaking, the status of animals in experiments perhaps most closely resembles that of the ignorant, uninformed subjects in Carl Stumpf's auditory experiments (see Kursell, this volume). Animals participate but remain uninvolved; the difference is that they often pay for their participation with their lives. In a scientific context, animals have a life of their own only insofar as they compel the researcher to make an extra effort to control them.

We may distinguish four concepts of control in the studies of Heß and Frisch on color vision in fish. The first is control as an activity to isolate and explore variables potentially significant to the research subject. The second is control as an activity to

---

[16] For a reflection on research materials, see the focus on "Materialgeschichten" in Hagner and Hoffmann (2018).

confirm or subvert experimental findings. Third is control as the activity of other scientists, who participate in reviewing results; here I am thinking of Frisch's public demonstrations of his experiments, for example. And the fourth is control as an activity concerned with the proper function of instruments, the care of the animals' living conditions, and the measures ensuring their cooperation in the experiments. I have emphasized this last point because it is rarely discussed within the broad notion of control as management of experimental situations (Schickore, this volume). Heß and Frisch took different approaches toward this goal, one based on environmental control and the other on behavioral control. Whether these strategies also evince disciplinary differences at the beginnings of the twentieth century, with Heß trained in physiology and Frisch trained in zoology, requires further investigation.

Finally, a question arises about the extent to which the basic problem of this research shapes experimentation: what can I learn about the sensations of a living being that is not myself, and with whom I cannot communicate? In this respect, the two methods for encouraging fish to cooperate constitute two different answers to this problem. Environmental control couples with the design of situations in which fish appear to respond spontaneously to the "question" posed by the researcher. When researchers, in turn, attempt to ensure cooperation by controlling behavior, it seems that the fish's ability or inability to respond unambiguously to a certain stimulus either directly confirms or directly disproves the existence of color vision. In both cases it appears that researchers favor observations, which seem to limit the scope for interpretation. But even then, the problem does not completely disappear. As we have seen, Heß and Frisch still felt that they must introduce additional plausible assumptions to strengthen their positions.

References

Ankeny, Rachel A., Sabina Leonelli, Nicole C. Nelson, and Edmund Ramsden. 2014. Making Organisms Model Human Behavior. Situated Models in North-American Alcohol Research, since 1950. *Science in Context* 27: 485–509.

Canguilhem, George. 2008. "Experimentation in Animal Biology." In *Knowledge of Life*. Trans. Stefanos Geroulanos and Daniela Ginsburg, 3–22. New York: Fordham University Press.

Dhein, Kelle. 2021. Karl von Frisch and the Discipline of Ethology. *Journal of the History of Biology* 54: 739–767.

———. 2022. From Karl von Frisch to Neuroethology: A Methodological Perspective on the Frischean Tradition's Expansion into Neuroethology. *Berichte zur Wissenschaftsgeschichte* 45: 30–54.

Doflein, Franz. 1914. Der angebliche Farbensinn der Insekten. *Die Naturwissenschaften* 2 (29): 708–710.

Frisch, Karl von. 1911. Über den Farbensinn der Fische. *Verhandlungen der Deutschen Zoologischen Gesellschaft auf der zwanzigsten und einundzwanzigsten Jahresversammlung zu Graz, am 19. August 1910, und zu Basel, vom 6. bis 9. Juni 1911*. Leipzig: Wilhelm Engelmann: 220–225.

———. 1912a. Ueber Färbung und Farbensinn der Tiere. *Sitzungsberichte der Gesellschaft für Morphologie und Physiologie in München* 28: 30–38.

————. 1912b. Über farbige Anpassung bei Fischen. *Zoologische Jahrbücher. Abteilung für allgemeine Zoologie und Physiologie der Tiere* 32: 171–230.

————. 1913a. Sind die Fische farbenblind? *Zoologische Jahrbücher. Abteilung für allgemeine Zoologie und Physiologie der Tiere* 33: 107–126.

————. 1913b. Über die Farbenanpassung des Crenilabrus. *Zoologische Jahrbücher. Abteilung für allgemeine Zoologie und Physiologie der Tiere* 33: 151–164.

————. 1914a. Weitere Untersuchungen über den Farbensinn der Fische. *Zoologische Jahrbücher. Abteilung für allgemeine Zoologie und Physiologie der Tiere* 34: 43–68.

————. 1914b. Demonstration von Versuchen zum Nachweis des Farbensinnes bei angeblich total farbenblinden Tieren. *Verhandlungen der Deutschen Zoologischen Gesellschaft auf der vierundzwanzigsten Jahresversammlung zu Freiburg i.Br., vom 2. bis 4. Juni 1914.* Berlin: W. Junk: 50–58.

————. 1918/19. Über den Farbensinn der Fische und der Bienen. *Schriften des Vereines zur Verbreitung Naturwissenschaftlicher Kenntnisse in Wien* 59: 1–22.

————. 1919. Zur Streitfrage nach dem Farbensinn der Bienen. *Biologisches Zentralblatt* 39: 122–139.

————. 1922/1932. Methoden sinnesphysiologischer und psychologischer Untersuchungen an Bienen. In *Handbuch der biologischen Arbeitsmethoden. Section VI: Methoden der experimentellen Psychologie. Part D: Methoden der vergleichenden Tierpsychologie*, ed. Emil Abderhalden, 121–178. Berlin: Urban & Schwarzenberg.

————. 1923. Das Problem des tierischen Farbensinnes. *Die Naturwissenschaften* 11 (24): 470–476.

————. 1957/1967. *A Biologist Remembers* (trans: Gombrich, L.). Oxford, London: Pergamon Press.

————. 1954/1966. *The Dancing Bees. An Account of the Life and Senses of the Honey Bee* (trans: Ilse, D. and N. Walker), 2nd ed. London: Methuen & Co.

Hacking, Ian. 1983. *Representing and Intervening. Introductory Topics in the Philosophy of Natural Science.* Cambridge, MA/New York: Cambridge University Press.

Haempel, Oskar, and Walter Kolmer. 1914. Ein Beitrag zur Helligkeits- und Farbenanpassung bei Fischen. *Biologisches Centralblatt* 34: 450–458.

Hagner, Michael, and Christoph Hoffmann, eds. 2018. "Materialgeschichten". *Nach Feierabend. Zürcher Jahrbuch für Wissensgeschichte* 14.

Hess, Carl. 1909. Untersuchungen über den Lichtsinn bei Fischen. *Archiv für Augenheilkunde* 64. supplementary issue: 1–38.

————. 1911. Experimentelle Untersuchungen zur vergleichenden Physiologie des Gesichtssinnes. *Pflüger's Archiv für die gesammte Physiologie des Menschen und der Tiere* 142: 405–446.

————. 1912a. *Vergleichende Physiologie des Gesichtssinnes.* Jena: Fischer.

————. 1912b. Untersuchungen zur Frage nach dem Vorkommen von Farbensinn bei Fischen. *Zoologische Jahrbücher. Abteilung für allgemeine Zoologie und Physiologie der Tiere* 31: 629–646.

————. 1913. Neue Untersuchungen zur vergleichenden Physiologie des Gesichtssinnes. *Zoologische Jahrbücher. Abteilung für allgemeine Zoologie und Physiologie der Tiere* 33: 387–440.

————. 1914. Untersuchungen zur Physiologie des Gesichtssinnes der Fische. *Zeitschrift für Biologie* 63: 245–274.

————. 1921/1937. Methoden zur Untersuchung des Licht- und Farbensinnes sowie des Pupillenspieles. In *Handbuch der biologischen Arbeitsmethoden. Section 5, Part 6, first half: Methoden zur Untersuchung der Sinnesorgane: Lichtsinn und Auge*, ed. Emil Abderhalden, 159–364. Berlin: Urban & Schwarzenberg.

————. 1922a. Farbenlehre. *Ergebnisse der Physiologie* 20: 1–107.

————. 1922b. Die Sehqualitäten der Insekten und Krebse. *Deutsche Medizinische Wochenschrift* 48 (37): 1238–1239.

Köchy, Kristian. 2018. Von der Ökologie der Forschung zu Forschungsumwelten. Bedingungen und Möglichkeiten einer erweiterten Lesart der Wechselbeziehung zwischen Forschenden und 'Forschungsgegenständen' in der Tierforschung. In *Philosophie der Tierforschung. Band 3: Milieus und Akteure*, ed. Matthias Wunsch, Martin Böhnert, and Kristian Köchy, 25–91. Freiburg: Verlag Karl Alber.

Köhler, Wolfgang. 1921/1932. Die Methoden der psychologischen Forschung an Affen. In *Handbuch der biologischen Arbeitsmethoden. Section VI: Methoden der experimentellen Psychologie. Part D: Methoden der vergleichenden Tierpsychologie*, ed. Emil Abderhalden, 69–120. Berlin: Urban & Schwarzenberg.

Kohler, Robert. 1993. *Drosophila*: A Life in the Laboratory. *Journal of the History of Biology* 26: 281–310.

———. 2002. Place and Practice in Field Biology. *History of Science* 40: 189–210.

Langlitz, Nicolas. 2017. Synthetic Primatology. What Humans and Chimpanzees do in a Japanese Laboratory and the African Field. *BJHS Themes* 2: 101–125.

McMullin, Ernan. 1987. Scientific Controversy and its Termination. In *Scientific Controversies. Case studies in the Resolution and Closure of Disputes in Science and Technology*, ed. H. Tristram Engelhardt Jr. and Arthur L. Caplan, 49–91. Cambridge: Cambridge University Press.

Munz, Tania. 2016. *The Dancing Bees. Karl von Frisch and the Discovery of the Honeybee Language*. Chicago and London: University of Chicago Press.

Pavlov, Ivan. 1911. Allgemeine Technik der physiologischen Versuche und Vivisektionen. In *Handbuch der physiologischen Methodik*, ed. Robert Tigerstedt , 1–64. Leipzig: S. Hirzel.Vol. 1, Part 1

Rheinberger, Hans-Jörg. 1997. *Towards a History of Epistemic Things. Synthesizing Proteins in the Test Tube*. Stanford, CA: Stanford University Press.

Star, Susan Leigh, and James Griesemer. 1989. Institutional Ecology, 'Translations' and Boundary Objects: Amateurs and Professionals in Berkeley's Museum of Vertebrate Zoology, 1907–39. *Social Studies of Science* 19: 387–420.

Todes, Daniel P. 2002. *Pavlov's Physiology Factory. Experiment, Interpretation, Laboratory Enterprise*. Baltimore, MD/London: Johns Hopkins University Press.

Christoph Hoffmann is Professor of Science Studies at the University of Lucerne. He is interested in animal experimentation, scholarly writing practices, and the epistemological enculturation of undergraduate students.

Printed in the United States
by Baker & Taylor Publisher Services